The Inaccessible Earth

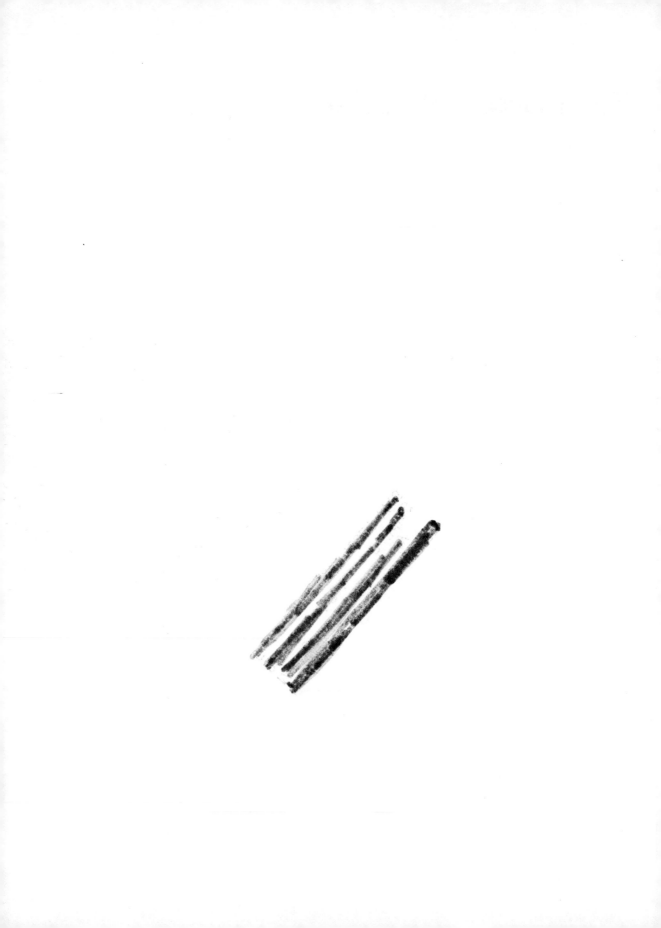

The Inaccessible Earth

An integrated view to its structure and composition

Second edition

G. C. Brown
Formerly of the Department of Earth Sciences, Open University, UK

A. E. Mussett
Department of Earth Sciences, Liverpool University, UK

 CHAPMAN & HALL
University and Professional Division
London · Glasgow · New York · Tokyo · Melbourne · Madras

Published by Chapman & Hall, 2-6 Boundary Row, London SE1 8HN, UK

Chapman & Hall, 2-6 Boundary Row, London SE1 8HN, UK

Blackie Academic & Professional, Wester Cleddens Road, Bishopbriggs, Glasgow G64 2NZ, UK

Chapman & Hall GmbH, Pappelallee 3, 69469 Weinheim, Germany

Chapman & Hall USA, One Penn Plaza, 41st Floor, New York, NY10119, USA

Chapman & Hall Japan, ITP - Japan, Kyowa Building, 3F, 2-2-1 Hirakawacho, Chiyoda-ku, Tokyo 102, Japan

Chapman & Hall Australia, Thomas Nelson Australia, 102 Dodds Street, South Melbourne, Victoria 3205, Australia

Chapman & Hall India, R. Seshadri, 32 Second Main Road, CIT East, Madras 600 035, India

First edition 1981
Second edition 1993
Reprinted 1995

© 1981, 1993 G.C. Brown and A.E. Mussett

Typeset in 10/12pt Times by Photoprint, Torquay
Printed in Great Britain by The Alden Press, Oxford

ISBN   0 412 48160 X

A Catalogue record for this book is available from the British Library

Library of Congress Cataloging-in-Publication Data available

∞ Printed on permanent acid-free text paper, manufactured in accordance with the proposed ANSI/NISO Z 39.48-199X and ANSI Z 39.48-1984

After the manuscript of this second edition had been submitted to the publishers – but before its publication – Professor Geoff Brown, with several other persons, was tragically killed by an eruption of Galeras volcano, Columbia, South America, on January 14th 1993. This occurred while he was taking measurements of the acceleration due to gravity, while carrying out research into the subterranean movement of magma. The volcano has since been declared sacred to the memory of those who died.

I take comfort from the fact that when Geoff died the second edition was far advanced and now has come to fruition. I hope that it will show that Geoff was as devoted to teaching as he was to research, and that it will help preserve his memory.

Alan Mussett

Contents

Preface to the Second Edition

In the dozen years since the first edition appeared, there has been a great advance in understanding of the Earth's deep interior. This is not because there have been breakthroughs in understanding, or even many changes of ideas, but largely because of many small advances, often the result of improved techniques. This has led to a complete revision of the book.

For instance, we have a much better idea of how the cloud of gas that formed the Solar Nebula evolved into the Sun and the planets, and of the chemical processes that accompanied its evolution and determined the mix of elements in the Earth. We have a better understanding of convection and how plates are an essential part of it, and how it is accompanied by chemical processes that have extracted the materials to build continents.

Although the major variation within the Earth is radial, improved geophysical and geochemical techniques have made progress in investigating and understanding the lateral heterogeneities, and it is encouraging that when geochemists and geophysicists talk about lateral heterogeneities they can sometimes be referring to the same thing. Plumes have become very fashionable as the cause of hot-spot magmatism and associated geochemical anomalies, probably originating at the base of the mantle (though clear evidence for their existence is lacking).

However, some old problems remain, notably the question of whether the 660 km discontinuity (formerly the 670-, and before that the 700 km discontinuity) is a barrier to whole-mantle convection. This, and the other controversial topics of plumes and heterogeneities and the nature of the D″ layer at the base of the mantle, have been quarantined in a new chapter (Chapter 9) of their own.

The huge growth in scientific literature has been reflected by an approximately 20% increase in the size of the book; without the encouragement of our publisher it would have been more – much more!

We are grateful to Chris Hawkesworth, Colin Pillinger, Nick Rogers and Dave Rothery for keeping us in touch with new scientific developments, and for their helpful comments on drafts; to Lynn Tilbury, who efficiently used her word-processor to erode the mountain of drafts, modified drafts and revised drafts; and to Una-Jane Winfield (of Chapman & Hall), without whose ultimata this project would never have been completed.

Preface to the First Edition

This book is about the Earth: its formation, evolution and, particularly, its present internal state and composition. It is here that the two great branches of the modern Earth sciences, geophysics and geochemistry, meet, though they are rarely combined in undergraduate texts to give a unified view of the Earth's interior.

Our understanding of the deep Earth has changed prodigiously in the past two decades and, although the future no doubt will see major changes in our understanding, there are signs that most advances will be a steady consolidation of present ideas, in a Kuhnian 'mopping up' phase. The problem in writing this book, then, has been one of amalgamating recent evidence, not just from geophysics and geochemistry, but also embracing parts of astronomy, meteoritics and so on. It is this breadth of input that makes the subject such a fascinating one but, because the available information is scattered through a wide range of books and papers – much of it is in specialist form – it is seldom read by the undergraduate.

Because our intended audience is generalist rather than specialist we have concentrated on getting clear the underlying physical and chemical principles, using mathematics only sparingly. For the less well-versed reader, some important basic groundwork, common to most undergraduates courses, is summarized in Notes at the end of the book. A 'Further reading' list appears at the end of each chapter. This includes non-specialist reviews or books or, when these are not available, recent papers that provide an entry into the literature. Those wanting to go more deeply into the subject can use the references towards the end of the book which include all but the oldest of the works cited. Two diagrams, in the front and back flaps, are included as overall summaries for easy reference.

If any of the views in the book do not find favour with our academic peers, then the responsibility is ours alone. But if the book achieves its aim of providing an accessible, generalist, undergraduate book, then our warmest thanks must go to all the friends and colleagues who helped with reviews and comments on draft chapters; in particular, Bill Fyfe, Ian Gass, Peter Harris, Aftab Khan, Richard Cooper, Peter Dagley, Peter Francis, Bob McConnell, Currie Palmer, Richard Thorpe, Rod Wilson and Brian Windley.

1 Introduction

1.1 AIMS AND OBJECTIVES

This book is about the interior of the Earth and the way in which internal processes influence surface morphology. It discusses the answers to questions such as: Of what is the Earth made? How does it behave? In other words, what chemical elements are present in different parts, and how are they combined to form the compounds that are built up into minerals and rocks? Also, what are their physical properties: are they liquid or solid, can they conduct electricity, and so on? The interior of the Earth is known to be a dynamic system because of its mobile surface features, expressed through the motions of lithospheric plates, for example, and this raises two other important questions: What forces and what processes are operating, and how have these and the Earth itself changed since our planet was formed about 4550 million years (Ma) ago?

We should like answers to these and many other questions, but the difficulty is that most of the Earth's interior is inaccessible to direct sampling or observation. It is true that rocks now at the surface, brought up in diamond-bearing kimberlite pipes and other volcanoes, have risen from depths of as much as 200 km, but, the deeper their source regions, the more likely it is that changes have occurred during their ascent to the surface. And even 200 km is but a small fraction of the Earth's radius (6370 km); so, valuable as such information is, it is far from sufficient to reveal the gross constitution of the Earth's interior.

There is no simple solution to this problem of inaccessibility. We have, instead, to combine information of many kinds, and it is the diversity of sources which makes the study so fascinating: astronomy, astrophysics, nuclear theory, planetary physics, as well as geophysics, geochemistry and geology all play a part. Often, each piece of information only narrows the possibilities, but if enough constraints are applied together we may arrive at a close approximation to the true constitution of the Earth. For example, the core is thought to be chiefly iron, for only iron is sufficiently dense, is an electrical conductor needed to produce the Earth's magnetic field, and is likely to be sufficiently abundant. No other material can satisfy all these requirements. But, before these techniques are discussed, our first object is to give an impression of the approach and general conclusions of the book.

1.2 'THE INACCESSIBLE EARTH': AN OUTLINE

To simplify matters, the following plan gives a logical development rather than the strict chronological evolution of knowledge. The first contribution comes from seismology, the most useful single technique for investigating the inside of the Earth. Earthquakes and explosions generate waves that may penetrate deep into the interior before eventually emerging far from their source. There are several sorts of seismic wave, which travel at different velocities, and the paths that the different kinds of wave follow depend upon the way the seismic wave

1

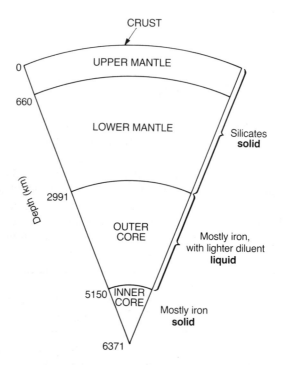

Figure 1.1 Sector of the Earth. Only the major features are shown. A fuller summary is given inside the front cover.

velocities vary within the Earth – which, in turn, depend upon the physical properties of the interior. Thus, by painstaking analysis of seismic records obtained at many recording stations set up around the world, it is possible to deduce how the different seismic velocities vary within the Earth. The picture that emerges is closely that of a concentrically layered Earth, in which the most striking division is the change, about half-way to the centre, from a solid (the **mantle**) to liquid (the **outer core**). However, there are many other discontinuities or rapid changes, some of which are shown in Figure 1.1, while a fuller summary is provided inside the front cover.

To find out whether these changes are due to changes of composition, temperature or other parameters, we turn to **density**, which is the most useful quantity that can be determined with reasonable precision. Density is deduced by combining information from seismology with a knowledge of the **mass** of the Earth, deduced from its gravitational attraction, and its **moment of inertia**, determined from the movement, or precession, of its axis of rotation. The resulting density variation with depth inside the Earth is not uniquely determined, but it is known within fairly close limits at most depths, so that next we can ask what material will match the density at any specified depth.

Obviously, a vast number of substances can have a particular density, so the purely physical description used so far has to be reinforced by other types of information that constrain the possible chemistry of the Earth. For example, current theories about the way in which the Solar System formed suggest that the Sun and planets condensed from a cloud, or **nebula**, of gases plus a little 'dust' of solid particles. **Meteorites** play a prominent role here, for they are believed to represent material dating from the time of planet formation. In particular, a few are thought to be close in composition to the original nebula, thereby providing an approximate bulk composition for the Earth.

It is well known from seismic studies that the Earth is not a homogeneous body, but that the chemical elements have been segregated, or differentiated into the layers mentioned earlier. Most of this segregation happened very early in the Earth's development, while it was forming or soon after, but some processes have operated over a long time, particularly the moulding of the continents, principally via the processes of plate tectonics. Given the constraints of temperature and also pressure variation in the Earth, the simple rules of geochemistry can be used to predict the combinations of chemical elements that can coexist stably as minerals at various depths. We are now in a position to apply these considerations to each region of the Earth in turn, together with any other relevant information.

The first region considered is the core. It is believed to be composed chiefly of iron, for the reasons given above, and the problems remaining are: to determine what small quantities of other elements must be present to bring the density into agreement with the calculated density variation throughout the core; to explain the existence of the solid inner core; and to account for the origin of the Earth's magnetic field.

Moving outwards, the mantle is thought to be far more complex. This is because it contains more elements in major amounts than does the core, and because it is solid and the various minerals formed from these elements change their *crystalline* form in response to the variations in pressure and temperature throughout the mantle. But though it is solid, all the mantle except the top few tens of kilometres is hot enough to be able to deform very slowly by solid-state creep. This allows it to circulate by thermal convection and permits the rigid, outermost part of the Earth, the **lithosphere**, to move, leading – among other results – to mountain building. Were it not for these movements, the Earth's surface would be almost featureless. In the upper mantle, beneath the lithosphere, the material is close to its melting temperature, and in some circumstances a partially molten, or liquid fraction, may be produced which, on rising towards the surface, leads to igneous activity.

The **crust**, the uppermost part of the lithosphere, is the most complex region of the Earth, because the cyclic phenomena of sedimentation, metamorphism and igneous activity, which process and reprocess its materials, can lead to extreme differentiation of the chemical elements. Principally through the techniques of stratigraphy and petrology, we shall be concerned with large-scale features of the continental crust, and the processes that exchange heat and material between the crust and upper mantle. The continental crust has been evolving irreversibly throughout geological time and also has grown in bulk at the expense of the upper mantle. The combination of crustal growth and thickening of the lithosphere on average, and the decreasing rate of heat production by radioactive decay, mean that mountain building and other major tectonic processes have changed in style as well as intensity during the Earth's history. Associated with the internal and surface evolution of the Earth have been changes in the composition of the atmosphere, the types of sedimentary rocks deposited, and life itself.

1.3 HISTORY OF IDEAS ABOUT THE EARTH

Throughout the history of science, increased understanding of the Earth's interior often had to await advances in physics, chemistry and astronomy, as well as in geology. But this should not be taken to mean that Earth science has been a fringe subject; often, it was at the forefront of science, giving rise to important controversies, such as those between the proponents of spontaneous creation and natural evolution, or between the literal adherents of the biblical age of the Earth and those who studied radioactivity, or the rates of slow geological processes.

It is customary in this kind of review to refer either to the Chinese or to the Greeks, or even, if erudition is excessive, to more remote civilizations. In this case, it is to the Greeks. During the period of 600–200 BC, their wealth of intellectual enquiry and speculation was amazing, and it had a major formative influence on our own ideas, particularly through the writings of Aristotle. Geological matters formed only a minor part of their interests, but they realized that land could become submerged, or form out of the sea. In addition, they understood that fossils represent organisms buried by past seas. On a larger scale, they knew that the Earth is a sphere and devised means of measuring its radius to within a few per cent. They also believed in a central fire, a belief that was to recur throughout history. One philosopher, Aristarchus (310–250 BC) even put forward the heliocentric (i.e. Sun-centred) theory of the planetary motions, but the idea did not find favour, and the geocentric (Earth-centred) theory was almost universally accepted until the 16th century AD.

Greek scientific reasoning had one serious weakness, which was its heavy reliance on elegant solutions that were deduced by reason, with little recourse to observation and experiment. For example, Socrates expelled a pupil from his logic class for suggesting that the best way to calculate the number of teeth a horse has was to open the horse's mouth and count them. As a result, they were unable to take many of their ideas beyond speculation.

The Greek body of thought was absorbed by the Romans, but with the fall of Rome in the 5th century AD it was almost lost, being preserved only in the Byzantine Empire. Fortunately, much of this knowledge percolated through the Islamic world and eventually was transmitted to the West, together with other knowledge, such as the decimal system, which originated in India. At first, medieval Europe regarded this pagan knowledge with suspicion, but gradually it was reconciled with Christian thought by the scholastic philosophers, prominent among whom (by girth as well as intellect) was St Thomas Aquinas (AD 1225–1274). As a result, the ancient works of Aristotle came to have an authority second only to that of the scriptures. This influx of knowledge into the West paved the way for the great upsurge of science, discovery and general enquiry that characterized the Renaissance.

But, before science as we know it could blossom, the reliance on authority to settle points at issue had to be challenged; it took a long time to accept that observation can overrule the stated opinions of eminent persons (practice has not quite caught up with this precept!). It is hard for us now to realize the true nature of this revolution in thought: an extreme act of faith must have been required to induce belief in the rational investigation of a world supposedly populated with witches, unicorns and hippogriffs, and a mineralogy that included jewels in toads' heads. But, gradually, observation became accepted as the final arbiter.

Copernicus (AD 1472–1553), a Pole, was one of the many dissatisfied with the geocentric theory that had the Earth stationary at the centre of the universe. The version of this theory then prevailing was due to Ptolemy who lived in Egypt from AD 90 to AD 168. To account for the fact that the planets occasionally appear

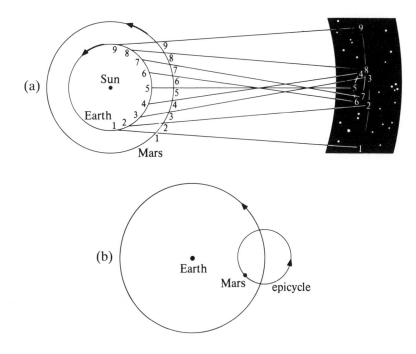

Figure 1.2 Illustration of epicycles. (a) Planets take different times to make one revolution about the Sun, and so a planet, viewed from the Earth, appears to form a loop in its motion. The numbers show the positions of the Earth and Mars at the same successive instants. (Redrawn from *Structure and Change* by G. S. Christiansen and P. M. Garrett © 1960 W. J. Freeman and Co.) (b) To explain such loops, the Ptolemaic system, which puts the Earth at the centre of the universe, made Mars move in a small circle (epicycle) which itself travels around the Earth. (Redrawn from *New Horizons in Astronomy*, 2nd edn, J. A. Brandt and S. P. Maran (eds.), © 1976 W. H. Freeman and Co.)

to reverse their motion across the sky (see Fig. 1.2), Ptolemy had proposed that they moved in epicycles. Copernicus probably took his idea of a heliocentric system directly from the ancient Greek philospher Aristarchus, for he admitted as much in a paragraph which he later deleted. To avoid criticism from the Church for demoting the Earth from the centre of the universe, a friend, who had been entrusted with publishing Copernicus' ideas, inserted a preface stating that the theory was merely a convenient method for simplifying calculations!

Tycho Brahé, a Dane who lived from 1546 to 1601, did not accept Copernicus' theory but, unlike Copernicus, he was a great observer and he produced greatly improved data that paved the way for Kepler (1571–1630), his German assistant, to eliminate epicycles by showing that planets move in elliptical orbits. Galileo (1564–1642), too, helped to disprove the Ptolemaic system; as a result he incurred the displeasure of the Italian Inquisition, who forced him to recant. On his death, he was denied a monument in the hope that he and his works would be forgotten, a vain hope, for the year of his death was that of Newton's birth in England.

Sir Isaac Newton (1642–1727) set the seal on these international advances with his theory of gravitation, and showed that if the force of attraction between two bodies was proportional to the product of their masses (m) and inversely proportional to the square of their separation (d), i.e.

$$F \propto \frac{m_1 m_2}{d^2}$$

elliptical orbits followed naturally. Newton was partly inspired to his theory by the work of William Gilbert, physician to Queen Elizabeth I, who had written the first geophysical treatise, a perceptive account of the Earth's magnetic field. Newton was unable to test his theory of gravitation directly by measuring the very small force between two masses in the laboratory. However, he showed theoretically that, under the action of gravitational and centrifugal forces, the Earth ought to have an equatorial bulge. Paradoxically, attempts by the French to measure the shape of the Earth seemed at first to show an elongation along its axis. To settle the matter, the French Academy organized expeditions to different latitudes in the decade 1735–1745, and these showed conclusively that an equatorial bulge exists.

Newton's formulation of the theory of gravitation and the laws of motion marks a watershed in the development of modern science from medieval belief and alchemy. It comes as something of a shock, therefore, to learn that Newton regarded his many scientific discoveries as a minor part of his work, and most of his 20 million words output concerned theology and such almost inconceivable topics as the topology of hell.

Early progress in geology was not so dramatic. As a descriptive science, many of its practitioners could only establish laws by the painstaking accumulation of many observations, and not by the elegant mathematical breakthroughs of astronomy or physics. However, systematic geology began with the observation of strata, which had practical applications in mining and engineering. Gradually, the idea of water-lain sediments became accepted, but fossils within them caused a lot of trouble. In medieval and later times, fossils included not only animal and plant remains but also curiously shaped minerals. It was also thought that fossils could form spontaneously within a rock. However, Leonardo da Vinci (1452–1519) understood clearly that fossils found deep within hills were the remains of long-dead creatures, and could not be attributed to the biblical deluge. Steno (1638–1686) documented certain of the rules of stratigraphy; for example, the law of superposition, which states that the lowest layer in a succession must be the oldest, and so paved the way for geological mapping by relative ages. William Smith applied these stratigraphical ideas in surveying and canal building, and also used fossils to correlate strata from different places. In 1815, he produced the first (hand-coloured) geological map of Britain, and in the following year published a book, *Strata Identified by Organized fossils*.

It has become a well-established principle of scientific method, known as Occam's razor, not to use two hypotheses in an explanation if one can be found to serve the purpose. But some geologists carried economy of hypotheses to fanatical lengths and claimed an all-embracing validity for their theories, forgetting the fact that, outside the carefully controlled conditions of the laboratory, many unrelated processes may be operating simultaneously. For example, the importance of water in forming rocks was seized upon by the Neptunists to account for nothing less than the formation of all the solid Earth. The theory was championed by Abraham Werner (1750–1817), Professor at Freiberg University, who believed that virtually all rocks were formed either by precipitation or crystallization in a universal sea. The theory was simple, but much of it was arbitrary and beyond test. Moreover, the sheer volume of water involved was a problem, for not only did it have to disappear after the initial formation but it had to return for succeeding epochs of deposition (the biblical deluge was only the last of many such epochs).

Neptunism was opposed by the Plutonists, whose cause was expressed by Sir James Hutton of Edinburgh (1726–1797) in his book *Theory of the Earth*, one of the most important geological texts ever written. The Plutonists did not dispute that sedimentary deposition occurred, but they did not believe that this could lead to hard, crystalline rocks without the action of subterranean heat (producing rocks 'formed in the underworld'). Their theory was less simple than that of the Neptunists, but it did offer a more dynamic Earth in which rock formations could be lifted and tilted and so oppose the levelling effects of erosion.

Disagreement between Plutonists and Neptunists sometimes reached a vigour not encountered in present-day controversies; in Edinburgh, a play written by an ardent Huttonian was hissed off the stage by an audience of Neptunists!

The Neptunist–Plutonist controversy merged within a generation into another: Catastrophism versus **Uniformitarianism**. This issue particularly concerned *time*: nowhere has the Bible been more restrictive to geology than concerning the age of the Earth. The computation in 1664 of Archbishop Ussher, a contemporary of Newton, that the Earth was created at 9.00 a.m., 26 October 4004 BC is well known. But this did not suit some geologists, who had begun to appreciate that slow geological processes could move mountains given sufficient time.

That geological change had occurred could no longer be denied. Catastrophism accounted for most of this change by a series of immense upheavals due to supernatural forces, while between these 'catastrophes', geological processes resumed normal operation, but produced only relatively small changes. The advantage of this theory was that a great deal of change could be accomplished in a very short time and so it could be reconciled with the literal biblical time-scale. It had the drawback, from the point of view of science, that it was beyond the reach of rational enquiry. By contrast, the theory of Uniformitarianism emphasized the continuity of geological processes. Hutton had observed geological unconformities and recognized that sedimentary successions could be built up, tilted and then eroded during a very long period of time, after which another cycle of deposition occurred. He enunciated the principle that processes in the past were the same as those observable today: 'The present is the key to the past'. This offered a rational means of explaining how rocks are formed, in terms of presently observable everyday processes that relate to the conditions existing during formation. For instance, the presence of marine fossils shows that the land was once below sea-level. A picture of some past environment can be built up and tested for self-consistency: the uniformitarian approach is therefore a very powerful method of investigation. The ideas of Uniformitarianism were developed and clarified by Playfair (1748–1819) and by Sir Charles Lyell of Edinburgh (1797–1875). Lyell, in his book *Principles of Geology*, published in 1833, showed that existing processes could account for observed geological changes, but vast periods of time were required. Lyell's arguments were persuasive, and the short biblical time-scale gave way to immense tracts of time. But the pendulum swung to the other extreme: Hutton's cautious statement that he could see no evidence of a beginning or an end to geological processes became, for some of his disciples, a world without end, with geological cycles of erosion and mountain building continuing indefinitely. Although we now believe the Earth formed 4550 Ma ago, opposition to such an apparently infinite time-scale was soon to come.

Physicists had been establishing the laws of thermodynamics, which demonstrated that processes could not continue indefinitely: the mechanism had to run down as the available energy was expended. Prominent among them was Lord Kelvin who, in the middle of the 19th century, calculated the time for the Earth to cool from a molten ball to its present temperature, which he estimated from the temperature gradient in mines, and the existence of volcanoes. He deduced a value of only about 100 Ma. This figure dismayed not only the geologists, but also Darwin, whose theory of evolution (published in 1859) owed much to Lyell's ideas, and was possible only if much longer intervals of time were available for evolution to bring about its changes. We know now that the

Earth gets much of its heat from radioactivity, but at that time Kelvin's case seemed unassailable; physical laws allied to mathematics seemed more than a match for semi-quantitative estimates of time based upon rates of geological or biological processes.

More and more geologists, over a period, came to accept the 100 Ma estimate, until it became almost a dogma for them. But the physicists, using progressively improved physical data, were refining their estimate of the age of the Earth, usually downwards, and in some cases to as little as 20 Ma. This, the geologists found progressively more constricting and it forced them to refine their calculations – based upon processes such as rates of sedimentation and the total thickness of sediments – until they were sufficiently confident to challenge the physicists. But then, at the end of the 19th century, radioactivity was discovered, and it was soon realized that the physicists' estimate of cooling time should be regarded as more like a minimum value for the age of the Earth, rather than a maximum. So another fruitful controversy ended, with the acceptance of a time-scale of hundreds of millions of years, which opened up new possibilities not only in geology but also in the life sciences and cosmology.

While the above ideas were developing, other people were thinking about the inside of the Earth. Buffon was the first to attempt to deduce the internal constitution of the Earth in a realistic manner. He believed that a molten interior is needed to allow the equatorial bulge to form due to rotation, and in 1776 published a theory whereby a molten Earth was formed when a comet collided with the Sun and knocked out material. He suggested that the subsequent evolution of the Earth required the more refractory materials to solidify first, leaving the volatile ones to form the oceans; later, the continents separated out and, finally, life emerged. Buffon performed crude experiments to measure the cooling rates of masses of various materials. By extrapolating the results to a mass the size of the Earth, he deduced a total cooling time of about 75 000 years, thereby foreshadowing Kelvin's method for estimating the age of the Earth.

In 1828, Cordier made measurements of the temperatures in mines and obtained a value for the geothermal gradient near the surface of the Earth of 30°C km^{-1}, in remarkable agreement with modern values. This was additional support for a hot interior of the Earth whose heat was attributed (since radioactivity was not known at that time) to an initially molten state.

A further inference from the cooling Earth model was that the solidified 'crust' or outer layer would buckle into mountains as the interior cooled and contracted. However, a new explanation of mountain belts was required once radioactive heat production was discovered around the turn of the century, for this gives the Earth a much longer active span. Nevertheless, the early geologists were right to see internal heat as the major source of energy that indirectly elevates the Earth's surface and so prevents the destructive forces of erosion, powered by solar energy, from reducing the relief to almost a plain.

Earthquakes, naturally, had long attracted attention. Explanations offered in classical times involved movement of internal water supporting the surface of the Earth, fire bursting out of the interior, or collapse of caverns. In medieval times, these mechanical, if vague, ideas were replaced by animalistic explanations; for example, the restless movement of a giant serpent in the depths of the sea was seriously proposed. Modern ideas began with Hooke, who recognized in 1705 that earthquakes are associated with land movements, and Mallet (1810–1881), in the middle of the 19th century, realized that most earthquake damage results, not from the gross movement of the land itself, but from waves that are generated by the movement and which spread out from the source. With great perspicacity, Mallet suggested that by timing the arrival of seismic waves which have travelled right through the Earth it should be possible to increase our knowledge of the Earth's interior. He was right, and seismology has proved to be a major tool for exploring the inside of the Earth.

Evidence for layering within the Earth came in 1906 when Oldham deduced that seismic waves travel more slowly in the very deep interior, and with refinement of his measurements this led to the recognition of the major divisions of core and mantle. There was much debate on the nature of the core and it was only in 1926 that Jeffreys conclusively showed that it was liquid, to account for the low rigidity of the Earth estimated from tidal and other measurements. This still left plenty of room for controversy about the composition of the core, which has not entirely ended, though iron is generally accepted as the main constituent.

The existence of earthquakes, volcanism and the growing body of evidence for geologically recent changes in sea-level, made it increasingly clear that the Earth had not only changed in the past but is still a dynamic planet. Yet, what forces were operating to cause vertical movements and how did the Earth respond to them? One clue came from an unexpected quarter. An attempt by Bouguer in 1735–1745, to 'weigh the Earth' by comparing the gravitational pulls of the Earth and the Andes, showed that the mountains apparently had considerably less mass than expected for their volume. Later, the effect was found elsewhere, particularly in the Himalayas during surveys of northern India by Sir George Everest and others. It was suggested that the deficit in mass is due to the rocks beneath the mountain being lighter than in adjacent areas. But how could this have come about? As the effect was too common to be assigned to chance, there had to be some mechanism that would bring it about naturally. In 1855, Pratt and Airy each published the concept of **isostasy**, the idea that surface rocks 'float' on a layer of denser but yielding rock.

This was only a partial explanation of mountain building, for it did not explain how the lighter rocks came to be accumulated into thick masses in the first place, a problem that was solved only with the advent of **plate tectonics**. This was a major development of the earlier idea of continental drift, first put forward in a coherent form in 1910 by Alfred Wegener, a German geophysicist. He argued, on the basis of different kinds of evidence (for example, matching of continental edges, rock types and faunal provinces across major ocean basins), that all the continents had been part of a single supercontinent several hundred million years ago, which since had broken up and 'drifted' apart. Acceptance of **continental drift** itself only became widespread with the blossoming of palaeo-magnetism, in the 1950s, which provided independent evidence for the movement of continents.

About the same time, techniques were being developed to investigate the two-thirds of the Earth's surface which is under water, and this led to the recognition that the ocean floors and the rocks beneath them are quite different from those of the continents. This difference was accounted for by plate tectonics which, unlike continental drift, embraced the whole surface of the Earth, and was formulated in the 1960s. According to plate tectonics, the rigid outer layer of the Earth, the lithosphere, is divided into huge pieces or plates which may comprise either or both continent and ocean. When plates move apart new oceanic lithosphere is created by sea-floor spreading, and where they move towards each other oceanic lithosphere is destroyed by diving into the mantle, in a subduction

9

zone. The continents, being thicker, stronger and less dense than the oceans are much less easily subducted and so are long-lived, whereas oceans are more ephemeral. Continental drift is simply the relative motion of the continental parts of the plates.

Tectonic activity is largely confined to the plate edges and gives rise to earthquakes, volcanism, departures from isostasy and other phenomena. But though plate tectonics is a surface process, there must be associated processes at depth, to provide the new material at spreading margins and remove it at subducting ones, and to permit the plates to move, and this has focused interest on the interior. And that, of course, is the subject of this book.

FURTHER READING

General books:
Adams, (1938); Gillespie (1951); Haber (1959): history of geological reasoning.
Burchfield (1975): the early controversy between geologists and geophysicists about the age of the Earth.
Huggett (1990): history of the role of catastrophe in geological thought.
Le Grand (1988): an investigation of how science operates, through an examination of geology, and particularly the development of the theory of plate tectonics.
Mather and Mason (1939): source book in geology.
General journals:
Badash (1989): the age of the Earth controversy.
Brush (1983): history of ideas about the Earth's core.

The contribution of seismology

2.1 SEISMIC WAVES

Seismology can mean the study of earthquakes, but it also means the study of the interior of the Earth by means of seismic waves, which can penetrate right through it. Much of the information has come from the times it takes the waves to travel along different paths through the interior.

Seismic waves can be generated by any sudden disturbance of the ground, but to be detectable on the far side of the Earth, only earthquakes and nuclear explosions are large enough **sources** of waves. The waves produced are of four kinds, as illustrated in Figure 2.1. They can be divided into body and surface waves: body waves travel within the Earth, while surface waves are largely confined to near the surface. Body waves, in turn, are of two kinds. **P-waves** are longitudinal waves, which means that individual particles of the material oscillate back and forth along the direction of wave propagation as the waves pass through, so that the material experiences successive compressions and rarefactions as the particles bunch and spread apart (Fig. 2.1(a)). P-waves are simply sound waves in the Earth, though usually of a frequency too low to be heard. **S-waves**, on the other hand, have their oscillations transverse to the direction of propagation (Fig. 2.1(b)), causing the material to undergo a shearing (or shaking) deformation that changes its shape.

A seismic wave is therefore the propagation of a deformation, or **strain**, through the material. In order that the material does not deform permanently, it must behave elastically during the deformations, which last only a fraction of a second. If any small volume within an elastic medium is distorted or strained in some way, a **stress** is developed which tends to restore the material to its undisturbed state. The ratio of stress to strain is known as the **elastic modulus** of the material. The speed of propagation increases with the value of the modulus but decreases with the density of the material. (Note that, in seismology, the vector term 'velocity', which implies a specified direction as well as a speed, is commonly used, rather than the more correct scalar term 'speed'.) The general expression for a wave velocity is:

$$\text{wave velocity} = \sqrt{\left(\frac{\text{elastic modulus}}{\text{density}}\right)}. \tag{2.1}$$

Because a material can often be deformed in different ways, it has more than one elastic modulus. The **compressibility** (or bulk) **modulus**, κ, is a measure of material's resistance to a change of volume – compression or dilation – while the **rigidity modulus**, μ, is its resistance to a change of shape caused by a shear stress (see Section 3.4 for expressions for these moduli). Examination of Figure 2.1 shows that the passage of an S-wave deforms the squares into lozenges, i.e. changes their shape but not volume, while P-waves deform the squares into rectangles, changing both their size and shape. The corresponding formulae for the velocities are:

$$V_{\mathrm{p}} = \sqrt{\left(\frac{\kappa + \frac{4}{3}\mu}{\rho}\right)} \qquad V_{\mathrm{s}} = \sqrt{\frac{\mu}{\rho}} \tag{2.2}$$

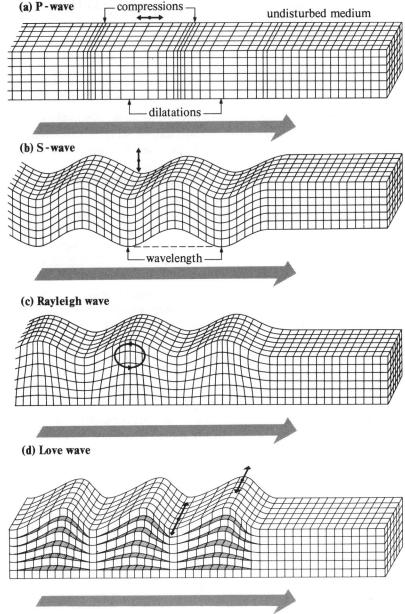

(a) P-wave — compressions — undisturbed medium

— dilatations —

(b) S-wave

— wavelength —

(c) Rayleigh wave

(d) Love wave

Figure 2.1 Four types of seismic wave. (a) In P-waves particles oscillate to and fro along the direction of propagation of the wave; (b) in S-waves particle motion is transverse. (c) The particle motion of Rayleigh waves is more complex, but near the surface is a backwards vertical ellipse; (d) in Love waves, it is transverse and horizontal. In both Rayleigh and Love waves, particle motion decreases with depth from the surface. (After Bolt, B. A., 1976.)

where ρ is the density.

Examination of the two formulae in Eqs (2.2) shows that V_p must always be greater than V_s and so P-waves from an earthquake will always arrive before S-waves at a seismograph recording station (see Fig. 2.2). Before their nature was known, these arrivals were simply called primary and secondary waves, and the initials P and S have remained; however, it may help to think of them as pressure and shear waves. A second deduction that we can make from the formulae is that S-waves cannot propagate in a material such as a liquid, which cannot resist a change of shape and so has no rigidity (i.e. $\mu = 0$). Moreover, while P-waves are propagated in a liquid, examination of the relevant equation (2.2) shows that setting $\mu = 0$ reduces the velocity considerably. This

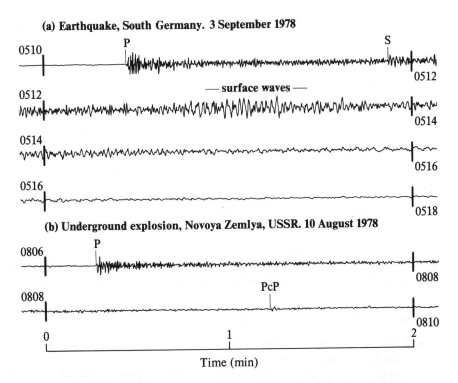

(a) Earthquake, South Germany. 3 September 1978

(b) Underground explosion, Novoya Zemlya, USSR. 10 August 1978

Time (min)

2.2 P- and S-wave velocities at depth

Figure 2.2 Seismic recordings or seismograms. The upper record is for an earthquake only 8° from the recording station and shows P-, S- and surface waves[1]. The lower record is of a Russian explosion which being 31° from the recording station, allows the PcP arrival to be seen clearly. Both were recorded at Leicester University's seismic station (CWF) in Charnwood Forest, UK, and were provided by courtesy of Dr M. A. Khan. (The PcP arrival is defined in Figure 2.6 and the angles are epicentral, subtended by source and recording station at the centre of the Earth.)

point is important because we shall use it to deduce that part of the Earth's core is liquid (Section 2.3).

The major object of this chapter is to deduce V_p and V_s at all depths in the Earth and so show that the interior can be divided into distinct concentric zones, separated by discontinuous, or rapid, changes in velocity. A further purpose is to introduce other seismic information, so that in Chapter 3 we can quantify how density varies within the Earth, for density provides a much better clue to the composition at depth than do the seismic velocities alone. This additional information comes from studying the other two types of wave shown in Figure 2.1: surface waves (Section 2.4), together with free oscillations of the Earth (Section 2.5), all of which depend on elastic moduli and density, but differently from body waves.

2.2 DEDUCTION OF P- AND S-WAVE VELOCITIES AT DEPTH

Firstly, we need to understand the rules that govern wave propagation within the Earth. A 'point' source of seismic waves, such as an explosion or an earthquake, will generate waves that spread out spherically, but when they encounter a region with different density or elastic properties – and hence a different seismic velocity – these spherical wave fronts will be distorted. This is closely analogous with the behaviour of light, and, as with light, it is usually more convenient to think in terms of rays rather than wave fronts, with a ray being the path followed by a small section of wave front, as illustrated by Figure 2.3. The section of wave

---

[1] One degree of epicentral angle equals about 111 km around the curved surface of the Earth.

Figure 2.3 Refraction and
reflection of seismic waves.
(a) shows equi-spaced wave
fronts spreading out from a
source. When they enter the
lower medium, with a faster
velocity they slew
anticlockwise, but also are
reflected (not shown). It also
shows two rays, which are
just the paths traced out by
points on a wave front; the
rays are refracted and
reflected at the interface. (b)
Detail at the interface. If the
two rays are close together
then they can be treated as
parallel over a short distance;
they are, of course,
perpendicular to the wave
fronts.

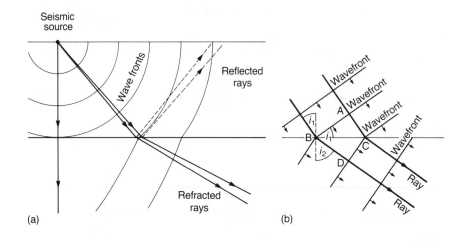

(a) (b)

front AB moves to CD, where AC and BD are in proportion to V_1 and V_2, the
seismic velocities of the two materials (in this case $V_1 > V_2$, so we can write:

$$\frac{\sin i_1}{\sin i_2} = \frac{AC/BC}{BD/BC} = \frac{AC}{BD} = \frac{V_1}{V_2} \quad \text{or} \quad \frac{\sin i_1}{V_1} = \frac{\sin i_2}{V_2} \qquad (2.3)$$

This is the basic rule governing refraction at an interface between two materials
and is equivalent to Snell's law in optics.

Equation (2.3) can be employed to trace rays through the Earth, with the aid
of some geometry. Figure 2.4 shows three concentric interfaces. Given the layers
and their velocities, we can start at an arbitrary interface, the *n*th, and an
arbitrary incident angle, i_{n-1}. Equation (2.3) is applied at A to give i_n

$$\frac{\sin i_{n-1}}{V_{n-1}} = \frac{\sin i_n}{V_n}. \qquad (2.4)$$

The distance AB, l_{AB}, is then given by:

$$l_{AB} \cos i_n = (r_n - r_{n+1}).$$

So the time to go A → B:

$$t_{AB} = \frac{l_{AB}}{V_n} = \frac{(r_n - r_{n+1})}{V_n \cos i_n}. \qquad (2.5)$$

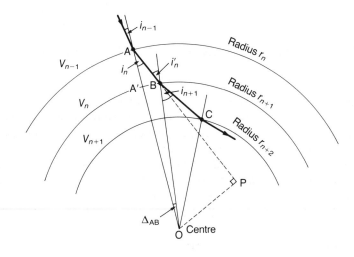

Figure 2.4 Ray tracing
through the Earth. This shows
the path of a ray refracted by
three interfaces. See text for
equations governing angles,
etc.

14

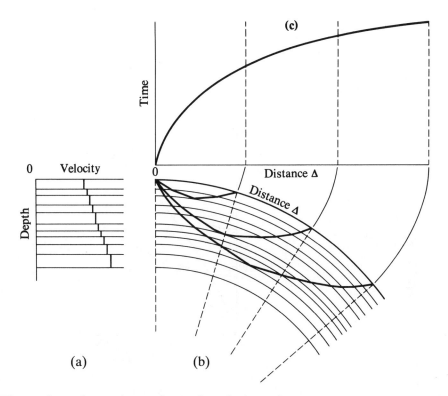

Figure 2.5 Schematic diagram of rays in a concentrically layered Earth. (a) Smooth changes in velocity can be approximated by many thin layers with small increments in velocity. (b) Velocity increasing downwards causes rays to curve back to the surface. (c) Diagram showing the corresponding time–distance or travel-time curve. Note that distances are measured by the angle subtended at the centre of the Earth, the epicentral angle, Δ.

The ray has advanced an epicentral angle Δ_{AB}. As:

$$A'B = r_{n+1}\,\Delta_{AB} \text{ and } \frac{A'B}{A'A} = \tan i_n$$

then:

$$\Delta_{AB} = \tan i_n \frac{(r_n - r_{n+1})}{r_{n+1}}. \tag{2.6}$$

We can repeat the above after advancing to B. Since the interfaces are arcs of circles, the perpendiculars at A and B converge at the centre of the Earth, so

$$AO \sin i_n = OP = BO \sin i_n'.$$

Therefore:

$$\sin i_n' = \frac{r_n}{r_{n+1}} \sin i_n. \tag{2.7}$$

This relates $\sin i_n'$ to $\sin i_n$, and hence to $\sin i_n'$, which we started with, and the above steps can then be repeated for $B \rightarrow C$.

By adding up the increments of distance and time, the ray path – and the time that it takes to travel along it – can be deduced, provided that the variation of velocity with depth is known. In the Earth, much change of velocity takes place continuously, rather than abruptly at sharp interfaces, but this can be dealt with by dividing such regions into a series of thin concentric shells (Fig. 2.5), which can be treated as above.

Of course, we do not know initially the velocity–depth profile (Fig. 2.5(a)) but only the end result, the **travel-time curve** (Fig. 2.5(c)). It is possible to invert the above mathematics, and before the advent of computers this was the only way to

15

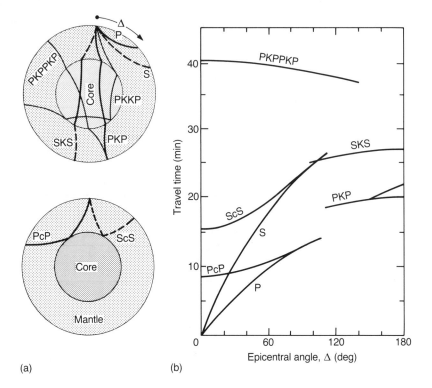

Figure 2.6 Selected rays and corresponding travel-time curves. (a) In the mantle, where the velocity, in general, increases with depth, the rays bend away from the vertical. Those reaching the core may either be reflected back into the mantle, or refracted into the core, which has a lower velocity. Capital letters denote both the type of wave and the part of the Earth through which it is travelling. K denotes a P-wave in the core; c indicates a reflection at the core. Rays can experience many more reflections and refractions than shown. (b) Corresponding travel times. In practice, of course, (a) is deduced from (b) and not vice versa. (From Richter, 1958 and Jeffreys, 1962.)

(a) (b)

deduce velocity–depth profiles. Now it is possible to employ the above equations, using powerful computers. But before this is described we shall first look at the way that accurate and detailed travel-time curves have been built up over the years from careful analysis of many earthquake records.

The process begins when an earthquake or large explosion occurs and waves from it are received at seismic stations scattered around the world. The next step is to identify the various arrivals at the different stations for, in addition to the first arrivals of P-, S- and surface waves so far mentioned, there are many other ray paths involving reflection, refraction, or both, within the Earth, and these will also provide discrete arrivals. Some of the more important ones are shown in Figure 2.6(a); they can be important for improving our knowledge of variations of velocity within the Earth.

As yet, neither the location nor time of the earthquake are known. These can be estimated by comparing records from the different stations, for those closest to the earthquake will record its arrival earliest and with the largest amplitude. In this way, rough travel-time curves for P- and S-waves could be constructed. Once this is done, we have a better method for estimating time and distance of earthquakes, using the *difference* in arrival time for the P- and S-waves. Figure 2.6(b) shows how this interval is related to the distance of the seismic station from the earthquake (expressed as the epicentral angle Δ, the angle subtended at the centre). Thus, the time and distance of the earthquake can be read off. One station is not enough to determine the location of the earthquake, only its distance, but by using several stations it is possible to pin-point its position.

A single earthquake only gives values on the travel-time curve corresponding to the specific distances of the various stations recording it. To build up almost continuous curves, as shown in Figure 2.6(b), requires the recording, over many years, of earthquakes occurring all over the world. This synthesis of a continuous

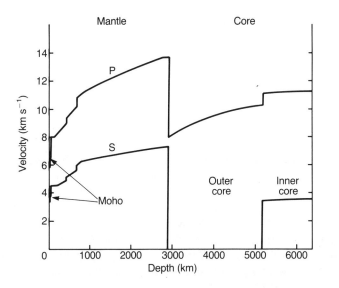

Figure 2.7 Seismic velocity versus depth curves for P- and S-waves, according to the iasp91 travel-time tables (Kennett and Engdahl, 1991).

travel-time curve from the records of many earthquakes received at many separate seismic stations is possible only because the Earth is spherically symmetrical to a high degree, after allowing for the equatorial bulge due to rotation (Section 3.3).

Having obtained an accurate travel-time curve, the next step, as mentioned earlier, was to use it to deduce the velocity–depth profile, having inverted the ray-tracing equations for that purpose. This was a laborious process but gave results so good that they have been in use until recently. However, it is not suitable for dealing with the very large amounts of data that have been accumulating. Nowadays, powerful computers are used to deduce velocity–depth profiles using **forward modelling**; that is, velocity–depth profiles are guessed and used to calculate travel times which are then compared with actual travel times following Figure 2.4 (a) to (c), rather than vice versa. The profiles are adjusted to improve the match, and this process is repeated until the profiles are as good as the data will allow. Of course, the starting point is not a 'blind guess' but is based upon the profiles obtained using the earlier methods. The details of the method are similar to those used to deduce the density–depth profile (Section 3.6).

The most recent of such profiles is **iasp91** (Fig. 2.7). This is intended to be the new standard, replacing tables dating from 1940 (Jeffreys and Bullen), which indicates that the changes, though significant, are actually quite small. In fact, for distances over 1000 km the differences are only about two seconds. These new tables are based on data accumulated over 20 years, comprising over six million arrival times recorded at over 3000 seismic stations.

It is important to realize that the profiles specify a model, being a spherically symmetrical Earth that best fits the observations (after allowance for the equatorial bulge). They average out lateral variations in velocity, and so form a **reference model**. This is a reasonable thing to do because the lateral variations are quite small (in fact, they are detected by the differences, found in some parts of the world, between actual travel times and those given by the reference model). Lateral variations do exist, for instance, between the regions beneath continents and oceans; and, because most seismic recording stations are on continents, iasp91 is biased towards the sub-continental structure, but that is not significant for the global features being considered in this chapter. The results will be discussed in the following section.

2.3 SEISMIC VELOCITIES AND THE STRUCTURE OF THE EARTH

Figure 2.7 shows a recent determination of the velocity–depth profile. There is a general trend of velocity increasing with depth, sometimes in jumps, termed **seismic discontinuities**. The major exception to this trend is the abrupt decrease in P-wave velocity about 2900 km down, accompanied by the disappearance of S-waves. It is this discontinuity which, by definition, divides the mantle from the core. We infer, from the absence of S-waves, that the outer core is liquid because, as we saw in Section 2.1, S-waves cannot propagate in a liquid.

The abrupt decrease in V_p at the **core–mantle boundary** causes rays entering the core to be refracted *towards* the vertical and this results in a shadow zone (Fig. 2.8): P-waves cannot directly reach the Earth's surface within a zone between 103° and 142° from the earthquake. However, it is not a perfect shadow zone, and because of some arrivals within it, it was suggested that there is an inner core with a sharp boundary that causes rays to reflect and so arrive in the shadow zone, as shown in Figure 2.8, rays E and F (these are called PKiKP rays, the small 'i' denoting a reflection at the inner core boundary). It used to be thought that the change from outer to inner core is complex, because the associated arrivals were not what would be expected from a simple discontinuity. However, there is now general agreement with the interpretation of Chang and Cleary (1978), that the complexity is due to scattering of rays, probably by bumps on a boundary which otherwise is a simple discontinuity (see Section 9.2.3).

From soon after its discovery, it was thought that the inner core is solid, because a non-zero value for the rigidity modulus, μ, would explain the abrupt increase in V_p compared to the liquid outer core. However, direct evidence for this, in the form of S-rays that have travelled through the inner core, has not yet been found definitely, despite claims made. The problem is that the S-ray in the

Figure 2.8 Core shadow zones. A ray that just grazes the core arrives at C, but a slightly steeper ray is refracted into the core to arrive at C′. Steeper rays move back towards the source, D, but not further than F. Thus C → F is a shadow zone for P-waves, except that the inner core can deflect a few rays into the shadow zone, E. As S-waves cannot propagate in the outer core their shadow zone is C → H. (After Gubbins, 1990.)

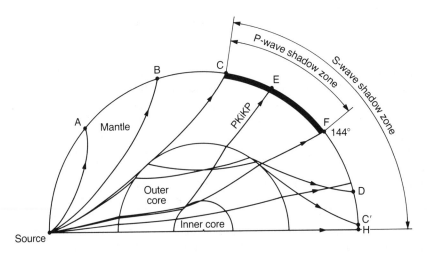

inner has to derive from a P-ray in the outer core, and then convert back to a P-wave to travel back to the surface, and these conversions mean there is little energy in the rays; moreover, a number of completely different ray paths can take nearly the same time to reach the receiving station. The best evidence that the inner core is solid comes from free oscillations (see Section 2.5). The existence of a solid inner core is important, for instance, to our understanding of convection in the core, which in turn is necessary to account for the existence of the Earth's magnetic field (Chapter 6).

Turning to the mantle, there are a number of discontinuties in the top several hundred kilometres. Just how many and at which depths they occur has been debated in recent years. There is little doubt that major discontinuities occur at about 410 and 660 km, with recent estimates placing them at 415 (±3) km and 659 (±8 km) (Shearer, 1991). They are usually called the 410–(or 400) and 660–(or 670) km discontinuities. Recently, good evidence has been found for another, though lesser, discontinuity at about 520 (519±4) km (Shearer, 1991). In contrast, the existence of the once widely accepted discontinuity at 220 km is not well supported. These discontinuities are probably due to phase changes of the mantle material into denser crystalline forms, as will be discussed in Chapters 3 and 7.

The mantle does not extend quite to the Earth's surface. The Moho (a convenient abbreviation of the Mohorovičić discontinuity, named after the Balkan seismologist who discovered it in 1909) is so close to the surface that it hardly shows on Figure 2.7. Although this is usually present as an abrupt increase of seismic velocity from about 7 km s^{-1} to about 8 km s^{-1}, there is considerable variability in the velocities immediately above and below the discontinuity. The Moho occurs nearly everywhere but its depth varies, being typically 35 km beneath the continents (with a range of 20–90 km), and 5 to 10 km beneath the ocean floor. It defines the boundary, in seismic properties, between **crust** and mantle, and the nature of the associated material change will be discussed in Section 7.4. The term 'crust' suggests a rigid layer over a yielding one, but the crust is just the top part of the rigid layer, the lithosphere, as described in Sections 1.2, 8.2.2 and 8.4.4.

In addition to the several discontinuites described above, there are two other interruptions to the smooth increase of velocity with depth in the mantle, but of a gradational nature. In the approximate depth range 60–220 km there is a decrease in velocity[1], called the **low-velocity zone, LVZ**. This is not regarded as such a prominent feature as it was a decade ago, and is thought to be largely or entirely confined to S-waves (Montagner and Anderson, 1989). The LVZ also has the property of partially absorbing S-waves as they pass through, so attenuating their energy, probably due to the presence of a small proportion of partial melt (Section 7.3). The LVZ does not have sharp boundaries, and so its top limit is taken as the depth at which the velocity begins to decrease, while the lower limit is arbitrarily taken as the depth at which the velocity has regained the value of the upper limit.

The remaining feature to be mentioned is at the base of the mantle, the bottommost 200 km or so where the velocity is nearly constant. This is called the D″ ('D double prime') layer, after a former system of naming the radial regions of the Earth alphabetically. Its nature will be discussed in Section 9.3.2.

These are only the more obvious features evident from the velocity–depth

---

[1] In Figure 2.7 apparent only as an approximately constant value of V_s.

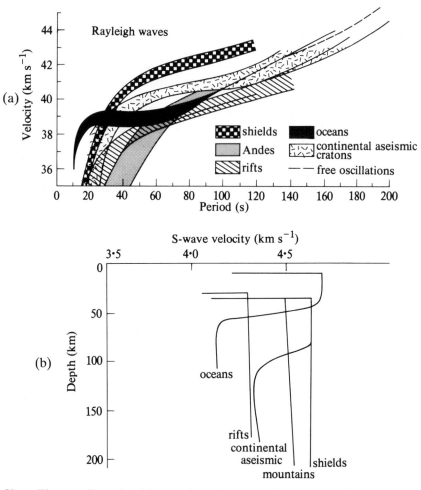

Figure 2.9 Dispersion of surface waves. (a) Dispersion of Rayleigh waves in regions with different structures. (The period of a wave is a measure of its wavelength, since it equals the velocity divided by the wavelength.) (b) S-wave velocity profiles consistent with (a). Love waves give similar results. (After Knopoff, 1972, by permission of Elsevier Scientific Publishing Co.)

profiles. The profiles should also be able to tell us something about the properties of the concentric zones between the discontinuities. However, this cannot be achieved without additional data, so we will turn to other types of seismic information.

2.4 SURFACE WAVES

Figures 2.1(c) and (d) illustrate the two types of seismic surface wave. **Rayleigh waves** have similarities with waves on water, for passage of the wave causes a point on the surface of the Earth to move in an ellipse in a vertical plane, as does a cork on water. Again like water waves, the amplitude of particle motion decreases with depth below the surface. This behaviour of water waves is an advantage to submarines for, if they submerge to 100 m or so, they are hardly disturbed by the waves produced by a storm. The amplitudes of waves of the same type, but with different wavelengths, fall off with depth at the same rate, *provided that the depth is measured in terms of wavelength*; for example, if the amplitude of a 10-m-long wave falls to half at a depth of 6 m, the amplitude of a 20-m-long wave would fall to half at 12 m depth. Thus the disturbance due to long wavelengths makes itself felt at greater depth than that due to short waves.

The velocity of Rayleigh waves depends chiefly on the rigidity modulus and on the density of rock, and usually may be taken to be about 0.92 of V_s, the velocity of S-waves. Their disturbance is not confined to the actual surface, and so their velocity is not determined entirely by the properties of the surface rocks, but also by those at depth, though in a progressively decreasing way. This dependence of velocity upon the properties at depth will be larger for longer waves, since their disturbance extends to greater depths. Consequently, in a region where V_s varies with depth – as is usually the case – the longer wavelengths will have a velocity different from that of shorter waves, and this variation of velocity with wavelength, called **dispersion**, can be used to determine how V_s varies with depth.

The velocity determined by measuring the time taken for surface waves of a given wavelength to travel between two places on the Earth's surface must be an average for the material along the path. This may be an advantage, compared with body wave results, if we wish to ignore small-scale variations such as those due to the great variability of crustal rocks. Another advantage is that this surface-wave dispersion yields data about areas, such as the oceans, where there are very few recording stations. Figure 2.9(a) shows results for some different types of area, and Figure 2.9(b) shows the variation of V_s with depth, which has been deduced from the surface-wave dispersion. It should be appreciated that, although these results demonstrate regional differences, the departures from spherical symmetry are quite small. The subject of lateral variations in the mantle will be considered further in Chapter 9.

So far, we have considered only Rayleigh surface waves. The second type of surface wave is the **Love wave**, which has motion entirely horizontally (Fig. 2.1(d)). The properties of Love waves are rather more complex than those of Rayleigh waves, but, as regards the type of information we shall derive from them, they can be considered as complementary to Rayleigh waves. The surface waves of interest here have a much longer period than body waves. Surface waves have revealed that the Earth is seismically anisotropic, that is, at some depths in the Earth seismic velocities vary with their direction of propagation (see also Section 3.6). Extremely long surface waves wrap around the Earth and may then form standing waves, resulting in free oscillations, as will be explained in the following section.

2.5 FREE OSCILLATIONS OF THE EARTH

Any elastic body can be set into characteristic oscillations by suitable disturbances, as for example a bell, a violin string or the column of air inside an organ pipe. The Earth, too, is elastic and can be set into its natural oscillations by a large earthquake, and may continue to vibrate or 'ring' for many hours, or even days.

The calculation of these frequencies is a formidable mathematical problem, and so this section attempts only to give a qualitative understanding of what is involved, starting with a general understanding of standing waves.

When any elastic body has waves of some sort propagated through it, or over its surface, then, in general, these waves will be reflected at the ends or boundaries of the body. This can be verified by watching water waves hitting a harbour wall or the end of the bath. For simplicity, if we consider parallel waves

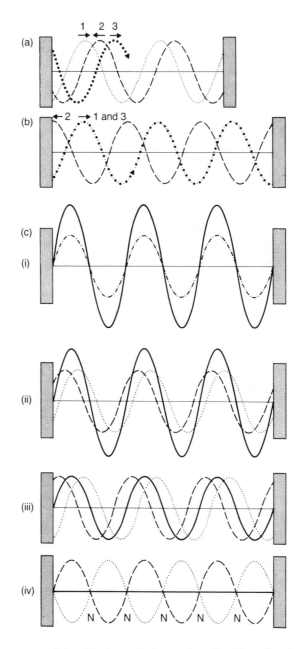

Figure 2.10 Standing waves. (a) A water wave travels outwards (1) and then is reflected back (2) by the right-hand wall. In general, after reflection at both walls, the wave (3) does not coincide exactly with the original wave, and so they partly cancel. This cancellation is complete after many reflections. (b) If an exact number of wavelengths fits into twice the distance between the walls, the doubly reflected wave coincides with the original one and enhances it. (c) These four diagrams show the addition of any wave and its reflection at different times, under condition (b). The amplitude of the combined waves (solid line) varies from nothing to twice that of either component wave, but the positions of its crests and troughs do not move, hence the term 'standing wave'.

moving across a straight-sided canal, then, after double reflection at the opposite sides in turn, they will be moving in the original direction (Fig. 2.10). They may then coincide with new waves being generated, and add together, in which case the wave amplitude will be increased, or they may be 'out of phase' and so partly cancel. The condition for reinforcement is that an exact number of wavelengths can be fitted into twice the width of the canal (Fig. 2.10).

If the condition for reinforcement is met, then there will be only two trains of waves, one moving across and the other returning in the opposite direction. At some instant their peaks will coincide, producing peaks of twice the height of either separately; between them will be troughs of double depth. A little later, the peaks and troughs of the two trains will have moved so that the peaks of the outward wave-train coincide with the troughs of the returning one; this gives

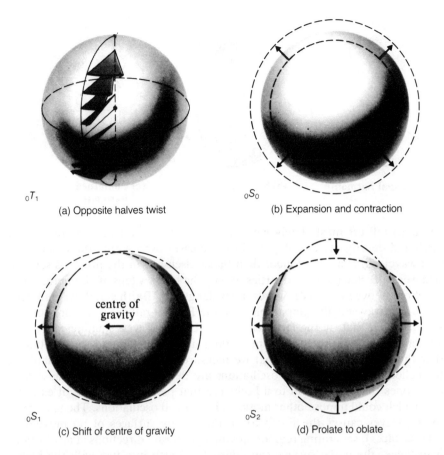

$_0T_1$

(a) Opposite halves twist

$_0S_0$

(b) Expansion and contraction

centre of gravity

$_0S_1$

(c) Shift of centre of gravity

$_0S_2$

(d) Prolate to oblate

Figure 2.11 Displacements and motion of some modes of oscillation. Movement may be entirely tangential to the Earth's surface, causing it to twist (toroidal modes, T), or predominantly or entirely radial (spheroidal modes, S). The subscripts refer to the number and type of nodal surface, following the terminology of spherical harmonic analysis, but will not be explained further. Note that the axis of symmetry relates to the position of the earthquake which excites the oscillation, and is not the Earth's axis of rotation.

complete cancellation, so momentarily the water will be at its undisturbed level. The same interval later, peaks will coincide again. Since the two wave trains are moving at the same speed but in opposite directions, the peaks of this set are in exactly the same position as those of the previous set. The peaks and troughs do not move along, but instead wax and wane in height, and therefore they are called **stationary** or **standing waves**. Between the points that wax and wane in amplitude are points that do not move at all, labelled N in Figure 2.10; these are nodal points.

The same reinforcement to form standing waves occurs in a two-dimensional body, such as a sheet of metal that is struck (e.g. a gong), except that the waves will reflect several times from different edges before regaining their original direction. Similarly, we can have standing waves in a three-dimensional body, which is why a bathroom singer finds certain notes enhanced, or resonant. In all cases, when an elastic body is struck, certain wavelengths persist because they reinforce themselves, whereas intermediate wavelengths die away. Many separate wavelengths can occur simultaneously, provided that they each meet the conditions for reinforcement. For example, wavelengths equal to twice the width of the canal, the width of the canal, one-half, one-third and one-quarter the width, will all reinforce and form standing waves. Normally, many will be found, though usually the longest ones predominate.

In the case of the Earth, surface waves, rather than body waves, are important in producing standing waves. The ones that most interest us are the very long ones, with wavelengths comparable to the radius of the Earth, because their

23

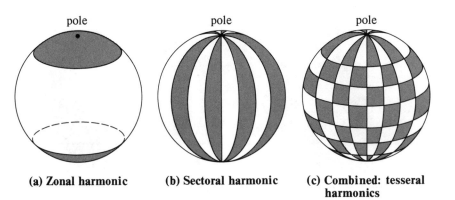

(a) Zonal harmonic **(b) Sectoral harmonic** **(c) Combined: tesseral harmonics**

Figure 2.12 Surface movements of some oscillations of the Earth. Light and dark areas have displacements which are in opposite senses at any instant, and are separated by nodal lines where there is never any movement (nodal lines are where nodal planes within the Earth intersect its surface). Movements corresponding with the left-hand diagram are shown in Figure 2.11 ($_0S_2$ mode).

amplitudes fall off most slowly with depth and so are affected by the elastic moduli and density at great depths, in some cases even within the inner core. These wavelengths are far longer than those discussed in the previous section, which were affected by the properties of only the top few tens of kilometres. The distinction between surface waves (wavelength $\ll$ Earth's radius) and free oscillations (wavelength comparable with the Earth's radius) is not only one of wavelength but that, in the latter case, we are dealing with standing waves.

In general, it is more convenient to think in terms of modes of oscillation, rather than wavelengths, and what we measure are the periods of oscillation, often called **eigenperiods**. The oscillations are described as 'free' because they are the ones that are natural to a body and that persist if it is free of external forces which could impose other motions, i.e. forced oscillations. The pattern of standing waves causes the Earth to be divided by surfaces of no movement (nodal surfaces), separating regions moving in opposite directions at any instant. For instance, the nodal surface can divide the Earth into two with the halves twisting in opposite senses (Fig. 2.11). The nodal surfaces are of several kinds:

Table 2.1 Periods of free oscillation (eigenperiods) of the Earth. This is a small selection of the modes observed. Errors are less than 0·1%. (From a compilation by Anderson and Hart, 1976.)

Spherical modes		Toroidal modes	
Mode	*Period* (min)	*Mode*	*Period* (min)
$_0S_0$	20·46		
$_0S_2$	53·83	$_0T_2$	43·94
$_0S_3$	35·56	$_0T_3$	28·37
$_0S_4$	25·76	$_0T_4$	21·72
$_0S_{10}$	9·67	$_0T_{10}$	10·31
$_0S_{20}$	5·792	$_0T_{20}$	5·993
$_0S_{40}$	3·538	$_0T_{40}$	3·333

The meaning of the symbols $_0S_0$ etc., is outlined in Figures 2.11 and 2.12.

concentric surfaces, diameters, and cones centred on the Earth's centre. Of course, we can only observe the motion of the surface of the Earth, and the cones and diameters intersect the Earth's surface like lines of latitude and longitude (Fig. 2.12). Motions of different modes may be predominantly radial i.e. up and down, or sideways, and are called spheroidal (S) and toroidal (T) oscillations respectively, corresponding to very long-period Rayleigh and Love waves. Figure 2.11 shows some of the slowest types of oscillations with few nodal lines. (The $_0S_1$ oscillation does not occur, for it would involve the *whole* Earth moving back and forth; for this to happen in an isolated body would be equivalent to picking oneself up by one's boot-laces!)

Table 2.1 gives the periods of the slowest oscillations. In general, the simpler the oscillation the longer the period ($_0S_0$ is an exception, for reasons that will not be discussed here) and hence the longer its equivalent wavelength, and therefore the more its period is affected by the physical properties at depth. It is because the different modes of oscillation 'penetrate' to different depths that they are a useful tool for investigating the Earth's interior. For instance, calculations show that the periods of these very long-period oscillations can only be accounted for if the inner core is solid.

In summary, the periods of oscillation of the Earth, equivalent to standing waves produced by long-wavelength surface waves, are determined by the variation of density and elastic moduli within the Earth. These are the same parameters that determine the body-wave velocities, but oscillations of the Earth are an independent way of investigating them, averaged over large volumes of the Earth, and so complement body-wave studies.

SUMMARY

1. Seismic body waves are studied chiefly by the time it takes them to travel through the Earth and re-emerge at different distances from their source, sometimes after one or more reflections or refractions at interfaces within the Earth. This information shows that to a very good approximation the Earth is spherically symmetrical (after allowance for the equatorial bulge).
2. Body-wave travel-time curves for P- and S-waves can be constructed and inverted to deduce the seismic velocities V_p and V_s at all depths. Major discontinuties define the boundaries between crust, mantle and core (divided into inner and outer cores), while lesser features are discontinuities in the mantle, the velocity inversion of the low-velocity zone (LVZ), and the D″ region, of reduced velocity gradient at the base of the mantle (Fig. 2.7).
3. Surface waves penetrate to a depth that increases with their wavelength, resulting in dispersion (variation of velocity with wavelength) when V_s varies with depth. Study of the dispersion yields average values of V_s that are independent of and complement body-wave results for shallow depths.
4. Free oscillations of the whole Earth – standing surface waves with wavelengths comparable to the radius of the Earth – depend upon the same physical parameters (compressibility and rigidity moduli, density) as do body-wave velocities, but are entirely independent. They are, therefore, a complementary source of information about the variation of these quantities within the Earth.
5. Although neither a knowledge of oscillation periods nor of body waves' velocities by themselves can be used to deduce density within the Earth, they are a vital constraint which, with other information, set close limits on the possible range of density at any depth. This will be the subject of the next chapter.

FURTHER READING

General books:

Fowler, C. M. R. (1990): contains a chapter on seismology and the Earth's interior.

Garland (1971): contains several chapters on aspects of seismology.

Gubbins (1990): an introduction to the theory of seismology, covering the topics of this chapter.

Press and Siever (1986): contains a chapter on seismology and the Earth's interior.

Advanced journals:

Kennett and Engdahl (1991): travel times.

Shearer (1991): mantle discontinuities.

3 The density within the Earth

3.1 INTRODUCTION

Seismology has told us much about the layered structure of the Earth, but little about the physical or chemical properties of the layers. Of course, the body wave seismic velocities depend upon the density and two elastic moduli, which, if known, would help determine the material, but because we have only two equations to solve for these three unknowns (Eqs (2.2)), it is not possible to deduce these quantities from the P- and S-wave velocities alone. The periods of free oscillation provide extra equations which, in principle, allow us to solve for ρ, κ and μ but, in practice, their use is limited, for reasons to be discussed in Section 3.6. To help solve this problem, we measure two other quantities which depend upon the density within the Earth, namely, the total mass and the moment of inertia.

3.2 THE MASS OF THE EARTH

We determine the Earth's mass, M_E, from the gravitational field that it produces. Newton's law of gravity states that the attractive force F between two point masses is:

$$F = G \frac{m_1 m_2}{r^2} \qquad (3.1)$$

where m_1 and m_2 are the two masses at a distance r apart, and G is the universal gravitational constant. If this equation is applied to all the particles in a perfectly spherical shell, it can be shown that the attraction the shell exerts *outside* itself is the same as if all its mass were concentrated at its centre. Thus, a body that consists of concentric layers each of uniform density – as does the Earth, to a good approximation – has an external attraction which is symmetrical and falls off as the inverse square of the distance from its centre. It is convenient to measure this attraction at any point by the force it would exert upon a unit mass at that point. If we put m_1 equal to unity and m_2 equal to M_E, the mass of the Earth, the force on the unit mass due to the Earth is:

$$F = G \frac{M_E}{r^2} .$$

A force accelerates a mass, m, according to $F = m \times$ acceleration, so the unit mass, if allowed to fall freely, would experience an acceleration g_r:

$$g_r = G \frac{M_E}{r^2} . \qquad (3.2)$$

(When referring to this acceleration at the surface of the Earth, the suffix r will be omitted.) One method of determining M_E is to measure g at the surface of the Earth, by timing the free fall of a mass in a vacuum. The radius of the Earth, R_E, 27

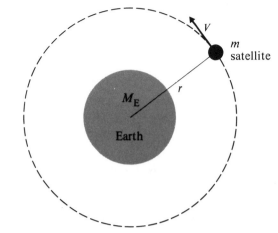

is determined by surveying. The constant G can be determined in the laboratory by measuring the tiny force of attraction between two masses. This was first done by Cavendish in 1798, who went on to calculate the mass of the Earth.

A second method of determining M_E is to measure the period of rotation of a satellite about the Earth. If, for simplicity, we suppose that it is in a circular orbit of radius r (Fig. 3.1), the centrifugal and gravitational forces are always equal and opposite,

$$G \frac{M_E m}{r^2} = \frac{mV^2}{r} \tag{3.3}$$

where m is the mass of the satellite and V is its speed. The mass m 'cancels', and since the time T that the satellite takes to make one revolution is:

$$T = \frac{2\pi r}{V} \tag{3.4}$$

the velocity V in Eq. (3.3) can be replaced and the equation rearranged to give:

$$M_E = \frac{r^3}{G} \left(\frac{2\pi}{T}\right)^2. \tag{3.5}$$

The period T is easily observed, and r can be measured by radar or by laser. (Eq. (3.5) still holds for the more general case of an elliptical orbit. It is also useful for finding the masses of planets that have satellites.)

The value found for M_E is $5 \cdot 98 \times 10^{21}$ Mg. Since we know the Earth's radius, we can calculate its volume, and we obtain a mean density of $5 \cdot 52$ Mg m^{-3}. As the densities of most near-surface rocks are less than $3 \cdot 2$ Mg m^{-3} ($3 \cdot 2$ times as dense as water), it follows that the inner parts of the Earth must be much denser than the outer parts. But because any arrangement of concentric shells of different densities, *having the same total mass*, produces the same external gravitational field, it is not possible to deduce from the gravitational field how the density varies radially within a spherical body. Therefore we turn to other constraints. (To the extent that the Earth is not perfectly symmetrical, it is possible to learn something of its interior inhomogeneities from the variation of gravity over the surface of the Earth. On a small scale, this can be used to interpret geological structures – a subject outside the remit of this book – and on a global scale will be discussed in Section 8.7. Nevertheless, here we can neglect the variations in g, since they are less than $0 \cdot 1\%$.)

3.3 THE MOMENT OF INERTIA OF THE EARTH

The Earth's **moment of inertia** (see Note 1 for an explanation of this and related terms) is another quantity that depends upon the mass in its interior, but unlike M_E, the moment of inertia depends upon the distribution of mass.

In the laboratory, the moment of inertia of a body can be determined by measuring the angular acceleration produced by a known torque, or couple; the greater the moment of inertia, the smaller the acceleration. In the case of the Earth we have to rely on naturally occurring torques, and these are produced by external bodies, principally the Moon and Sun. Since these exert a torque only because the Earth is not quite spherical, we need to consider the shape of the Earth. The main deviation of the Earth from sphericity is the equatorial bulge, which results from its rotation. Its cross-section is close to an ellipse so that an ellipsoid of revolution (the shape swept out by rotating an ellipse about a diameter) can be fitted to the Earth, with discrepancies of no more than 100 metres (the significance of these discrepancies will be discussed in Section 8.7). The ellipticity itself is small, only one part in 298, with the equatorial radius of 6387 km being about 22 km more than the polar one.

The ellipticity can be found without viewing the Earth from space. Suppose you travelled from the equator to the pole, measuring the angle between the local vertical defined by a plumb-bob and some fixed astronomical direction, such as the position of the distant Pole Star (Fig. 3.2), the angle between them would steadily decrease. The distance travelled for each degree of change of this angle would increase slightly, because the curvature of the Earth is less near the poles than near the equator.

Because of its equatorial bulge, we can think of the Earth as consisting of a sphere with a belt of extra material round its equator. Further, the belt can be thought of as made up of many pairs of masses at opposite ends of a diameter, such as m_1 and m_2 in Figure 3.3. Next, consider a satellite in a circular orbit about the Earth. It is kept in its orbit by the balance of the centrifugal force and

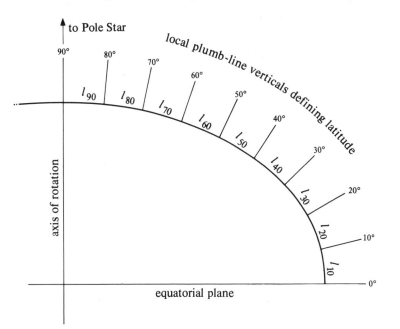

Figure 3.2 Determination of the shape of the Earth. As the Earth is slightly flattened, the lines of latitude – deduced from the angle between the local vertical and the Pole Star or other convenient star – are not an equal distance apart, i.e. $l_{90} > l_{80} > l_{70}$, etc. Note that verticals in general do not point to the centre of the Earth. The cross-section is very close to an ellipse.

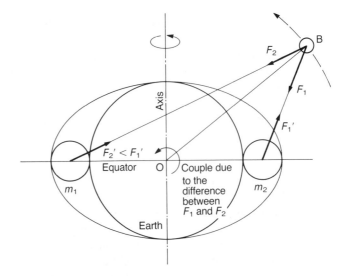

Figure 3.3 Torque on the Earth's equatorial bulge (much exaggerated). For details see text.

the gravitational force of attraction acting along the line of centres of the two masses. Although the equatorial bulge contributes to this attraction, it also produces a torque: this is because the forces F_1 and F_2 due to m_1 and m_2 are slightly unequal in size and also at a slight angle, because of the different positions of m_1 and m_2 with respect to the satellite.

From Newton's third law of motion, that action and reaction are equal and opposite, there must be corresponding forces F_1 and F_2 acting upon the Earth; because they are unequal they will produce a small counter-clockwise torque, as shown. If this argument is repeated for all the pairs of masses that make up the equatorial bulge, it is seen that they add to produce a net torque. Note that the satellite is not in a symmetrical position, i.e. not in the equatorial plane or on the Earth's axis; if it were, then the torque would be zero.

In the case of an artificial satellite the torque upon the Earth is negligible, but significant torques are produced by the Moon and, to a lesser extent, the Sun (though the Sun is far more massive, its extra distance more than outweighs this).

It would seem obvious that the effect of the torque of the Moon (or Sun) upon the Earth's equatorial bulge would be to move both bodies until the Moon lay in the Earth's equatorial plane, where the torque is zero. This, indeed, would be the case if the Earth were not rotating, but because it is rotating it behaves like a spinning top, or one of those toy gyroscopes that wobble around on their stands, apparently in defiance of gravity. In Figure 3.4, the weight of a top and the reaction at its tip produce a torque that would cause the top, if *not* rotating, to fall on its side. But we know that its axis traces out a cone; similarly, because of the torque due to the Moon, the Earth's axis is not fixed in space but precesses slowly, with the axis of the cone perpendicular to the plane of the Earth's orbit and passing through the centre of the Earth. The rate is small because of the smallness of the bulge and the great distance of the Moon, and it takes about 26 000 years to complete one revolution. This can be detected because, though the Earth's axis at present points almost at the Pole Star (Fig. 3.5), its direction is slowly changing.

If the above considerations are followed through quantitatively, then a mathematical expression for the rate of precession can be derived (see e.g. Garland, 1971; Stacey, 1977). It contains quantities such as the mass and

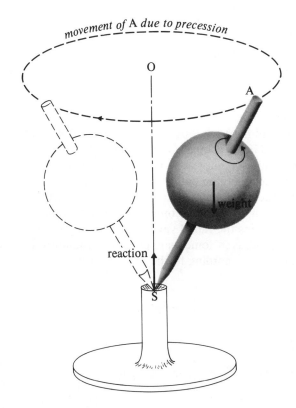

movement of A *due to precession*

O

A

weight

reaction

S

Figure 3.4 Precessing top. Due to the torque formed by weight and reaction, the top precesses, its axis moving in an inverted cone whose axis is OS. The direction of precession is opposite to the spin of the top.

Figure 3.5 Time exposure of the night sky. In the course of several hours, each star traces an arc about a point. This point is the celestial pole and lies on the Earth's axis of rotation, when extended. It is, therefore, exactly above the North Pole. The Pole Star is close to the celestial pole and so traces out only a very small arc.

distance of the Moon (or other body), plus two additional terms. One of these terms shows how the torque, and hence rate of precession, depends upon the deviation of the Earth's gravitational field from perfect symmetry, and we know this from the motions of satellites. The other term shows that the rate of precession is inversely proportional to C, the Earth's moment of inertia about its rotation axis. Thus, the Earth's moment of inertia can be calculated from its measured rate of precession.

The Earth's moment of inertia is 8.07×10^{34} Mg m^{-2}. The significance of this value can be appreciated if it is compared to that of an Earth of uniform density, with the same radius and total mass. The measured value is only 83% of such a uniform Earth. To reduce the moment of inertia of a sphere, while maintaining its mass and radius, the mass must be concentrated towards the centre, i.e. density must increase with depth. This confirms the result of the average density (Section 3.2), but constrains the radial variation of density more specifically. If the density varies with radius in some way, $\rho(r)$, then the mass and moment of inertia will depend upon it according to the expressions:

$$Mass \qquad M_E = \int_0^{R_E} 4\pi r^2 \rho(r) dr. \qquad (3.6)$$

$$Moment\ of\ inertia: \qquad C_E = \int_0^{R_E} 4\pi r^2 \rho(r)\frac{r^2}{3} dr. \qquad (3.7)$$

Thus, though neither measurement can define the radial variation of density, $\rho(r)$, each is able to rule out many distributions, and together they rule out more than they do separately. They are, therefore, constraints on the density variation, and will be used in the next section.

3.4 THE SIMPLE SELF-COMPRESSION MODEL: THE ADAMS–WILLIAMSON EQUATION

In this section we describe an early method used to estimate the density–depth profile (or radial variation of density). Although it does not give a fully correct solution, it is instructive, and valid for parts of the Earth, and is still useful. We start with the equations for the velocities of P- and S-waves met earlier in Eq. (2.2), namely:

$$V_p = \sqrt{\frac{\kappa + 4/3\mu}{\rho}} \qquad V_s = \sqrt{\frac{\mu}{\rho}}. \qquad (3.8)$$

These cannot be solved for κ, μ and ρ because there are three unknowns but only two equations. A third independent relationship between the variables is needed but, unfortunately, no rigorous one exists. Early attempts to deduce the density tried to get round this obstacle either by assuming a relationship or by trying to discover one empirically.

In 1923, Adams and Williamson (Williamson and Adams, 1923) suggested that density increases with depth only because of compression due to the weight of material above and not, for instance, because of a change of composition. The

32

quantity that defines the change of volume, and hence density, due to pressure is the compressibility modulus, κ (Section 2.1), given by:

$$\kappa = \frac{\text{compressional stress}}{\text{volumetric strain}} = \frac{\text{increase in pressure}}{\text{resulting proportional decrease in volume}} = \frac{dP}{dv/v}. \quad (3.9)$$

Since volume and density are inversely related, we can write:

$$\frac{d\rho}{\rho} = -\frac{dv}{v}. \tag{3.10}$$

(e.g. decreasing the volume of a given mass of material by 1% increases the density by 1%). Therefore, the compressibility modulus κ is:

$$\kappa = -\rho \frac{dP}{d\rho}. \tag{3.11}$$

The extra pressure on descending through a spherical shell of thickness dr (Fig. 3.6), due to the extra weight above, is:

$$dP = -\rho_r g_r dr \tag{3.12}$$

(the negative sign is because P increases as r decreases). Of course, ρ_r and g_r are not absolutely uniform through the thickness of the shell, but they change much less rapidly than P; by making the shell infinitesimally thick (the calculus limit), the error in assuming ρ_r and g_r to be uniform can be made negligibly small. The density increase, produced by the increase in pressure, is found by substituting Eq. (3.12) into Eq. (3.11). The result is rearranged to give:

$$\left(\frac{\kappa}{\rho}\right)_r = g_r \rho_r \frac{dr}{d\rho}. \tag{3.13}$$

The seismic velocity information in Eq. (3.8) has not yet been used. These two equations can be combined to give:

$$V_p^2 - \frac{4}{3} V_s^2 = \kappa/\rho \tag{3.14}$$

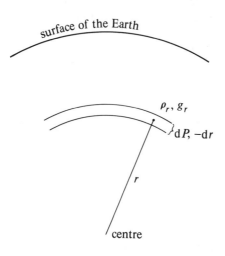

surface of the Earth

ρ_r, g_r

$dP, -dr$

r

centre

Figure 3.6 Pressure and density within the Earth. Increasing depth by dr increases pressure by dP due to the extra weight above. See text for further details.

33

(μ has been eliminated). Hence Eq. (3.13) can be rewritten as

$$(V_p^2 - 4/3V_s^2)_r = g_r\rho_r \frac{dr}{d\rho}.$$
(3.15)

The suffix r denotes that the values of V_p, V_s, etc. are those occurring at the radius r, for it should be appreciated that the elastic moduli, as well as the density, increase with depth. This must be so because a density increase alone would produce a *decrease* of V_p and V_s with depth (see Eqs (3.8)), whereas we know they increase. It may seem a contradiction that the elastic moduli, usually regarded as constants, can vary with depth. In fact, the ratio of stress to strain is constant only provided the *variation* of stress is relatively small, such as is encountered in engineering. It is not constant over the huge range of pressures within the Earth. However, provided that the variation remains small, the ratio of stress to strain will be constant even at very large pressures, but the constant of proportionality, i.e. the modulus, will depend upon the value of the pressure. Thus Eq. 3.8. for instance, will still hold.

Equation (3.15) still contains g_r which is unknown, and so must be eliminated. g_r depends upon the radial density variation. We have already seen that *outside* a shell the gravitational attraction is the same as if its mass were concentrated at its centre. It can be shown from Eq. (3.1) that anywhere *inside* the shell there is no attraction. Thus, the acceleration due to gravity at a radius r is the same as if all the mass inside r were concentrated at the centre, the mass outside being ignored, i.e.

$$g_r = \frac{G}{r^2} \times \text{(sum of masses of all shells inside } r\text{)}$$

or:

$$g_r = \frac{G}{r^2} \int_{r_1 = 0}^{r_1 = r} 4\pi\rho_1 r_1^2 dr_1$$
(3.16)

where $4\pi r_1^2 dr_1$ is the volume of a shell at distance r_1 from the Earth's core, and ρ_1 is the density at the same distance; these are summed from the centre out to r.

Finally, substituting Eq. (3.16) into (3.15) and rearranging gives:

$$\frac{d\rho}{dr} = \frac{G}{r^2} \frac{\rho_r}{(V_p^2 - \frac{4}{3}V_s^2)_r} \int_{r_1 = 0}^{r_1 = r} 4\pi\rho_1 r_1^2 dr_1.$$
(3.17)

This untidy expression is the **Adams–Williamson equation**. The left-hand side is the rate of change of density with depth, at some particular radius, r, which, as the right-hand side shows, relates to the seismic velocities and the value of density at all depths within the radius r.

How does it help us? First, we see that since V_p and V_s are known at all values of r (Chapter 2), the only unknown is how ρ varies with r. Secondly, at the surface of the Earth, we also know ρ, r and the integral, which is just the mass of the Earth (Eq. 3.6). Therefore, we can evaluate the right-hand side of the equation *at the surface of the Earth*. Then, since the left-hand side gives the rate at which density increases with depth, we can calculate the density at a small depth below the surface, i.e. at the base of a thin surface shell. The right-hand side can now be evaluated at the new, slightly smaller, value of r since the integral will now equal the mass of the Earth less the mass of the shell, whose density we know. And so on, for progressively greater depths.

In practice, the integration is begun at the top of the mantle, because the crust is a layer of variable thickness and density (its mass is allowed for). A density of about 3.2 Mg m^{-3} is chosen, based upon samples derived from the mantle, and then the density is deduced at all depths down to the base of the mantle, beyond which it would be absurd to go because clearly a major change occurs there that cannot be due to a simple compression (we shall come to the other discontinuities of Fig. 2.7 in due course). How, then, can the core density be deduced? This is done by guessing a density for the top of the core and then using the Adams–Williamson equation to deduce the density down the centre of the Earth. But this density profile must be such that the total mass, adding together crust, mantle and core, is equal to the known mass of the Earth (this is the constraint provided by our knowledge of the mass of the Earth, Section 3.2); the density of the top of the core is adjusted until this is so. The result is shown in Figure 3.7.

To check whether this density distribution is correct, it is used to calculate the moment of inertia of the Earth, which is compared with the known value (this is the second constraint, Section 3.3). They are found to differ significantly. It can be shown that the discrepancy is not within the core, as follows. The density distribution of the mantle (plus crust) is used to calculate both the mass and moment of inertia of the mantle, and, thence by subtracting them from the known values for the whole Earth, the values for the core. It turns out that the resulting ratio of the moment of inertia of the core to its mass is 1.4 times that of a uniform sphere, which would imply that the mass of the core is concentrated towards its surface. As it is highly implausible that the density of the core markedly *decreases* inwards, the only conclusion is that there must be more mass in the mantle than the self-compression model predicts.

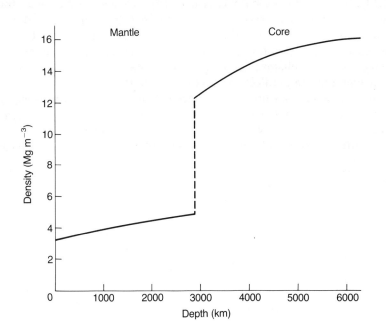

Figure 3.7 Density–depth profile in the Earth, according to the self-compression model. It is assumed that the Adams–Williamson equation holds within both mantle and core, but not across the core–mantle boundary. (After Bullard, 1957.)

The possible reasons for the mantle not obeying the Adams–Williamson equation will be discussed in the following section.

3.5   DEFECTS OF THE SELF-COMPRESSION MODEL

As something is wrong with the model applied to the mantle, the assumptions upon which it is based must be examined carefully. There are several assumptions concealed within the derivation of the Adams–Williamson equation. One assumption is that the pressure at a given depth is due simply to the weight of material above. Clearly, this is not true in a cave, because the strength of the rock supports the weight above, but at a depth of only a few kilometres the weight becomes too great for empty cavities to exist. In fact, the strength of the whole Earth with respect to *long-continued* forces is negligible on a global scale, and the assumption that it behaves as a liquid, implicit in Eq. (3.12), is closely correct on this time scale (Section 8.3). Corrections for departures from spherical symmetry due to inhomogeneities in the densities of rocks and the equatorial bulge also are minor.

Another factor is temperature. Temperature does not appear explicitly in the Adams–Williamson equation, but if the temperature rises with depth, it will act in the opposite sense to pressure, tending to expand the material and so reduce its density. It might, therefore, be thought that the Adams–Williamson equation implies a constant temperature, but this is not so because of the compressibility modulus used. There are two major definitions of this modulus, depending upon what happens to the heat that is produced when a material is compressed. In the case of the isothermal modulus, this heat is removed so that the temperature remains constant; in the adiabatic modulus the heat remains in the material, and as heat tends to expand the material, a greater pressure is required to produce a given compression; therefore, this compressibility modulus is the greater. The compressibility modulus used in the Adams–Williamson equation derives, via Eq. (3.14), from the equation for V_p (Eq. (3.8)). When a P-wave passes through a material, a compression lasts so short a time that the heat produced does not have time to exchange with the surrounding material before the ensuing rarefaction cools it. Therefore, the appropriate modulus is the adiabatic one, and so the Adams–Williamson equation implies that inside the Earth there is an **adiabatic temperature gradient,** or **adiabat** for short.

The adiabat is such an important concept that it is necessary to explain its meaning at some length. Consider a compressible liquid in which both temperature and pressure increase with depth. If the two equal small masses δm_1 and δm_2 in Figure 3.8 were to be exchanged instantaneously in some magical way

Figure 3.8 Adiabatic temperature gradient. Two equal masses δm_1, δm_2 initially at the locations shown, are instantaneously interchanged in a hypothetical experiment. For further details see text.

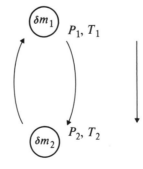

36

without change of temperature or pressure δm_2 initially would be at both a higher temperature and a higher pressure than its new surroundings (and vice versa for δm_1). If then, the pressure were allowed to equilibrate instantaneously with the new surroundings, δm_2 would expand, and, in so doing, cool. If, after pressure equilibration, the temperature were equal to that of the surroundings, the temperature gradient is adiabatic. If δm_2 were hotter than its new surroundings, the gradient is superadiabatic, and if cooler subadiabatic. In the superadiabatic case, the excess temperature after pressure equilibration would mean that the material would expand, to a density below that of its surroundings, so giving it an upward or positive buoyancy. On a large scale, a superadiabatic temperature gradient in a liquid tends to lead to convection, because if any small disturbance in the liquid displaces material upwards, the resulting buoyancy will tend to cause it to rise further. This result will be used in the sections on convection, in Chapters 6 and 8.

If temperature were the cause of the density in the mantle increasing with depth more rapidly than predicted by the Adams–Williamson equation, it would require a subadiabatic temperature gradient, since cooling generally causes contraction and hence an increase in density. But the gradient is almost certainly at least slightly superadiabatic (Section 8.5) everywhere in the mantle, so temperature variations cannot be the explanation.

This leaves only two other factors: compositional changes or changes of state. Since materials of different composition can have different densities at the same pressure, it is evident that the Adams–Williamson equation cannot take into account a change of material within the Earth, either abrupt or gradational. Changes of state include the change from liquid to solid, which has been deduced to occur at the core–mantle boundary (Section 2.3), but also include **phase changes**, in which a solid material changes its crystalline structure by a repositioning of its atoms. An example is the change of pure carbon from graphite, with a density of 2 Mg m^{-3}, to diamond, density $3\cdot5$ Mg m^{-3}, at high pressures. The increase of pressure with depth in the Earth will favour more compact forms and hence higher densities of the common silicate minerals of which the Earth is made (see Chapter 7 for details). Since crystals exist only as regular arrays of atoms, the transition from one form to another must be abrupt, at least on a local scale, and such changes are outside the scope of the Adams–Williamson equation, for use of the bulk modulus assumes that a smooth increase in pressure produces a smooth decrease in volume. The simple self-compression model therefore, in its essentials, assumes a homogeneous, liquid Earth with an adiabatic temperature gradient. The extent to which the departures of the mantle from the self-compression model are due to phase changes or to changes of composition will be discussed in Chapters 7 and 9.

Although the Adams–Williamson equation is not rigorously applicable, there are regions where it applies fairly well. Density variation in the outer core is deduced using this approach because, being liquid and probably in convective motion, the material should be well mixed and with a closely adiabatic temperature gradient (see Chapter 6). The equation probably also holds fairly well throughout most of the mantle between phase changes. Where the equation does not hold, the density–depth gradient that it predicts will generally be less than the actual one, because both compositional and phase changes tend to result in a more rapid increase of density with depth, and so it provides a lower limit on the density. The only likely exceptions to this are regions of very high temperature gradient, which are most likely to occur at the top and base of the mantle, leading to lower density gradients. Thus the result that the Adams–Williamson equation deduced too small a mass in the mantle is accounted for mainly by the presence of phase changes in the upper parts of the mantle, each of which adds a discontinuous jump in density, superimposed on the smooth increase predicted by the equation; however, compositional changes are not ruled out (Section 7.6). Together they account for more of the Earth's mass being in the mantle than is deduced by the Adams–Williamson equation.

3.6 MORE SOPHISTICATED MODELS AND A CONTEMPORARY RESULT FOR THE EARTH'S DENSITY VARIATION

Following the realization that the simple self-compression model has only limited applicability, other attempts were made to find a third relationship linking density and elastic moduli, or, alternatively, seismic velocities. From experiment, Birch (1961) proposed the relation:

$$V_p = a\bar{m} + b\rho \tag{3.18}$$

where a and b are empirically determined constants and $\bar{m}$ is the **mean atomic weight** of a mineral, i.e. the sum of the atomic weights of all the atoms in a mineral, divided by the number of atoms. The physical reasons underlying this law are not fully understood, and there are doubts whether it holds at the high pressure of the deep mantle. (For further details, see Note 2 and Section 7.5.)

Another approach used the assumption, on the basis of the limited information available, that the compressibility modulus changed in a particular way with depth. This led to a family of models by Bullen (1963) which steadily evolved over the years to take account of improved data, and which had wide use. A further approach was to deduce the density from proposed petrological models of the Earth (this is the reverse of the course adopted in this book). Other models were a patchwork with different assumptions and sources of data for different depths. Although such models have the merit of trying to use the best available information, the choice is subjective and the resulting model may not be consistent with all data relevant to all depths.

The periods of free oscillations of the Earth offer a way around this problem, since the period of each mode depends upon ρ, μ and κ. Matching a finite number of modes does not give a unique solution, but the more modes matched, the more tightly the radial variations in these variables are constrained. For instance, the period of the $_0S_0$ (the fundamental spherical) mode could be matched by a uniform Earth, but to model the $_0S_2$ as well would require at least two layers with different values of density and the elastic moduli, and so on. Thus by considering many modes a multi-layer model can be deduced. However, because the periods of the modes are not sensitive to small-scale variations in the way that body-wave travel times are, this method does not define well the positions of the boundaries between regions.

To combine the huge amount of information available on arrival times of P- and S-waves, due to many earthquakes and received at many different epicentral angles, plus surface-wave dispersion data and free oscillation periods, requires the power of modern computers. This is done using **forward modelling** as already described for the velocity–depth profiles (Section 2.2); that is, a structure of the Earth is guessed and used to calculate observable quantities, such as travel times and free oscillation periods; these calculated values are compared with the actual values and the model is modified iteratively to produce a good agreement. The starting point is based on existing models, as described above.

Like the iasp91 velocity–depth profiles described in Section 2.2, the values so deduced are **reference models**; that is, they specify the *spherically symmetrical* Earth (after allowing for the equatorial bulge) that best fits the observations, with lateral variations averaged out.

One of the best known of the recent reference models is **PREM**, the **P**reliminary **R**eference **E**arth **M**odel (Dziewonski and Anderson, 1981). The variables 'guessed' were V_p, V_s, ρ, q_κ and q_μ (the significance of the last two quantities will be discussed shortly).

A problem with this approach is how to specify the variables at all depths. If, for instance, values were assigned independently at every kilometre of depth, the resulting computation would be far beyond the power of even modern computers. Instead, the radial variation is partly specified by simple mathematical functions. The Earth is divided into a number of radial divisions, such as inner and outer core, and between the regions there is either a jump in the value of the variable (first-order discontinuity) or a change in the rate of variation of the variable with depth (second-order discontinuity). Within each region the variable changes smoothly according to a polynomial equation of the form:

$$\text{value of variable at } r = c_0 + c_1 r + c_2 r^2 + c_3 r^3 \qquad (3.19)$$

where r is the distance from the Earth's centre. The values of the coefficients have to be found for each of the variables and many turn out to be zero. This way of describing how the value of a variable depends upon depth in the Earth is called parameterization.

The observed data that the model has to satisfy are values for about two million V_p and a quarter of a million V_s travel times, resulting from 26 000 earthquakes or explosions, about 900 periods of free oscillation, data on surface wave dispersion, and also the mass, radius and moment of inertia of the Earth. The process starts with an existing model and then all variables are systematically changed, including the positions of the discontinuities and the coefficients of the polynomials for V_p, V_s, ρ, q_μ, and q_κ, to find the combination that best matches the observed data.

In 1989, PREM was revised using an new data set for free oscillation periods (Montagner and Anderson, 1989), resulting in some changes in the model for the upper mantle. The results are shown in Figure 3.9. In addition to V_p and V_s, already met in Chapter 2, the density, pressure and gravity, g, are shown.

The main result, so far as we are concerned, is the density–depth profile. This does not show any features not already indicated by the velocity–depth curves, but density is a more useful property to know when we come to discussing the composition of various parts of the Earth, starting in Chapter 6. In particular, the increases in density at the various discontinuities are a considerable constraint upon composition, for any postulated composition profile has to match, not only the densities at any selected depth, but also the abrupt increases. It is now possible to submit samples in the laboratory to pressures and temperatures corresponding with most depths in the Earth, and so to test whether a postulated composition matches the density–depth profile.

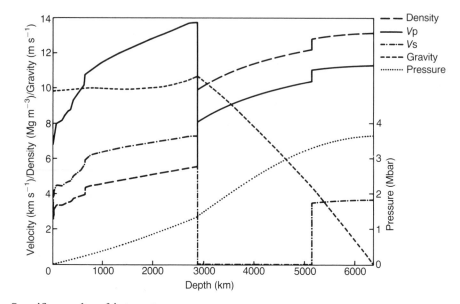

Figure 3.9 Density (ρ), V_p, V_s and gravity (g) according to PREM (Preliminary Reference Earth Model, Dziewonski and Anderson, 1981); density in top 1000 km modified according to Montagner and Anderson, 1989). (Figure drawn from tabulated values.)

Specific results of interest are:

1. There is decrease in velocity of V_s but not V_p in the top 200 km; i.e. there is a low-velocity zone for S-waves, but probably not universally so for P-waves.

2. The topmost mantle is anisotropic in seismic velocity, and requires that both P- and S-waves travel faster in a horizontal direction than in the vertical, up to about 2% for P-waves and 6% for S-waves (because it is a reference model, assuming spherical symmetry, any variations of velocity in different horizontal directions are averaged out). Anisotropy extends down to at least 200 km and probably, though to a lesser extent, to 410 km. The existence of anisotropy was detected by the discrepancies found between models based separately upon Rayleigh and Love waves; in Section 2.3 it was explained that the particle motion of Rayleigh waves is in the vertical plane whereas Love wave particle motion is entirely horizontal, and so the two waves respond differently when there is anisotropy.

3. There is significant attenuation of seismic energy, particularly of S-waves, in the LVZ, the inner core and – to a lesser extent – in the mantle transition zone, 410–660 km. (Attenuation is total, of course, for S-waves in the liquid outer core.) Attenuation is detectable because it increases the periods of free oscillations, just as a pendulum swinging with its tip in water would have a longer period than one swinging in air. The other effect of the water, of course, would be to decrease progressively the amplitude of the swings. The attenuation is measured by the fraction of the energy of the swing lost at each oscillation, q. Similarly, seismic waves decrease in amplitude as they pass through an attenuating medium, and q_μ and q_κ refer to S- and P-waves respectively. (The reciprocal of q is Q the quality factor, widely used in electrical circuits, for instance.) Attenuation is likely to be high if a material is close to its melting temperature.

4. Finally, the regions where PREM deviates from the density–depth gradient predicted by the Adams–Williamson equation can be deduced. By combining Eqs (3.11), (3.12) and (3.14) from the derivation of the Adams–Williamson equation, a new parameter η_B can be defined and shown to equal 1:

40

$$\eta_B = \frac{d\kappa}{dP} + \frac{1}{g}\frac{d}{dr}\left(V_p^2 - \frac{4}{3}V_s^2\right) = 1.$$

The term after the first equals sign is the dependence of the bulk modulus upon pressure (and hence depth), while the second is the radial (and hence depth) dependence of an expression involving the seismic body-wave velocities, divided by the acceleration due to gravity. But η_B equals unity only if the Adams–Williamson equation is valid, and this can be tested because all the terms in the expression for η_B can be evaluated directly or indirectly from the values of the variables selected by PREM. It is found that η_B is close to unity throughout most of the core and lower mantle, while in the upper mantle it is also close to unity between the discontinuities, except for the low-velocity zone.

Although PREM is an advanced model, with anisotropy and attenuation included, it is important to appreciate its limitations. Firstly, the parameterization of variables is a simplification, and small but significant departures may well occur. To study particular features in detail – such as the sharpness of the 660 km discontinuity – it is still necessary to use separate body-wave studies. Secondly, lateral inhomogeneities or variation of anisotropy within the horizontal direction have been averaged to give a model that varies only radially. However, the fact that such a good fit to observed data can be achieved with a radial model is confirmation that the Earth to a very good approximation is concentrically layered. We have therefore a good estimate of how density varies with depth in the Earth and this is a vital step to deducing the composition.

SUMMARY

1. The measurement of g, the acceleration due to gravity, shows that the average density of the Earth is $5{\cdot}52$ Mg m^{-3}. Since near-surface rocks seldom exceed a density of $3{\cdot}2$ Mg m^{-3}, this shows that the density must be much greater at depth, but a knowledge of g alone cannot constrain the radial density distribution further.
2. The moment of inertia of the Earth, determined from the precession of the equinoxes, also shows that density increases with depth in the Earth. As moment of inertia depends upon the radial distribution of density, and not only upon the total mass, the result provides an additional constraint on the density–depth variation.
3. The *two* seismic velocities V_p and V_s are known with considerable precision at most depths in the Earth (as was shown in the previous chapter), but because they depend upon *three* variables, the density ρ and the two elastic moduli κ and μ, it is not possible to deduce how density varies with depth, in the absence of a third relationship.
4. An attempt to provide a third relationship, by making the simplifying assumption that density increases with depth only because of the compression due to the weight of the overlying layers, leads to the Adams–Williamson equation. The density–depth profile it predicts fails to satisfy the constraints of the mass and moment of inertia even after allowing for a density jump at the core–mantle boundary.
5. The most likely causes for this failure are phase and/or compositional changes – leading to abrupt increases in density at the depth where there are velocity

41

discontinuities. Departure from an adiabatic temperature gradient is very unlikely to contribute to the failure.

6. The addition of surface-wave data and periods of free oscillations of the Earth offer further constraints on the density–depth distribution. By using powerful computers to generate radial models that can be tested against a large number of observations of travel times, surface-wave data and periods of free oscillations, very well-constrained density–depth profiles can be deduced.

7. The most comprehensive of such models is PREM (Preliminary Reference Earth Model), with later modifications for the upper mantle. It describes the Earth in terms of a number of shells separated by – from the surface downwards – the Moho, the low-velocity zone, upper mantle discontinuities (at about 410, 520 and 660 km depth), the D″ layer at the base of the mantle, the core–mantle boundary and the inner core boundary. Within each shell the density follows a smooth curve given by an equation.

8. Additionally, PREM has been so constrained by the observational data that it has to incorporate anisotropy in at least the top 200 km, plus attenuation at various depth ranges.

9. PREM is a *reference model* because it has only radial variations of density, etc.; it is the spherically symmetrical model that best matches the observations.

FURTHER READING

Advanced journals:
Dziewonski and Anderson (1981); Montagner and Anderson (1989): current methods for deducing density, etc. with depth in the Earth.

4 The formation of the Solar System and the abundances of the elements

4.1 WHY IT IS NECESSARY TO LOOK OUTSIDE THE EARTH

In the two previous chapters, the density and other physical quantities within the Earth were deduced with considerable precision. Next, we want to know what sort of material is there, i.e. what is the chemical composition of the Earth's interior, and of what minerals is it made.

It has already been established that the Earth is layered, that the density of crustal rocks is much less than the density at depth, and that some of the boundaries within the Earth, such as the core–mantle boundary, are probably compositional. Therefore, though we must take into account the composition of the surface rocks, they cannot be taken as representative of the mantle and core. Admittedly, some of the crustal rocks have derived directly from the mantle, such as the basalts of the ocean floors, but the process of partial melting that produces them (Section 7.3) ensures that their composition is different from that of their source region. In any case, they come from relatively shallow depths. For these reasons, we need more than samples of crustal rocks to deduce the composition at depth.

We are forced to ask from what materials the Earth formed, but to answer this we have to determine how the Earth itself formed. We believe that it formed together with the rest of the Solar System – the Sun, planets and lesser bodies – which pushes the enquiry back to how the Solar System formed, and from what materials. This is inextricably linked to the formation of stars from clouds of gas and dust. Thus we have to embark on a widening investigation, outward in space and back in time.

This and the next chapter are concerned with establishing the bulk composition of the original gas-and-dust cloud, and the processes that formed it into the Solar System and particularly into the Earth. This chapter is about the physical processes that operated, while the next chapter examines the associated chemical processes.

4.2 INTRODUCING THE SOLAR SYSTEM AND THE GALAXY

4.2.1 The Solar System

The Solar System consists of the Sun, and all the bodies that orbit about it, of which the most massive are the planets. Table 4.1 lists many of their properties, while Figure 4.1 shows the relative sizes of the planets and their orbits.

Most of the mass of the Solar System is in the Sun, nearly 99·9% of it, with the largest planet, Jupiter, having little more than a thousandth of the Sun's

Table 4.1 Chief properties of the Sun, the planets and the Moon.

Line	Property	Sun	Terrestrial planets					Major planets				
			Mercury	Venus	Earth	(Moon)	Mars	Jupiter	Saturn	Uranus	Neptune	Pluto
(a)	Distance from Sun (mean value) (units of 10^6 km)	—	58	108	150	—	228	778	1427	2870	4497	5900
(b)	(Earth = 1)	—	0·39	0·72	1	—	1·52	5·20	9·54	19·2	30·1	39·4
(c)	Mass (Earth = 1)	343 000	0·055	0·815	1	0·012	0·108	318	95	14·6	17·2	$2·6 \times 10^{-5}$
(d)	Mean density (water = 1)	1·4	5·4	5·2	5·5	3·3	3·9	1·3	0·7	1·2	1·7	1·5
(e)	Radius (km)	696 000	2440	6052	6378	1738	3394	71 400	60 000	25 900	24 750	1145
(f)	Year, i.e. period of revolution about Sun (Earth years)	—	0·24	0·62	1	—	1·88	11·9	29·5	84·0	164	248
(g)	Spin period, i.e. rotation about axis (days)	27	59	−243*	1	27·3	1·03	0·40	0·43	−0·75*	0·53	6·4
(h)	Eccentricity of orbit	—	0·206	0·007	0·017	0·055	0·093	0·043	0·056	0·047	0·009	0·25
(i)	Inclination or orbit, with respect to the Earth's (deg)	—	7	3·4	0	23†	1·9	1·3	2·5	0·8	1·8	17·2
(j)	Inclination of axis to orbit (deg)	7§	<28	3	23	23†	24	3	27	82	29	?
(k)	Number of moons known	—	0	0	1	—	2	16	18	5	2	1
(l)	Atmosphere, chief constituents	—	none	CO_2	N_2, O_2	none	CO_2	H_2, He	H_2, He	H_2, He, CH_4	H_2, He, CH_4	CH_4
(m)	Magnetic field, dipole moment‡ (Earth = 1)	3×10^6	$6·6 \times 10^{-4}$	$<10^{-4}$	1	$<2 \times 10^{-6}$	3×10^{-4}	$1·9 \times 10^4$	?	?	?	?

* Minus sign denotes rotation is retrograde, i.e. opposite to majority direction.
† That is, orbit is in plane of Earth's equator.
‡ That is, strength of equivalent bar magnet (but some planetary fields are poorly represented by a dipole).
§ Inclination to the mean axis of rotation of planets.

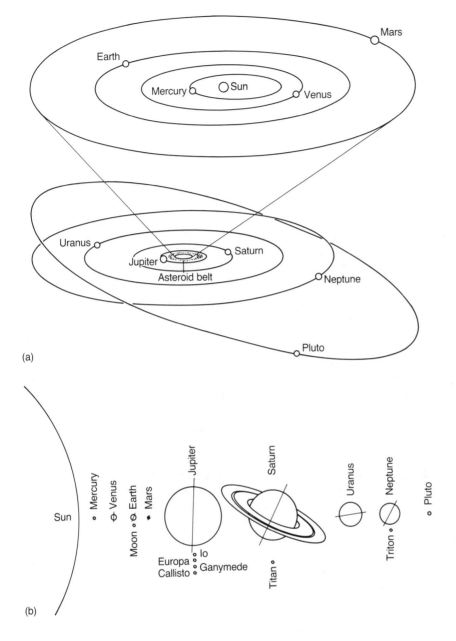

(a)

(b)

Figure 4.1 Orbits and sizes of the planets. (a) Shows the relative sizes of the orbits of the planets, with those of the terrestrial planets enlarged five times (Murray, 1983: the planets – readings from *Scientific American*). (b) Shows their relative sizes (and those of the seven largest moons), and the inclinations of their axes of rotation.

mass, yet Jupiter's mass is greater than that of the remaining bodies added together. All the planets rotate around the Sun in the same sense, and nearly in a plane (Table 4.1, row (i)), close to the equatorial plane of the Sun; this common axis of rotation is one reason for believing that the members of the Solar System formed together. But though the Sun has nearly all the mass of the Solar System, the orbital motions of the planets have about 99·5% of the angular momentum (see Note 1 for an explanation of this and related terms). This separation of mass from angular momentum once seemed a major obstacle to what is now the preferred theory of formation (Section 4.3).

So far, we have dwelt on the common features, but closer inspection reveals differences. The planets may be grouped as follows: the terrestrial planets – Mercury, Venus, Earth (and also the Moon), and Mars – have small masses but

45

high densities, whereas the giant planets – Jupiter, Saturn, Uranus and Neptune – have large masses but low densities. The giant planets therefore must have a much higher proportion of light elements, which we shall learn are predominantly hydrogen, helium and some other gases. In addition, though the orbits progressively increase in size across Table 4.1, there is a jump between Mars and Jupiter. The tiny planet Pluto has an elongated orbit which comes within Neptune's, and it is thought to be an escaped satellite of its big neighbour.

The spin axes of the planets are not generally parallel to their orbital axis (Table 4.1, row j) and they have a wide range of spin periods (row g). For instance, the Earth's is inclined at 23° (giving us summer and winter); Uranus is roughly 'lying on its side', while Venus is rotating very slowly in the opposite (retrograde) sense, as indicated by the minus sign.

In addition to the planets, there are many smaller bodies, mostly too small to be appear in Table 4.1 or Figure 4.1. Most of the planets, particularly the giant ones, have moons, or satellites, and our own Moon is one of the largest. The formation of the Moon is believed to have an important bearing on the Earth's composition (Sections 4.3.1 and 5.4). Generally smaller than the satellites are the asteroids, most of which orbit in the 'gap' between Mars and Jupiter (Fig. 4.1(a)), though a few cross Mars' and even the Earth's orbit. The largest is Ceres, 760 km in diameter, while many of the 20 000 known asteroids are only a few kilometres or even less across and most are irregular lumps, like the Martian satellite, Phobos, which is probably a captured asteroid. Their total mass is less than a tenth of one per cent of the Earth's mass. Asteroids are important to us because they are believed to be the major source of meteorites – extraterrestrial samples that arrive at the Earth's surface – which are mostly relics of early stages in the evolution of the Solar System. For a long time, meteorites were the only extraterrestrial material that could be examined in the laboratory. The launchings of planetary probes have not eliminated their importance, for they provide much information that can be obtained in no other way (Section 4.5). Comets are occasional but sometimes spectacular celestial objects, some having a tail of very diffuse gas and dust stretching across much of the sky. The head from which the tail derives is only a few kilometres across, and composed of frozen water plus ices of frozen gases and some dust; they are often referred to as dirty snowballs. They are probably primitive, having formed very early in the development of the Solar System. They have very elongated orbits and are thought to derive from the Oort Cloud, a reservoir of a vast number of comets that orbit beyond Neptune, mostly at many times the distance of Pluto, from which they are occasionally deflected into the inner part of the Solar System by the influence of passing stars or interstellar clouds.

In summary, the Solar System consists of the Sun and a huge number of bodies orbiting about it, with an enormous range of masses, and a considerable range of densities (Table 4.1, row (d)) and hence compositions. The largest bodies have a common axis of rotation – indicating a common origin – but at a more detailed level there are many irregularities. In Section 4.3 we explain how such diverse bodies were formed, but first we describe the context in which the Solar System formed, in relation to other stars.

4.2.2   The Galaxy

The night sky is filled with innumerable stars, and the daytime sky would be too if it were not for the brightness of the nearest star, the Sun. Stars vary greatly in

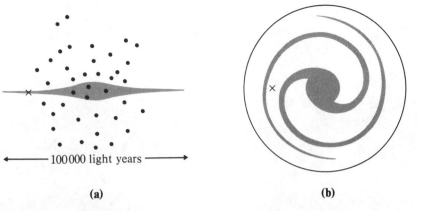

Figure 4.2 The Galaxy in
(a) elevation and (b) plan.
Most of the stars are in the
centre or the spiral arms. The
approximate position of the
Sun is shown by the cross.
See Figure 4.3(a).

100 000 light years

(a)

(b)

mass, brightness and colour, but all orbit in the same sense within a disc, with a
central concentration of stars (Fig. 4.2); the whole assemblage of about 10^{11}
stars forms the Galaxy. Many of the stars outside the central concentration lie in
two spiral arms, so that, if we could look at the Galaxy from a great distance –
millions of light-years away (a light-year is nearly 10^{13} km) – it would appear as a
spiral nebula, similar to many others that we can view through telescopes (Fig.
4.3(a)).

The Solar System lies close to one of the spiral arms (Fig. 4.2), so that if we
look into space along the plane of the disc we see the concentration of stars as a
belt across the sky, the Milky Way. In addition to the stars, there are many
'clouds' composed of gas plus a small proportion of solid particles, i.e. dust;
some of these can be seen as dark, starless patches, but others can only be
detected by more subtle observations. It is out of such clouds that stars are
formed.

4.3 FORMATION OF THE SOLAR SYSTEM

No other solar system similar to our own is known. This is probably not because
none exists, but because at present we have no way of detecting such a solar
system. The light reflected from a planet is too feeble in comparison with that of
its sun to be detectable, and attempts using other methods have not yet
unambiguously proved the presence of a planet even as large as Jupiter. (Planets
have probably been detected orbiting pulsars, but these, being remnants of
supernovae (Section 4.4.2), are not comparable with the Sun.) Therefore, our
ideas about the formation of the solar systems are largely theoretical and only
indirectly supported by observational evidence. But stars in the course of
formation have been observed, and it is believed that a proportion of them have
planets. That the following account is necessarily schematic should be borne in
mind.

The evolution of celestial bodies is largely determined by the interplay of the
inward force of gravity and several possible outward forces. Gravity is important
because every particle of matter attracts every other particle, with a force
expressed by Newton's Law of gravity (Eq. (3.1)). Since this force operating
alone would cause all matter to collapse together, it must be opposed by other
forces, of which there are three mechanical ones:

(a)

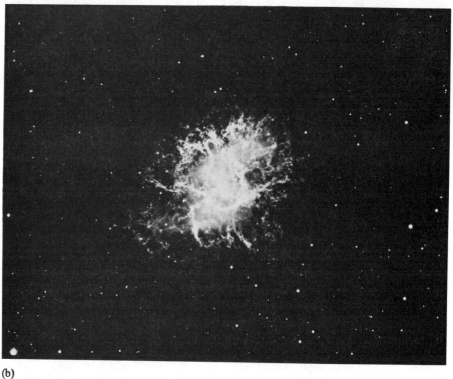

Figure 4.3 (a) A spiral nebula
or galaxy, composed not of
glowing gas but of about 10^{11}
stars. It is similar to our own
Galaxy, a few of whose stars
are seen dotted over the
picture. (b) The Crab Nebula,
the expanding remains of a
supernova which exploded in
AD 1054 (see Section 4.4.2).
(Photographs from the Hale
Observatories, reproduced by
permission of the Royal
Astronomical Society,
London.) (b)

48

1. internal pressure of a body, like the pressure that keeps a balloon inflated;
2. 'centrifugal force', which spins objects outwards, like a weight swung round on the end of a string, or the Earth orbiting about the Sun; and
3. outward velocity, like a rocket attempting to leave the Earth.

All three forces are important to aspects of this chapter.

In addition to these mechanical outward forces there is growing recognition of the importance of electromagnetic forces. As explained in Note 3, the presence of a magnetic field in a conductor effectively stiffens it. Gases in space are usually conductive either because cosmic rays, very energetic particles that permeate space, collide with some of the gas molecules and eject electrons, or the temperature is high enough to ionize it; the resulting plasma is a good conductor. As magnetic fields are found everywhere in space, this is an important force.

We start our account of the formation of the Solar System with a giant molecular cloud, similar to those where today stars like the Sun can be seen in the act of formation. These clouds consist mostly of molecular hydrogen, H_2, but contain other gases and some 'dust' of solid particles too, and they are permeated by weak magnetic fields. They may have enough mass to form 100 000 stars. Although they have a very low density, they are so vast and so cool – only about 10 K – that their self-gravity slightly exceeds their internal pressure, and collapse is only prevented by electromagnetic stiffening forces. Within the cloud, a volume of slightly higher density began slowly to collapse by self-gravity. If this smaller volume were not rotating it would have tended to collapse into a sphere but, like all bodies in galaxies, it was rotating. As a consequence, as it contracted it also flattened because, whereas gravity pulls radially towards the centre of mass of the sphere, the centrifugal force acts only perpendicularly to the axis of rotation (Fig. 4.4). As a result, individual portions of the cloud moved both

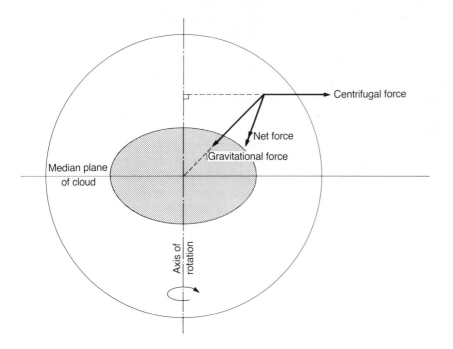

Figure 4.4 Contraction of a rotating, spherical cloud. The gravitational and centrifugal forces are not antiparallel, and therefore there is a net force towards the median plane, so the cloud tends to become flattened into a disc as it contracts; the inner shaded area shows an intermediate stage.

49

radially towards the axis of rotation and 'vertically' towards the median plane, to form a disc. The cloud had begun its development into a star – the Sun – and the Solar System.

However, before the bulk of the mass could contract to the centre to form the Sun, it had to lose much of its angular momentum. To understand this, suppose that as the cloud contracts all parts rotate with the same angular velocity. Conservation of angular momentum would then ensure that as the cloud contracts it speeds up – just as an ice-skater pirouetting on one toe speeds up if outstretched arms are moved inwards. The equations (Note 1) show that the angular velocity, ω, increases as $1/r^2$, so that if the cloud contracted to half its radius it would rotate in a quarter of the time. The centrifugal force, proportional to $r\omega^2$, would therefore increase by a factor of eight. The pull of gravity would also increase, as $1/r^2$, and so the halving of radius would quadruple the pull. Therefore, during contraction, *the inward pull of gravity increases less rapidly than the outward centrifugal force*, and there comes a point where centrifugal force is equal to gravity and the contraction ceases. If the assumption that all parts of the disc rotate at the same angular velocity is relaxed, we would expect the matter to arrange itself radially so that gravity and centrifugal force everywhere were in balance, which would result in the inner parts of the disc spinning faster than the outer parts, just as planets closer to the Sun have a shorter 'year' than those further out (Table 4.1, row (f)). We would also expect the Sun to follow the same rule, rotating at its centrifugal limit; at its present radius this would be one rotation in under three hours, whereas the Sun actually rotates only once every 27 days.

Although it is clear from observation that stars do slow down, the mechanism by which they do so is not agreed, though two are suggested. One is turbulence, swirls in the disc which mix more rapidly rotating inner material with slower rotating outer material and vice versa, tending to make the disc rotate at a uniform angular velocity. The second and more favoured mechanism relies on magnetic stiffening, as described in Section 4.3.2.

This rotating disc of gas and dust, with a central concentration, destined to form the Solar System, was the **Solar Nebula** (also called the pre-Solar Nebula). This concept replaces earlier theories of a filament of hot gas drawn out by gravitational attraction when a star passed close to the Sun. Although this filament rotating about the Sun could account for the angular momenta of the planets, it had difficulty explaining their fast spin rates, and theories of this type fell out of favour when it was realized that the filament, being initially very hot, would disperse.

The Solar Nebula consisted of two parts that evolved rather differently. The matter that lost most of its angular momentum fell to the centre of the disc. Owing to its lack of angular momentum, it formed a spherical mass, and also because of the same lack it could not balance the inward gravitational pull with centrifugal force. Instead, it compressed and heated up, and this raised the internal pressure until it balanced the pull of gravity. This allowed it to avoid further contraction except that heat inevitably flowed from the hot interior to the cool surface, whence it was lost by radiation into space. This loss had to be made good by release of more heat by further contraction, but a more compact sphere has a higher inward gravitational pull (since g increases as $1/r^2$, Eq. (3.2)); in turn, this requires a higher central pressure and hence temperature to balance it. Thus we have the apparently paradoxical result that loss of heat resulted in a higher internal temperature. The **proto-Sun** had formed at the centre of the disc and entered into a stage of relatively slow contraction, with rising central temperature.

This stage continued until the temperature reached about 10 million degrees, hot enough to allow nuclear reactions to cause hydrogen 'burning' and release heat (Section 4.4.2). This new source of heat replaced contraction as the main mechanism to make good the surface heat loss. The proto-Sun had become the Sun, and it had entered upon a long, stable phase in its development, as a **main-sequence star,** which will last about as long again as its present age of 4550 Ma.

4.3.1 Accretion of the terrestrial planets

Accretion of dust particles to form planetesimals. Although the material in the disc was prevented by its angular momentum from falling into the centre, it had enough mass to compress itself towards the median plane, and this released gravitational energy sufficient to vaporize most of the dust grains. When the disc stabilized under its internal pressure it cooled and the some of the material recondensed as solid particles, commencing with the most refractory, within a period of no more than 10^5 years.

There is an important difference between the gas molecules and the solid dust particles of the disc. When gas molecules collide, they do so perfectly elastically, so that, on average, they never slow down. It is this bouncing back that creates the pressure of a gas (with the average speed increasing with the temperature). This gas pressure prevented the gas cloud contracting down to a very thin disc. But particles collide inelastically, and if their velocity is not large they will tend to stick together (as found in laboratory experiments) and this reduces their average speed. As a result, the dust particles tended to fall towards the median plane.

At first, the rate of fall was very slow, because the particles were so small – much less than one micrometre (millionth of a metre) across – that they tended to be carried along with the gas. However, as they collided they grew, and so fell faster, since their surface areas increased less rapidly than their masses (the drag of the gas was proportional to the surface area of the particles, whereas the pull of gravity was proportional to their mass). The size of grains increased exponentially with time, and resulted in a wide range of sizes, since the bigger a grain the more of the smaller ones it would hit. At one AU (astronomical unit: the distance of the Earth from the Sun), the largest grains grew to several metres in diameter and reached the median plane within a few tens of thousands of years, a tiny fraction of the total duration of formation of the Solar System.

Once grains had grown to a few metres in diameter the impacts tended to compact the grains and reduce their surface area. They were now large enough for gas drag to be quite small, and they moved in orbits largely determined by their velocities and distance from the Sun. Such masses, which are large enough to move independently of the gas, are called **planetesimals.**

Accretion of planetesimals to form planetary embryos. The huge number of planetesimals in the median plane were orbiting about the Sun, but as their orbits were not circular their orbits intersected and collisions were frequent. Two bodies that collide will merge rather than fragment if their velocity of collision is less than about half their escape velocity. (The escape velocity is that needed for the two planetesimals, initially in contact, to escape from their

51

mutual gravitational pull by moving directly away from each other, and it increases as the bodies grow.)

It is believed that there was runaway growth, for several reasons. Firstly, larger planetesimals make more collisions simply because of their larger size. Secondly, as they grow, their gravitational attraction allows them to attract bodies at a distance and so collide with bodies with which they would not have made contact otherwise. A third effect is that the collision of two planetesimals in non-circular orbits tends to produce a body with a more circular orbit, and as orbits become more circular, their relative velocities decrease on average, and so are more likely to be less than the escape velocity.

These effects greatly speeded up the growth of planetesimals, and after about 10^5 years the original planetesimals in a narrow annulus of the disc, such as 0·99 to 1·01 AU, have evolved through collisions into a wide range of masses but with one very large one exceeding 10^{23} kg (the Earth's mass is about 6×10^{24} kg), as shown in Figure 4.5. This dominant mass is called a **planetary embryo**.

Accretion by this mechanism ended when each of these embryos had swept up the bulk of the matter within an annulus equal to its range of gravitational attraction. There were approaching 100 planetary embryos in the region now

Figure 4.5 Calculated evolution in the early Solar System of planetesimals between 0·99 and 1·01 AU. Each curve should be read from left to right. (a) Shows the starting conditions, assumed to be just over 10^8 bodies with a mass range 10^{15} to nearly 10^{17} kg. (b) After 30 000 years these have evolved to 10 bodies exceeding 10^{20} kg in mass, 100 exceeding 10^{19} kg, and so on; notice there are now nearly 10^8 bodies *smaller* than the original planetesimals, formed by fragmentation. (c) After 50 000 years there is a single large body of nearly 10^{24} kg, plus many, much smaller bodies. (After Wetherill, 1989).

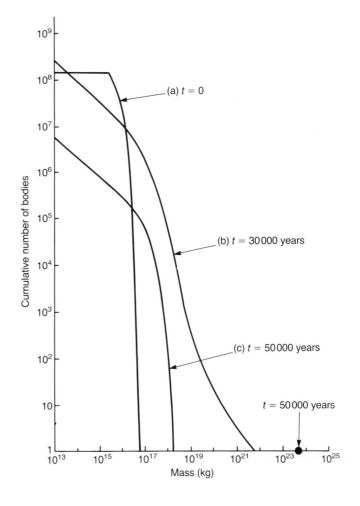

occupied by the terrestrial planets, with masses ranging from about 1 to 10% of the Earth's, moving in nearly circular orbits. It took less than 10^5 years to progress through this stage of accretion, at the distance of the terrestrial planets.

Accretion of planetary embryos to form the terrestrial planets. The embryos were large enough to account for the Moon, Mercury and even Mars but not the Earth, Venus or the giant planets, so a further stage of accretion is needed.

The embryos had swept up most of the material in their annuli, and could grow little by attracting the remainder, but they were now large enough to perturb each other's orbits slightly and this was enough to cause, over a period much longer than the time of their formation, further, catastrophic collisions, which led on average to just a few bodies, the planets. This final stage was between only a relatively few – less than a hundred – quite large bodies, and so there were two consequences: firstly, the effects of chance no longer averaged out, and their energies were sufficient to cause great heating. The randomness of collisions explains how Venus can have such a slow rotation, when the angular momenta of the colliding bodies happened to cancel, and it also offers an explanation for the Earth–Moon system. This is believed to be the result of a collision of a Mars-sized embryo with the nearly formed Earth: the impact disrupted the smaller body, which left some of its mass with the Earth, while the motion of the remainder was slowed enough for it to go into orbit around the Earth. Chemical consequences are discussed in Chapter 5.

Calculations show that this process seems adequate to account for the planets Mercury to Earth, but predicts a Mars larger than the Earth and a single body in the present asteroid belt. But this is only in the absence of Jupiter; Jupiter was probably already so massive that its gravitational force was able to prevent the bodies in what is now the asteroid belt from forming into larger bodies, even today it is able to eject bodies from this belt when they have certain configurations with it. At the greater distance of Mars the effect was less, but sufficient to stunt growth.

In summary, the successive stages of accretion take progressively longer, with very approximate times being:

1. grains sticking together, growing to become planetesimals and falling into the median plane in 10^4–10^5 years;
2. planetesimals merging to produce a few large planetary embryos – plus many smaller bodies – in less than 10^5 years, followed by:
3. growth of planets from embryos taking about 10 million years to reach half their mass, and 100 million to reach full size.

4.3.2   The proto-Sun and the T-Tauri wind

While the disc of the Solar Nebula was evolving into planets, the central mass – the proto-Sun – was evolving, too. This evolution influenced the planet-forming process.

The infall of material and its compression released much gravitational energy, which was converted to heat. Although some of this heat was lost by radiation into space, enough remained within to raise the surface temperature to a high value, far hotter than the present Sun. The proto-Sun that formed was very vigorous, and produced a strong solar wind, which is a stream of charged particles, such as is still produced today on a much reduced scale. The outflow of material was so large that the proto-Sun may have lost about a millionth part of its mass each year. As the wind mostly 'blew' out from the poles it did not interfere with the infall, but it caused a significant loss of mass in the 10^5 years it lasted.

When infall was practically complete the proto-Sun entered a T-Tauri stage, named after the type star. Although not as vigorous as the previous stage it still produced a strong solar wind, no longer confined to the poles, and this had two important effects. The solar wind is a plasma, and so it experienced magnetic stiffening. Therefore, instead of describing spiral outward paths, as do the jets of a rotating lawn sprinkler, the particles of the solar wind tended to move radially outwards, as if they were moving along the spokes of a wheel attached to the Sun. To do this they had to increase their angular momentum ($mr\omega^2$), because of their increasing distance from the axis of rotation. They did this at the expense of the Sun, so that, as the wind blew outwards, it braked the Sun's rotation, and over a period of millions of years reduced it to its present low value.

A second important effect is that the solar wind slowly removed all the remaining unaccreted gas and dust, bringing an end to planetary growth. The T-Tauri stage lasted roughly 10 Ma. Stars in the T-Tauri stage have been observed to have a disc containing dust and extending up to 30 times the diameter of Pluto's orbit, and which seems to thin as the star ages, consistent with loss of the dust. There is also some evidence that T-Tauri stars may experience brief outbursts, during which luminosities increase by a factor of hundreds or thousands. This would raise the temperature of the inner part of the Solar Nebula to over 1500 K, and may help explain the existence of chondrules (Section 4.5.2).

4.3.3 Formation of the giant planets and other bodies

The main reason why the giant planets are so much larger than the terrestrial ones is because they contain a large proportion of the light elements, particularly hydrogen and helium, as revealed by their low densities (Table 4.1, row (d)). The question then arises why the giant planets accreted large amounts of these elements whereas the terrestrial planets did not. Part of the answer is that the Earth even today is too small to retain hydrogen and helium indefinitely in its atmosphere, while in the past the conditions were even less favourable, with higher temperatures due to the proto-Sun and the heat released by impacts. At the distances of the major planets the lower temperature favoured retention of volatile elements, but the main reason is that they lay beyond the 'snow line'; the distance at which it was cool enough for water and methane to form solid particles. As we have seen above, solid particles can accrete where gases cannot, and this, together with the larger area of an annulus at this a larger radius, provided far more material from which the planets could grow. And having reached a certain mass they would be able to attract the gaseous hydrogen and helium as well.

There are problems in accounting for sufficient growth before the T-Tauri wind dispersed unaccreted material, for the mechanisms described above for the terrestrial planets take much longer at larger distances from the Sun, but it is supposed that some form of runaway growth occurred.

The early Jupiter itself probably removed some of the residual material, its strong gravitational field flinging small planetesimals far outwards, to form the

Oort Cloud of comets, as today it is used to boost space probes into the further parts of the Solar System. It also probably prevented the growth of a planet between it and Mars, the planetesimals remaining as the asteroids, and stunted the growth of Mars (see Section 4.3.1).

4.3.4 Summary

Starting as a cold, contracting and rotating cloud of gas plus solid particles, the Solar System probably formed from the Solar Nebula as a result of a number of processes which, for convenience, are treated as if they occur separately and successively, though they probably blended one into another, but also were diachronous across the Solar System.

First, a cloud contracted into a disc with a central concentration. The disc was initially stabilized by its rotation, centrifugal force approximately balancing gravitational attraction, but loss of angular momentum through the mechanism of magnetic stiffening or other process permitted much of the matter to fall into the centre, to form the proto-Sun. This infall produced a very vigorous stage in the development of the Sun, lasting for only about 10^5 years, followed by the somewhat less vigorous T-Tauri phase.

The presence of the early Sun raised the temperature of the material remaining in the disc, so partly determining which chemical elements were in solid particles and which in gases. The solar wind of the T-Tauri phase blew away any unaccreted gas and dust.

In the disc, contraction towards the median plane gas compressed the material sufficiently to volatilize briefly most of it, after which it soon cooled, and the more refractory elements condensed as dust. The dust grains settled towards the median plane, growing by 'sticky' collisions. At first, settling was slow, but as the grains grew in size, settling speeded up, and larger planetesimals were formed, largely unaffected by gas drag.

Collisions between planetesimals orbiting in the median plane sometimes resulted in destruction, sometimes in amalgamation, but on average they led to growth with the largest having runaway growth due to a combination of mechanisms. This stage culminated with most of the solid matter of each of a succession of narrow annular zones being concentrated into a planetary embryo. These embryos perturbed each other's orbits and the resulting collisions were massive, with their outcome partly dependent on chance. On average, however, in the region of the terrestrial planets, four or five planets, with sizes similar to those of the actual terrestrial planets were a likely outcome. The Earth–Moon system is probably the result of a massive collision.

4.4 ABUNDANCES OF THE ELEMENTS

Sections 4.4.1 and 4.4.2 examine how the elements in the Solar System formed, and how they came to be incorporated into it. The Sun is not only the largest part of the Solar System but it escaped the processes of segregation that occurred in the disc of the Solar Nebula. Therefore, the Sun can provide a good estimate of the initial composition of the Solar Nebula, assuming that it has not been affected by other processes.

55

Figure 4.6 Elements and the
s- and r-processes of
nucleosynthesis (unstable
isotopes are not shown). Each
dot represents a stable
isotope. Isotopes of an
element, determined by the
value of *Z*, are spread out
'horizontally' due to their
different values of *N*, the
number of neutrons. The
continuous line of the upper
'staircase' shows how the slow
neutron process builds up
from ^{56}Fe. The rapid neutron
process builds along the lower
'staircase'. Note the large
steps at the 'magic' neutron
numbers of 50, 82 and 126.
When rapid irradiation of
neutrons ceases, isotopes
spontaneously convert
neutrons to protons, moving
diagonally up to the left (after
Burbidge *et al.*, 1957;
Campbell, 1969).

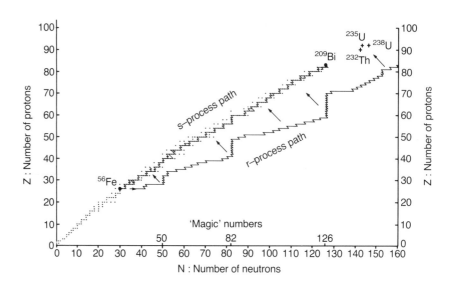

4.4.1 Nucleosynthesis

The chemical elements may be arranged in a series from the simplest, hydrogen
(just one proton), through helium (two protons) right up to uranium (92
protons) and beyond (the atom and nucleus are discussed in Note 5). In addition
to the protons, the nucleus usually contains neutrons, but since they have no
charge they do not determine the chemical properties of atoms. The most stable
isotopes of each element, except the lightest, have slightly more neutrons than
protons (Fig. 4.6). We believe that all the elements were built up from hydrogen
or, more correctly, their nuclei were built up from protons. To build up from
hydrogen it is necessary to force the protons together, despite the repulsion of
their electric charge, and this can be done only by giving them a high velocity
towards each other. This is achieved by very high temperatures – in excess of a
million degrees – or by particle accelerators, allowing the reactions to be studied
in the laboratory.

To build up all the elements from hydrogen requires several reactions. The
first is hydrogen fusion or 'burning', essentially the progressive addition of four
protons, two of which spontaneously convert to neutrons, with emission of a
positive electron, or positron, to conserve electrical charge. This results in one
α-particle, or helium nucleus:

$$4\,p \rightarrow {}^{4}_{2}He$$

where the superscript is the number of nucleons, i.e. protons plus neutrons, and
the subscript the number of protons.

The addition of one further proton to an α-particle is blocked because ${}^{5}_{3}Li$ is
unstable. Instead, the fusion of two helium nuclei produces ${}^{8}_{4}Be$, but because it is
unstable, having slightly higher energy than two α-particles (Fig. 4.7, upper
curve), a third must be added quickly, to produce ${}^{12}_{6}C$. The likelihood of there
being three nuclei with the correct velocities to fuse is much less than having just
two, and acts as a bottle-neck to the synthesis of heavier elements. Once ${}^{12}_{6}C$ has
been formed, further α-particles can be added, to form with increasing
temperature and in turn, ${}^{16}_{8}O$, ${}^{20}_{10}Ne$, ${}^{24}_{12}Mg$, ${}^{28}_{14}Si$, as carbon-, oxygen-, etc.
burning, exhausting the supply of α-particles.

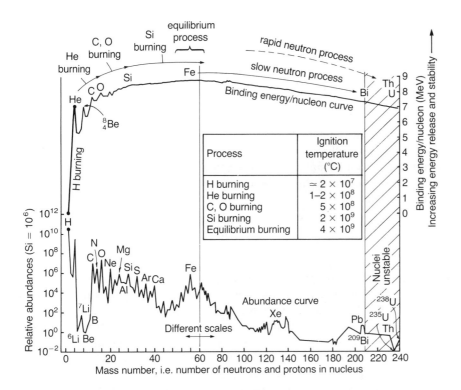

Figure 4.7 Nucleosynthesis and abundance of the elements. The upper curve shows the binding energy per nucleon curve: conversion of a nucleus to another higher up the curve releases energy equal to the vertical difference multiplied by the number of nucleons involved. The most important reactions are indicated, and the temperatures at which they commence are shown in the box below. The lower curve shows relative abundances of elements in the Sun plotted on a log scale with silicon set at 10^6. Superimposed on the general decrease with increasing nuclear mass are peaks corresponding with locally more stable elements, which tally with the peaks on the upper curve. (The abundance curve has been drawn from data in Table 3 of Anders and Grevesse, 1989. The binding energy/nucleon curve has been calculated from data in Appendix IV of Serrat, 1954.)

Before the temperature can rise high enough to be able to force two silicon nuclei together, another process comes into operation. The large energies of collision cause the nuclei partly to break up, but the resulting particles combine with other nuclei, and the net result is to build heavier nuclei, since they are more stable. This type of reaction culminates with the e-process (equilibrium process) in which particles are freely released and re-absorbed, tending to form the most stable nuclei. The upper curve of Figure 4.7 shows the binding energy/ nucleon curve, the amount of energy released per nucleon in fusing hydrogen into different nuclei. The e-process produces isotopes at the highest or most stable part of the curve, which is occupied by the iron-group elements, various isotopes of iron (Fe), cobalt (Co) and nickel (Ni). Just what proportions of each isotope are formed depends upon the detailed properties of the nuclei and the rates at which they react, but a combination of them is the culmination of fusion resulting from increasing temperature, which has reached 4×10^9 K.

Higher temperatures will not build beyond the iron-group elements because any nuclei formed would simply revert to the iron group. Instead, neutron reactions are needed; as neutrons have no charge they do not need a high velocity and hence temperature to react. There are two neutron processes. In the s-process (slow neutron process) neutrons are added at what in nuclear terms are long intervals, several minutes apart. At first this merely increases the neutron:proton ratio, but when this becomes too large for stability a neutron converts spontaneously into a proton (with emission of an electron or β-particle), so forming the next element (Fig. 4.7, upper curve). One by one, neutrons are added until ^{209}Bi is reached. This is the limit for the s-process, for the next nucleus is unstable and spontaneously decays to a lighter element. To build beyond this obstacle needs the r-process (rapid neutron process). Neutrons are added so rapidly that there is no time for the resulting nuclei to decay, and large excesses of neutrons over protons are developed (Fig. 4.6). When rapid addition of neutrons ceases, the excess neutrons begin to decay into protons, but some of the resulting nuclei, particularly ^{238}U, ^{235}U and ^{232}Th (Fig. 4.6), are nearly stable and decay with half-lives so long that significant amounts still remain in the Earth, providing a source of heat (see Sections 6.3, 8.3 and 11.1), and being the basis for important radiometric dating schemes (Note 4).

The s-process also helps to 'fill in' some of the gaps, such as between ^{12}C and ^{16}O, but there are still some isotopes not yet accounted for, such as those with a low rather than a high stable proton:neutron ratio. These are formed by the p-(proton) process, the protons being by-products of some of the reactions described above.

4.4.2   Synthesis of the elements and stellar evolution

We go back to the very beginning, the formation of the Universe itself. It began, according to current understanding, with all its matter and energy concentrated into a point or singularity, at unimaginable pressure and temperature. From that point it expanded in a vast explosion – the Big Bang – and, in doing so, cooled. This allowed the initial exotic matter to convert to more familiar protons and neutrons, and continued cooling allowed them to react together to form multi-nucleon nuclei, such as helium. But because of the bottle-neck beyond helium, described in the previous section, very little of the elements heavier than helium were formed before the temperature dropped too low – a few million degrees – for reactions to continue. Even today, despite subsequent synthesis of heavier elements, hydrogen and helium greatly predominate over other elements in the Universe. Further reactions that produced elements making up most of the materials we are familiar with had to wait a long time, until stars were formed.

Within the expanding Universe, there were variations in density in the form of huge swirling clouds. Although their density was extremely low, they were so vast that their gravitational self-attraction was sufficient to produce contraction, while their angular momentum prevented rapid implosion. As they contracted and the density rose, smaller volumes of higher density became self-contracting and the cloud divided into separate clouds orbiting about one another. Further contraction allowed the process to repeat, resulting in a hierarchy which we recognize today as galaxies belonging to groups, which in turn belong to supergroups.

The story continues with volumes of material within the protogalaxy locally having sufficiently high density to contract under self-gravity. As described earlier, such a volume loses angular momentum and contracts to a sphere, becoming a star. Thereafter, it contracts further to compensate for surface heat loss, developing progressively higher central temperatures. When these reach 10^7 K, hydrogen burning commences. This is such a potent source of energy – several million times greater than the chemical burning of hydrogen – that the star can continue to burn hydrogen for millions of years. The star has entered the most stable phase of its life, as a **main-sequence star.**

Hydrogen burning ignites at the centre of the star, where it is hottest. The heat produced flows outwards by radiation, rather than convection, so that the resulting helium remains where it is formed; only in the outer part of the star does convection occur. The accumulating helium forms a core that grows, with hydrogen burning confined to a shell at its surface. As the core grows, the outer parts of the star expand enormously and the surface cools so that it appears red, and the star has become a red giant. A red giant is so large that when the Sun reaches this stage, in about another 5000 million years, it will engulf Mercury, and more massive stars are larger still. Concurrently, the core contracts very slowly and the central temperature rises until helium burning commences, in turn leading to a carbon–oxygen core. This process of forming a core of burnt fuel which later ignites by a further reaction repeats until the ageing star consists of a series of shells, with the equilibrium process occurring at the centre, as shown in Figure 4.8.

Figure 4.7, upper curve, shows that each successive stage of burning from hydrogen towards the iron group releases less energy than the previous one, with hydrogen burning releasing several times as much as the remainder together. This diminishing of the energy supply coincides with an increasing need for energy by the evolving star, for its central temperature rises as it evolves and the heat loss by surface irradiation increases rapidly. So by far the longest stage is hydrogen burning, and most stars are therefore at this main-sequence stage. Secondly, the variable that has most influence upon the evolution of a star is its mass. A star can ignite hydrogen only if its mass is about ten times that of Jupiter; only stars with a mass approaching that of the Sun are able to ignite helium, and only stars with at least eight solar masses evolve beyond carbon–oxygen burning. Counter-intuitively, the bigger the star, the faster it burns fuel,

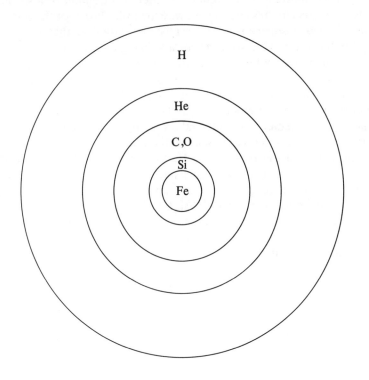

Figure 4.8 Cross-section of a massive star late in its evolution. The major products of each stage of nuclear 'burning' are shown; burning proceeds in a shell at the base of each layer. Having synthesized iron at its centre, the star has exhausted its fuel there and it will soon explode as a supernova.

59

for it has to support higher internal pressures, which require higher temperatures, so that massive stars have comparatively short lifetimes. For example, a main-sequence star 50 times as massive the Sun radiates energy at a million times the Sun's rate, and as a result uses up its fuel in a few million years, whereas the Sun, about 4550 Ma old (see Section 4.5.4), is only about half-way through its main-sequence stage.

After the main-sequence stage, the diminishing energy supply coupled with increasing demand ensures that a massive star moves through the succeeding stages ever more rapidly, so that a star that has existed for millions of years passes through the e-process stage in only seconds, to find itself bankrupt of fuel. Once the centre has exhausted its nuclear fuel, it has to revert to contraction to make good the heat loss, and the temperature rises even higher. When it reaches about 10^{10} K, the collisions between nuclei are so violent that they begin to break apart the nuclei. Since forming the nuclei released energy, breaking them apart must absorb energy, presenting the star with a predicament: now contraction results in a net loss of heat. The central pressure can no longer support the weight of the overlying layers and the star implodes, in a few tenths of a second, compressing the electrons and protons into neutrons, and so ultimately forming a neutron star with a diameter of only 10 kilometres.

But the star has exhausted its nuclear fuel only at its centre; in the layers outside there is plenty of unburnt hydrogen, helium, etc. The implosion of the centre causes the outer parts to fall in, at a very high speed, encounter the collapsed core and then 'bounce back' as a shock wave, that, sweeping through the outer layers, rapidly compresses them and raises their temperature. Nuclear reactions are very sensitive to temperature, so the temperature rise causes a huge rise in the rate of release of nuclear reactions. So rapid is the release that the star is not able to adjust by expansion, and a runaway stage ensues that consumes a large amount of the remaining fuel in a fraction of a second. The consequent huge rise in pressure blasts away the outer layers of the star in a vast explosion, termed a supernova. So vast is the explosion that a supernova's output is briefly comparable with that of a whole galaxy of stars!

The material of the outer layers, with its synthesized elements, disperses throughout space, contributing to the background of gas and dust. Since the galaxy is at least twice as old as the Solar System, there has been time for many massive stars to evolve into supernovae and then spew out their contents, so that the proportion of elements in the Galaxy heavier than helium has steadily increased. Stars that formed later, such as the Sun, incorporated some of these elements, which form the bulk of the Earth and of ourselves. While the star synthesizes up to iron at its centre, other nuclear reactions go on, such as the s-process. Slow neutrons reacting with the small amounts of iron, etc. deriving from earlier supernovae build up to ^{209}Bi, but the r-process only operates in the final moments of the star's life, when a flood of neutrons produces the unstable and semi-stable heaviest elements. The minute amounts of these heavy elements are dispersed by the supernova, together with the far more abundant lighter elements.

This account has described synthesis only in a stars massive enough to evolve all the way to the e-process and thence a supernova; elements are also synthesized in a variety of other types of stars, but will not be described here. Lastly, we come to the light elements, Li, Be and B, which are not produced inside stars; in fact, they tend to be destroyed, which is why lithium is less abundant in the Sun than the Earth and meteorites (Fig. 4.12). Instead, they are formed as fragments broken off ('spallated') nuclei in dust particles in space when hit by cosmic rays.

4.4.3   Solar abundances of the chemical elements

The importance of the solar composition is that it is the best sample we have of the original composition of the Solar Nebula, because it has been little affected by the segregating processes that operated upon the material destined to become the planets, etc., and neither have its outer parts, that we see, been affected by nuclear reactions which are confined to its centre, with the exception of lithium destruction.

The visible surface of the Sun, the photosphere, has a temperature of about 6000 K. It produces the continuous range of wavelengths recognized as the colours of the rainbow, but the spectrum is interrupted by dark lines caused by absorption by specific elements in the cooler outer parts. The strength of an absorption line depends not only upon the amount of the element present, but also upon the temperature, the pressure and the intrinsic efficiency of the element in producing or absorbing light ('oscillator strength'), all of which have to be measured or estimated.

The results are presented in Figure 4.7, lower curve. The general shape is a decrease of abundance with increasing atomic number, by a factor of about 10^{12} overall, but abundances lie above this curve if they are more stable than adjacent isotopes, such as multiples of α-particles (^{12}C, ^{16}O. . .), the iron group and nuclei having 'magic numbers' (see Note 5). The 'zig-zag' appearance of the curve is caused by the fact that nuclei with an even number of nucleons are more stable than those with an odd number.

4.5   METEORITES AND ASTEROIDS

4.5.1   Classification of meteorites

Meteorites are important to us because they are believed to represent intermediate stages in the development of the Solar Nebula into the Solar System, some of them very early on in the process.

Meteorites are defined as extraterrestial samples that have survived their passage through the atmosphere to land on the surface of the Earth. (Meteors, or shooting stars, are specks of dust that become incandescent and vaporize at a height of 60 km or so. They probably derive from comets.) The sudden arrival of a meteorite, often accompanied by an impressive fireball, sometimes led to them being treated with awe, and fanciful stories surrounding them perhaps provoked scientists to undue scepticism about their origin. Thus, Chladni was reluctant, in 1794, to publish his reasons for believing them to be of extraterrestrial origin, but his ideas were soon accepted. (A notable disbeliever was Thomas Jefferson, he who helped draft the American constitution and later became President. He remarked of a fall reported by two professors of Yale: 'It is easier to believe that Yankee professors would lie, than that stones would fall from heaven.' But as Jefferson lived in Virginia, this may tell us more about his views on Yankees than on meteorites.)

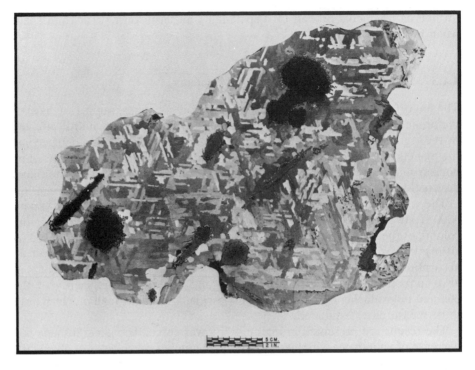

(a)

Figure 4.9 Polished sections of meteorites. (a) Iron: the intersecting lines are Widmanstätten patterns; the dark areas are troilite (FeS) inclusions. (b) Stony-iron: the metal matrix encloses olivine nodules. (c) Chondrite (Allende carbonaceous chondrite). The many small spheres are chondrules, while the two prominent irregular areas are 'high-temperature inclusions' or CAIs. (Photographs from the Smithsonian Institution.)

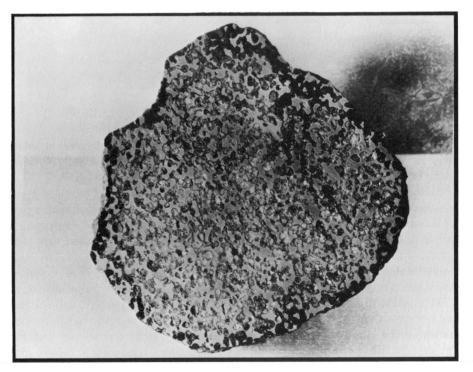

(b)

(c)

Figure 4.9 continued

Meteorites are evidently fragments of larger bodies (Fig. 4.9), believed to derive by collisions from much larger bodies, the **meteorite parent bodies**. All meteorites have experienced changes due to secondary 'processing' such as heating, aqueous alteration and changes caused by impacts. Although these changes potentially provide further information about conditions in the Solar Nebula and early Solar System, they complicate interpretation. What follows is a very simplified account, and further details – and qualifications – can be found in Kerridge and Matthews (1988), or other, less-advanced books listed in 'Further reading'.

Meteorites may be classified in different ways (Table 4.2). Based on major compositions and structures, they divide into **irons** (Fig. 4.9(a)), predominantly of iron with some nickel; **stones**, which are composed of silicate minerals; and **stony-irons** (Fig. 4.9(b)), about half metal and half silicate minerals. The stones may be subdivided into **chondrites**, containing chondrules: roughly spherical inclusions up to a millimetre across, in a matrix (Fig. 4.9(c)), and **achondrites**, lacking chondrules. Achondrites are not dissimilar to terrestrial rocks, some being like basalts or peridotites, but the structure of chondrites is unique to meteorites. These main divisions may be subdivided according to the types of minerals and structures present. The compositions of chondrites are much closer to the solar abundances than are other types of meteorites.

A second classification is based upon how the meteorites apparently were formed. **Differentiated meteorites** – irons, stony-irons and achondrites – acquired their composition and structure as the result of having differentiated

63

Table 4.2 Classification of meteorites (based on data in Table 1.1.1, Sears and Dodd, 1988).

Division				Falls	
Type			Class	Number	%
Differ-entiated	Irons		All	42	5
	Stony-irons		All	9	1
		Achondrites	SNCs	4	8
			Others	65	
Undiffer-entiated		Chondrites	Carbonaceous	35	86
			Ordinary	661	
			Enstatite	16	

into layers of different composition from a molten or partially molten material. In contrast, the **undifferentiated meteorites** – the chondrites – could not have been formed in this way, because melting would have destroyed the structure of chondrules in a matrix. The chondrites are undifferentiated, and also have compositions close to solar abundances, apart from the most volatile elements; they are therefore regarded as the most 'primitive' of meteorites – that is, closest to the original composition of the Solar Nebula.

The number of known meteorites has been greatly increased by large numbers recently found in Antarctica, where they may have been carried along deep in ice-sheets for periods sometimes reaching a million years, while others have been found in deserts. Table 4.2 gives the major types and their relative numbers. However, the proportions of the different types are estimated from observed falls, for iron meteorites are over-represented in museum collections as they are conspicuously different from most terrestrial rocks and so more easily recognized.

Even the meteorite falls probably do not represent the proportions of the parent bodies, because almost all known meteorites landed very recently in the Earth's history, and so may not represent the long-term average.

4.5.2 Chondrites

The chondrites are divided into several groups: the 'carbonaceous chondrites' have relatively high carbon contents, the 'enstatite chondrites' have enstatite – a magnesium-rich orthopyroxene – as their dominant mineral, while the 'ordinary chondrites' – by far the most numerous – differ from the others only in their amounts of iron (Fig. 4.10).

In 1967, Van Schmus and Wood devised a classification of the chondrites (Fig. 4.11) based upon both composition (vertically) and petrographic type (horizontally). Progression to a higher petrographic number involves a number of changes, including minerals becoming more homogeneous, amounts of volatiles

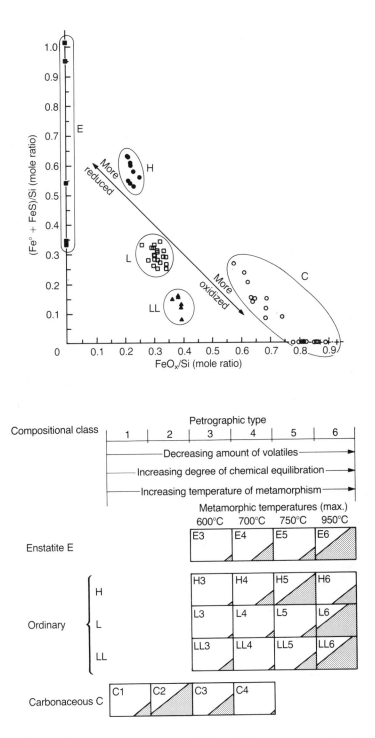

Figure 4.10 Plot of iron in the reduced versus oxidized states, for chondritic meteorites. The meteorite classes are denoted by capital letters: C, carbonaceous; E, enstatite; H, L and LL, ordinary chondrites with high, low and very low proportions of iron. They plot as discrete groups (after Larimer and Wasson, 1988).

Figure 4.11 Classification of chondritic meteorites. The shaded areas show the percentage of petrographic types within each compositional class. (After McSween, 1987).

decreasing, primary textures progressively obliterated, and with the distinction between chondrule and matrix becoming less distinct. Not all classes of chondrite are found in all petrographic numbers.

The accepted explanation of these trends, particularly the decrease of volatiles with increasing petrographic number, is an increased degree of heating of the original material, ranging from 400 to 1000°C, and therefore petrographic number is sometimes called degree of metamorphism. For this reason some C

65

chondrites, which have the largest proportions of volatiles and particularly water, were taken to be the most primitive of all, but it is now recognized that they have acquired their large amounts of volatiles by secondary aqueous alteration in the parent body, so 'degree of (thermal) metamorphism' really applies only to numbers 3 to 6. Instead, the least equilibrated chondrites of type 3 are believed to be closest to the composition of the Solar Nebula.

Chondrules have evidently been molten, which would have required temperatures up to about 2100°C, and the minerals present show that they cooled at rates of hundreds to thousands of degrees per hour. One possible cause of such intense but brief heating is a violent flare-up of the proto-Sun, as mentioned in Section 4.3.2.

Another kind of inclusion is the calcium–aluminium-rich inclusion, or CAIs (see Fig. 4.9(c)), irregular in shape and up to a centimetre or more across. CAIs consist predominantly of silicates and oxides of Ca and Al, notable for having very high vaporization temperatures (2100–2400°C), so they are called, alternatively, high-temperature inclusions. They are believed to have been the first condensates of the Solar Nebula, and will be discussed in Section 5.2.1.

Figure 4.12 compares the relative elemental abundances in chondrites with the solar abundances. The abundances range over eight orders of magnitude, but there is excellent agreement, apart from the most volatile elements and lithium. The volatile elements were only partially condensed in the accretion processes that formed the meteorite parent bodies (Section 4.3.1); lithium is less abundant in the Sun than in meteorites (and the Earth) because most has been destroyed by nuclear reactions in the Sun, as explained in Section 4.4.2.

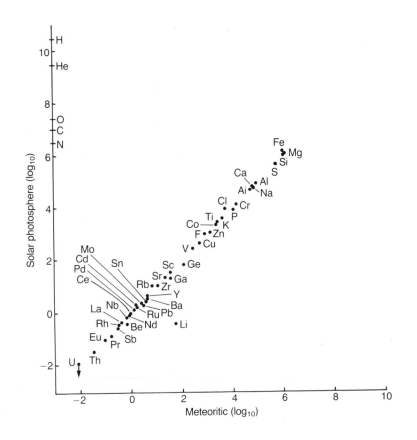

Figure 4.12 Comparison of solar and chondritic abundances (silicon has been set at 10^6). Note that logs have been used because the range of abundances is so large. Hydrogen, He, O, C and N are not compared because they have mostly been lost from meteorites, because of their great volatility, but they have been shown on the solar abundance axis to demonstrate their high abundances in the Sun. The reason for the anomalous position of Li is explained in the text. The meteorites analysed were C1 carbonaceous chondrites and are generally accurate to ± 5–10% (drawn from data in Anders and Grevasse, 1989).

66

4.5.3 Iron meteorites

These are composed predominantly of nickel–iron alloy, plus some non-metallic inclusions, of which troilite (iron sulphide) is the most common. Many irons, when sectioned and polished, reveal Widmanstätten patterns (Fig. 4.9(a)). These are intergrowths of two alloys of iron–nickel. Such patterns arise when two mineral or alloy components in solid solution cease to be fully miscible as the temperature falls. Suppose the atoms of two elements are similar but not identical – such as iron and nickel – so that separately they would form lattices that differ slightly. The two types of atom are freely interchangeable in a crystal at high temperature, because of the looseness of the expanded lattice. But as the temperature is reduced, their difference becomes more significant, and there comes a point when the energy of the whole system can be lowered by segregating the atoms into two separate lattices, each richer in one of the elements; then a second lattice grows within the original one. There is a mismatch where the two lattices abut, but this is minimized if the second lattice grows along certain directions of the host lattice, as **exsolution lamellae**. An example familiar to petrologists is the perthite structure of alkali feldspars.

Figure 4.13(a) shows the phase diagram for solid nickel–iron. Consider a mixture containing, say, 10% nickel, initially at 800°C (point A in the figure). At this temperature the two elements are fully miscible and form taenite (face-centred cubic structure) but when the temperature has fallen to B this ceases to be so. Below B, kamacite (body-centred cubic structure) of composition B_1 forms within the host lattice of taenite. Further cooling increases the disparity between the two lattices, and their compositions are given where a horizontal line at the appropriate temperature intersects the two heavy lines, for example C_1 and C_2 at 600°C. The proportions of C_1 and C_2 must be such that the bulk composition remains constant, 10% Ni, 9% Fe in this example. The kamacite forms along planes in the taenite that correspond with the surfaces of an octahedron (two pyramids base to base) (Fig. 4.13(a), inset); hence the name octahedrite is sometimes given to this class of meteorite. An arbitrary section through an octahedrite reveals the kamacite lamellae as intersecting lines, giving rise to the Widmanstätten patterns.

The compositions of the lamellae and host adjust to the required values by diffusion of iron and nickel atoms across the boundaries between the two minerals. But since diffusion rates slow with temperature, cooling will reach a point where diffusion cannot keep up and compositions will tend to become 'frozen', particularly where atoms have further to travel, i.e. away from boundaries. Figure 4.13(b) shows how Ni/Fe ratios vary across a crystal in such a case. From a knowledge of diffusion rates, such variations can be used to deduce cooling rates which, for typical iron meteorites, range from <1 to 20°C per thousand years (Saikumar and Goldstein, 1988). These are the rates expected for cooling of bodies of asteroidal size. The same method applied to tiny metal nuggets in CAIs reveal that they cooled over periods of 10 days to 10 years.

4.5.4 Ages of meteorites

Three types of age can be determined radiometrically. These are the **crystallization age**, which is the age when the minerals were formed or were last severely disturbed; the **exposure age**, the time since the sample was at or very near the 67

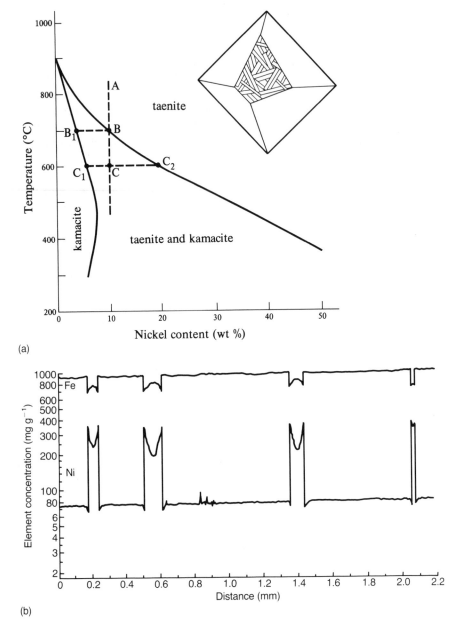

Figure 4.13 (a) Iron–nickel phase diagram at atmospheric pressure; at lower temperatures, kamacite exsolves along octahedral planes in the host taenite, and the inset shows the Widmanstätten patterns resulting from an arbitrary section through the lattice (after Goldstein and Short, 1967). (b) Variation of iron and nickel across lamellae; the variation within the wider lamellae is because equilibrium is not being maintained (based on Fig. IV–4 of Wasson, 1985).

surface of a body in space, and the **formation interval**, the time between the synthesis of some of its elements and their incorporation into the sample.

Crystallization ages. These may be determined by a variety of methods, particularly the Rb–Sr and Pb–Pb isochron methods (see Note 4), and the majority of ages fall in the range 4500 to 4600 Ma. An isochron results only if the parent/daughter ratios of all the samples contributing to it diverged at the same time from a single system with a uniform parent/daughter ratio; conversely, the existence of an isochron means that the samples had a common origin; the slope of the isochron then gives the time since the divergence. Samples of many different ordinary chondrites lie close to a straight line on an Rb–Sr isochron plot (Fig. 4.14(a)), showing that they had a common origin, which was

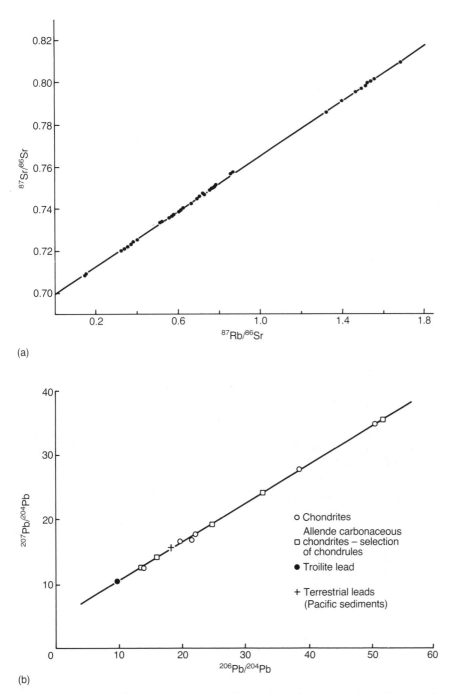

(a)

(b)

Figure 4.14 (a) Rb–Sr isochron plot of samples of E, H and LL chondrites. The line has a slope corresponding with an age of 4·555 ± 0·01 Ma (reproduced from Minster *et al.*, 1982). (b) Pb–Pb isochron plot of iron and stony meteorites. Its slope corresponds with an age of 4550 ± 70 Ma. Note that leads from terrestrial sediments fall on the same isochron. (Based on Murty and Patterson, 1962.)

4498±15 Ma ago (the error quoted is analytical; uncertainty in the decay constant of ^{87}Rb means that the date could be over 4500 Ma). Isochrons also result for other decay schemes, with dates of about 4550 Ma. A similar date has been found using the potassium–argon method for minerals from two iron meteorites.

The lead–lead dating method (Note 4) also provides convincing evidence that most meteorites are closely related, and the Earth, too. Chondrites yield a good line on a Pb–Pb isochron plot (Fig. 4.14(b)), with an age of 4550 ± 70 Ma. Iron meteorites often have troilite (iron sulphide) inclusions which contain significant amounts of Pb but not U, and the absence of U means that the Pb-isotopic composition has not evolved since the troilite formed. As the Pb-isotopic composition is almost identical for samples from different iron meteorites, and because it has the lowest ratios, troilite lead is the most primitive of all known leads, and is termed primeval lead. Further, it plots on the chondritic isochron extended down to the left (Fig. 4.14(b)); this shows that the chondritic leads evolved from this initial isotopic composition. (The Earth is related too: though we know of no rocks that have survived as closed systems since the Earth formed, the mantle as a whole can be regarded as such a closed system. Representative samples are provided by recent oceanic sediments and sea-floor basalts, which have derived from the mantle, and these plot close to the meteorite isochron, as shown in Figure 4.14(b).)

Collectively, all of the above radiometric data, plus other data not discussed, show that most of the meteorites were formed from a common, well-mixed reservoir of elements and crystallized within a time that has not yet been resolved; and neither have systematic age differences between different meteorite classes been found. Thus the meteorites formed within a few tens of millions of years, about 4550 Ma ago (Dalrymple, 1991).

A further inference is that chemical processes operated at this time to differentiate the chemical elements, for data will be distributed along an isochron only if the initial parent/daughter element ratios were different at the time of closure; if they were not, then a single point would result. Differentiation of the elements will be discussed in the next chapter.

There are many dates considerably younger than 4550 Ma. Some of these reflect collisions in the meteorite parent bodies, while the dates of the SNC and lunar meteorites (see Section 4.5.5) variously record the much later times of crystallization in these bodies, which have had a more active history, as discussed in Chapter 5.

Exposure ages. Exposure ages can be determined because cosmic rays react with the nuclei of atoms they encounter, to form new isotopes by spallation. As cosmic rays penetrate only a metre or so into a solid, the accumulation of spallation products in a meteorite occurs only after it is at or near the surface of a celestial body. As production rates are known only poorly, ages obtained using different isotopes can differ by a factor of two, but the *relative* values of dates using a single isotope are in correct relation. Exposure ages tend to cluster, believed to reflect times of breakup of the parent bodies. For example, many dates for H-group ordinary chondrites are about 7 Ma. Ages for irons are generally much older, hundreds of millions of years. In some meteorites there is evidence that more than one breakup occurred, surface material being buried in the rubble produced by an impact, to be re-exposed by a later impact.

Formation ages. From our understanding of nucleosynthesis, we know that many radioactive nuclei with half-lives much shorter than the age of the Earth must have been formed, but they have decayed to negligible proportions by the present time. However, it is sometimes possible to detect the former existence of such an isotope, from an 'excess' of its daughter product, as an example illustrates.

The isotope ^{129}I decays to form ^{129}Xe. Since xenon has many isotopes, a sample containing xenon with the common isotopic proportions, *except* for a

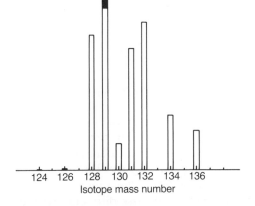

Isotope mass number

Figure 4.15 Xenon isotopic anomaly. The proportions of the various isotopes of xenon in a meteoritic sample are the same as in air, except that ^{129}Xe is larger by the shaded amount.

higher, or anomalous content of ^{129}Xe, is evidence that it once contained ^{129}I (Fig. 4.15). We would like to know how long it was from the synthesis of ^{129}I and its incorporation into the sample: the formation age. The ratio in which ^{129}I and the stable ^{127}I were synthesized can be estimated from nucleosynthesis theory, so that measuring the amount of ^{127}I in the sample gives the amount of ^{129}I synthesized. Then the ratio of the amount of ^{129}I present in excess in the sample, divided by the amount synthesized, can be substituted in the radioactive decay equation (Note 4) to yield the formation age. The half-life of ^{129}I is only 15·7 Ma, and the intervals deduced range from a few million to a few tens of millions of years.

Many other extinct isotopes have been claimed. For some, there is ambiguity whether the grains containing the isotopic anomaly were formed within the Solar Nebula, so that the interval refers to the time before the Solar Nebula formed that the isotope was synthesized, or whether some of the original dust from which the Solar Nebula formed already contained the isotopic anomalies and escaped volatilization. There are also sometimes doubts whether the anomalies could have been produced in other ways. But there is little doubt that ^{244}Pu, half-life 82 Ma, was present when the Solar Nebula formed, and it is likely that even ^{26}Al, with a half-life of only 0·75 Ma, also was present. They would have generated heat, which in the case of ^{26}Al could have been considerable for a few million years, and could have helped the planetary embryos to melt. When stars form from a cloud, many are formed, and there is recent evidence that much of the production of the heavier elements, including radioactive ones, occurred in massive stars that formed first. These stars evolved into supernovae and enriched the cloud before other stars, such as the Sun, formed. This could explain why traces of such short-lived isotopes as ^{26}Al are found in meteorites.

4.5.5   Parent bodies: the asteroids

Meteorites derive from a variety of sources. A few have come from the Moon, presumably ejected by collisions. Another small group are the SNC meteorites (standing for shergottites, nakhlites and Chassigny types of meteorites) are recognized to have come from Mars, ejected by impacts late in its history. The nakhlites, for instance, give a crystallization age of about 1300 Ma.

But most meteorites are believed to derive from the asteroids. This is because their orbits are elongated, having their nearest approach to the Sun (perihelion) a little inside the Earth's orbit, and furthest point (aphelion) between Mars and Jupiter. The aphelion indicates the position of the original nearly circular orbit from which they were probably perturbed. Different meteorite classes derive from different distances from the Sun, with the ordinary chondrites from a distance of about 2·5 AU, while the basaltic achondrites come from only a little over 2 AU.

How were the meteorites formed and perturbed into Earth-crossing orbits? Asteroids in resonance with Jupiter, particularly the 3:1 resonance at about 2·5 AU, are perturbed into elliptical orbits. In their repeated passages through the asteroid belt they suffer further collisions, the times of some of which are known from cosmic ray exposure ages. Some of the debris from these collisions chance into Earth-crossing orbits, and may hit the Earth, before the gravitational influence of planets they pass by deflect them into other orbits.

The next step is to attempt to identify different types of meteorites with different types of asteroid. Our knowledge of asteroids is mostly limited to what we can learn from the spectra of sunlight reflected from them, though their reflectance of radar beamed from the Earth also helps as this depends upon their metal content. The main drawbacks of this approach are that – in contrast to the sharp lines found in the spectra of gases – the spectra have only broad features (Fig. 4.16(a)); in addition, only the surface is analysed.

Figure 4.16(b) shows the spectra of some common minerals. The presence of, for example, pyroxene can be recognized in some of the asteroids from the dip just to the left of 1 μm wavelength. The asteroid spectra have been classified into various types, and it is found that the proportions of the various types varies with heliocentric distance. Some of the spectral types have been identified with meteorite types, but the proportions do not match; in particular, there are few known asteroids that match the abundant ordinary chondrites. A number of

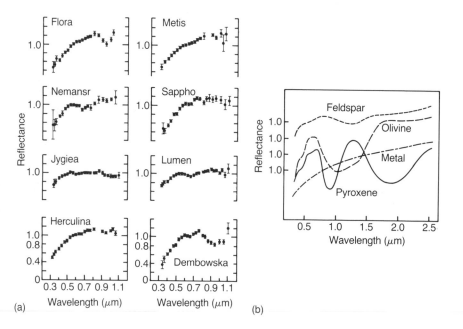

Figure 4.16 (a) Examples of asteroid reflectance spectra, showing the fraction of sunlight reflected back at various wavelengths. (b) Reflectance spectra of some common meteorite constituents. (Adapted from McSween, 1987, Figs 3.4 and 3.5.)

possible explanations for this lack of correlation have been offered, such as surfaces not being representative of the interior, or that the relevant asteroids are among the many yet to be examined, but none are widely accepted.

Despite the poor match of asteroids to meteorites, there is little doubt that many asteroids have melted. Probably the heat to melt such small bodies was not primarily due to impact energies, and other possible energy sources were perhaps decay of the short-lived ^{26}Al, or – more favoured nowadays – heating by electric currents induced by the intense magnetic fields of the Sun in its T-Tauri stage (Note 3).

In summary, the meteorites are a diverse group of objects from which a wide range of data has been obtained. Their parent bodies – the asteroids for most of them – were formed within a few tens of millions of years, about 4550 Ma ago, from which the meteorites derived, and still derive, by collisions. The chondrites are undifferentiated and are the most primitive, with compositions closest to that of the Sun and hence the Solar Nebula. Even so, all have been altered to some degree, and so record processes that presumably also affected material subsequently incorporated in the Earth. These processes are likely to have depended upon position in the Solar Nebula. The differentiated meteorites demonstrate that bodies, probably asteroid-sized, could melt and differentiate before the growth of the Earth was complete.

The implications of these conclusions to our understanding of the composition of the Earth will be explored in the next chapter.

SUMMARY

1. The elements were synthesized by a large number of nuclear reactions, mostly within massive stars, commencing with the hydrogen and helium produced by the Big Bang. When a massive star has exhausted the nuclear fuel at its centre, it becomes unstable and explodes as a supernova, blasting the synthesized elements into space. They are incorporated into later stars, such as the Sun.
2. The Solar System probably formed from a rotating cloud containing gas and dust, which flattened under gravity. Much of the matter contracted to the centre as it transferred its angular momentum outwards; it was destined to become the Sun, while the remaining matter remained in a disc rotating about the proto-Sun, and from it the planets, etc. formed.
3. The material of the disc moved towards the median plane, heating itself by compression and briefly volatilizing most of the dust. As it cooled, dust recondensed, the most refractory first. The dust particles accreted in several stages: first, they fell towards the median plane, coagulating by 'sticky impact', and becoming large enough to form planetesimals, a few metres in diameter and largely unaffected by the drag of the gas. Secondly, the planetesimals grew by collision when their orbits intersected, leading by runaway growth to planetary embryos, ranging in mass from 1 to 10% of the Earth's present mass. Thirdly, by mutual perturbation, the embryos collided and amalgamated into the planets. Owing to the relatively small number of bodies involved, the outcome of this third stage partly depended on chance, accounting for the different spins of these planets. The Earth–Moon system is probably the result of such a collision.

For the terrestrial planets, the three stages successively took approximately a few tens of thousands, approaching a million, and about 100 million years.

4. While the material of the disc was accreting the proto-Sun was evolving. Infall of material produced a hot body with a strong solar wind, but before 10^5 years had elapsed the Sun entered the less vigorous T-Tauri stage, lasting about 10 Ma. Its still considerable solar wind removed any unaccreted gas and dust. During this stage, there may have been very brief episodes of greatly increased luminosity.

   At the end of the T-Tauri stage the central temperature had reached about 10 million degrees and hydrogen burning commenced. The Sun then entered a still-continuing phase as a stable main-sequence star.

5. The giant planets grew large enough to attract a large amount of the light elements as ices before the T-Tauri stage dispersed them.

   Jupiter influenced the growth of the terrestrial planets, particularly Mars; prevented the asteroids growing into a planet; and slung the comets outwards to form the Oort Cloud.

6. The Sun's composition is our best estimate of the composition of the Solar Nebula (excepting lithium). Solar abundances are determined from spectroscopic measurements of the solar spectrum. Abundances vary over a range of 10^{12}, with elements heavier than helium together forming about 2% of the Sun. The proportions of the elements are much as predicted by nucleosynthesis theory: generally decreasing abundances with increasing mass, with local peaks of the more stable nuclei.

7. Meteorites are samples of extraterrestrial material, derived mainly from the asteroid belt, that have reached the Earth's surface. They may be divided into differentiated – formed from a melt – and undifferentiated – which contain minerals and structures which have not equilibrated. The differentiated meteorites comprise irons, stony-irons and achondrites; the undifferentiated are the chondrites.

8. Most meteorites are 'fossils' of the various stages of the accretion process. The chondrites are more primitive than differentiated meteorites, with carbonaceous chondrites being the most primitive of all, and have abundances close to those of the Sun, apart from the most volatile elements.

FURTHER READING

General journals:
   Binzel *et al* (1991): asteroids.
   Bethe and Brown (1985): how a supernova explodes.
   Henbest (1992): about the 1987 supernova.
   Stahler (1991): formation and early evolution of stars.
   Woosley and Weaver (1989): about the 1987 supernova.
General books:
   Dalrymple (1991): the ages of meteorites and the Earth.
   McSween (1987): meteorites.
   Murray (1983): the planets.
   Seeds (1990): an introduction to astronomy that includes stellar evolution, supernovae, galaxies, etc.
   Wasson (1985): meteorites.
   Wood (1979): the Solar System.

Advanced journals:
Chevalier (1992): Supernova 1987A.
Lada and Shu (1990): the formation of Sun-like stars.
Advanced books:
Kerridge and Matthews (1988): meteorites and the early Solar System.

5 The accretion and layering of the terrestrial planets

5.1 THE FIRST 100 MILLION YEARS

Our attention now focuses on the processes that determined the present compositions and internal structures of the Earth and its near neighbours. We have seen that the wide diversity of chemical elements found in the Solar System was produced by nucleosynthesis processes, mostly inside stars. Moreover, the chemical composition of the small, dense terrestrial planets is dominated by heavy elements that comprised just 2% of the Solar Nebula. So what processes were responsible for selectively concentrating these elements into the terrestrial planets, how did they operate and at what stage during the progressive accumulation of dust grains into planetesimals, planetary embryos and finally planets? Meteorites provide us with a good starting point (Section 4.5) for they show that similar chemical selection, and in some cases internal differentiation processes, operated as far out from the Sun as the asteroid belt. But beyond the asteroids, planetary densities are much lower (Table 4.1) implying compositions much closer to the Solar Nebula. Moreover, there are variations in density among even the terrestrial planets, and this may be due to chemical differences between them (Section 5.2.2). So there must have been a primary differentiation of the elements across the nebula.

We now know that there is a close similarity in the abundances of the elements heavier than oxygen in chondritic meteorites and the solar atmosphere (Fig. 4.12). It is this striking similarity that led to the suggestion that all of the inner Solar System planetary material might also have a chondritic bulk (i.e. total average) composition. The **Chondritic Earth Model** (CEM) proposes that this is true for our own planet, but we shall see that while this provides a good preliminary model, in detail there are important differences between the Earth and chondrites. Planetary and lunar exploration have allowed other chemical differences to be discovered by acquiring data from Mars, Venus and the Moon. In turn, this information provides vital clues about the chemistry of planetary accretion from the Solar Nebula.

To build bridges from the physics (Chapter 4) to the chemistry of planetary accretion we need to know more about the evolution of temperatures and material densities in the Solar Nebula. This is a complex subject because these physical constraints depend on the details of the model; the rates of processes are critical so we need a time-scale for planetary accretion. Among the most useful data available are the excellent radiometric ages that give a time-scale for the early history of the meteorites and terrestrial planets. The oldest known grains of millimetre size, unmodified since Solar System formation, are the so-called 'high-temperature inclusions' from the Allende meteorite (Fig. 4.10(c)) which are dated at 4559 ± 4 Ma, preceding by up to 30 Ma the ages for other types of meteorites (Section 4.5.4). A date of 4550 Ma is commonly accepted as the beginning of the Solar System, in the sense of dating the oldest known solid material. There are very few *terrestrial* rocks older than 3800 Ma (see Chapter

11), reflecting the lack of a stable crust, but ages of 4440 ± 20 Ma were obtained from lunar crust samples returned by various space missions. Given that the Moon was molten during its formation (discussed later), this age records the crystallization of the lunar crust. Assuming for now that the Earth and Moon formed together, this gives an upper limit of *c.* 100 Ma between the ages of the oldest known small particles and the existence of the Earth–Moon system. Physical models of planetary accretion (Section 4.3) also show that this is the time interval needed for the growth of Earth-like planets. Thus pre-planetary chemical segregation processes must have been well advanced by 100 Ma after the first solid grains accumulated in the Solar Nebula.

Physical models of the nature and timing of events within this first 100 Ma have preoccupied many astronomers and space scientists. Most now agree with calculations by Hyashi and co-workers at Kyoto University (e.g. Hayashi *et al.*, 1985) that the initial *cold* stage of dust and gas contraction (Section 4.3) to form the Solar Nebula, was a rapid event taking around 10^4 years. The next stage, the accumulation of submicron-sized particles in the mid-plane of the developing nebula was nearly as rapid, taking 10^4–10^5 years and involving strong heating due to the release of gravitational energy. While the outer, lower-density parts of the nebular disc may have remained quite cold, closer to the Sun all the dust grains are likely to have been completely vaporized. Indeed, the existence of a strong temperature gradient across the nebular disc during the vaporization phase is thought to be a key factor in fractionating the elements across the pre-planetary Solar System. Complete vaporization requires temperatures in the 1500–2000 K range, and the likelihood that the whole of the inner Solar System passed through this gaseous phase is consistent with geochemical evidence from meteorites and their inclusions. A note of caution, however: pre-solar grains of silicon carbide, diamond and some organic compounds occur in undifferentiated meteorites, indicating that temperatures may not have exceeded a few hundred degrees in parts of the asteroid belt and beyond. Radiative cooling caused recondensation of the gas to form micron-sized grains which coagulated and gradually developed into gravitationally stable planetesimals (upwards of a few metres across, see Section 4.3). Accumulation of planetesimals into planetary embryos would have taken a further 10^5 years in the inner Solar System, but probably 10^7 years further out. The final stages of planet assembly probably occupied much of the next 100 Ma.

The first aim of this chapter is to consider how planetary chemistry was affected by processes operating in the Solar Nebula, the pre-planetary stage. The second question that it addresses is the origin of large-scale stratification, or layering in planets, particularly the development of iron-rich cores and silicate mantles such as are believed to characterize all the terrestrial planets. This involves the geochemical segregation of elements inside embryos and planets by melting processes during and after accretion. In a nutshell, the processes are:

1. segregation of elements across the Solar System before and during the early stages of accretion, which was a rapid process influenced by pressure and temperature gradients away from the Sun;
2. planetary accretion over a longer time-scale, which involved the progressive accumulation of planetary embryos, and then planets, which became hot internally, enabling them to develop layered structures.

5.2 PRE-PLANETARY CHEMICAL PROCESSES

5.2.1 Condensation processes in the Solar Nebula

What was the likely order in which dust grains formed by condensation? We start with a hot gaseous mixture of all the elements present in the Solar Nebula, including the light elements (H, He, C, N, O, etc.) observed in solar spectra (Fig. 4.7 lower curve). The exact order in which elements condensed depended on factors such as the mineral compounds that formed, the temperature, the pressure and other factors that will be introduced where relevant. Owing to the low pressures and their small range (10^{-3}–10^{-5} bars) the condensation order will have been little affected by change of pressure, and the sequence of compounds determined for 10^{-4} bars is taken as typical (Table 5.1). (This is based on

Table 5.1 Approximate condensation temperatures for grains forming in the cooling Solar Nebula at 10^{-4} bars (after Grossman and Larimer, 1974).

Mineral phase	Composition	Condensation temperature (K)	
Corundum	Al_2O_3	1680	↑ more
Perovskite	$CaTiO_3$	1560	
Melilite	$Ca_2Al_2SiO_7$–$Ca_2MgSi_2O_7$	1470	Refractory
Diopside (pyroxene)	$CaMgSi_2O_6$	1410	
Spinel	$MgAl_2O_4$	1390	
Metallic iron	Fe(Ni)	1380	
Forsterite (olivine)	Mg_2SiO_4 ($-Fe_2SiO_4$)	1370	
Enstatite (pyroxene)	$Mg_2Si_2O_6$ ($-Fe_2Si_2O_6$)	1360	↓ less
Anorthite (feldspar)	$CaAl_2Si_2O_8$	1230	
Alkali feldspars	$(Na,K)AlSi_3O_8$	1060	↑ more
Troilite	FeS	650	
Magnetite	Fe_3O_4	410	
Hydrated silicates	various	300	Volatile
Water ice	H_2O	240	
Ammonia ice	$NH_3.H_2O$	130	
Methane ice	$CH_4.6H_2O$	90	
Nitrogen ice	$N_2.6H_2O$	90	↓ less

thermodynamic calculations rather than experiment, and is highly simplified.) Different minor elements can substitute in these mineral structures, causing small changes in the condensation temperatures. At this stage, it is convenient to distinguish compounds that condense above 1200 K, known as **refractory,** from those that condense at lower temperatures and are **volatile.** Further subdivision is often employed, and bearing in mind how elements occur naturally in different compounds (Table 5.1), we can recognize the following categories: highly refractory (Al, Ca, Ti), moderately refractory (Mg, Fe, Si), moderately volatile (Na, K, S) and highly volatile or gaseous (H, C, N) elements.

Armed with this information, we can now appreciate the significance of the Allende inclusions. More than two tonnes of this carbonaceous chondrite fell at Pueblito de Allende, Mexico, in 1969, and it has since been the subject of endless research and debate. About 8% of this meteorite consists of tiny white Ca- and Al-rich inclusions (**CAIs**) containing **perovskite, melilite** and **spinel** as important minerals (see Table 5.1). Their compositions are therefore strikingly close to the predicted *first* refractory condensates from the Solar Nebula. Perhaps even more important, these inclusions contain tiny metal nuggets of highly refractory trace metals (Re, Mo, W and platinum-group elements) and, in some cases, less-refractory Fe–Ni alloys with exsolution textures (see Fig. 4.13). The rather rapid cooling over a period of 10 days to 10 years (derived from these textures) is taken to indicate that the refractory Allende inclusions, up to a centimetre in size, must have formed as small particles within the hot, but cooling nebula. They had not become incorporated within well-insulated planetesimals, which would have led to slow cooling rates such as observed from iron meteorites (Section 4.5.3).

It is important to distinguish the timing of refractory inclusion production from the subsequent period of meteorite chondrule generation. Although this, too, was an early event, in which condensed silicate grains were fused into molten droplets, perhaps by solar flares (see Section 4.5.2), their estimated cooling times (<1 hour) are so fast that the nebula must have been cooler when they formed. Again, chondrules are tiny, just a few millimetres across, so they must have formed before grains had accumulated to planetesimal size. So it seems that chondrules may be a fossil remnant of a *cool pre-planetesimal stage which followed condensation*. Differences in iron metal/silicate ratios observed in the various chondrite groups (Fig. 4.10) also provide evidence about the pre-planetesimal stage, but of a different kind. Going back to the condensation sequence, one important change with pressure is the order in which iron and silicate form (Fig. 5.1). At high pressures, perhaps close to the centre of the nebula, metallic iron condenses first and so may be more common. At lower pressures, further out, primary iron silicate condensation may prevent the formation of much metal because the available iron is used up in producing silicate. (The process is almost certainly more complex than this, for example, because the *partial* pressure of oxygen in the nebular gas would influence condensation temperatures.) In general terms, we can see that the local pressure of condensation influenced the nature of the grains then available for planetesimal growth in any part of the nebula. This might account for the observation (Section 4.5.5) that the most-oxidized bodies tend to occur towards the outer perimeter of the asteroid belt where pressures would have been lower.

Whatever the details of the Fe–Si fractionation process, it may well have affected not just chondrites but also the terrestrial planets, for the anomalously high density of Mercury (Table 4.1) must imply a higher Fe/Si ratio than for any other planet, including the Earth.

A final question to consider about condensation processes is why so little of the highly volatile compounds in Table 5.1 appears in the terrestrial planets. After all, these compounds comprised over 98% of the Solar Nebula, so does their absence from the inner planets mean that temperatures never fell below those necessary for large amounts of water ice, ammonia ice, etc. to form? And what about the most abundant species in the Solar Nebula: H and He – where have they gone? As discussed in Section 4.3.2, the T-Tauri wind and associated brief outbursts would have been capable of sweeping much of the Solar System clear of uncondensed gases and, certainly in the terrestrial planet zone, of any sub-planetesimal material smaller than about 1 metre in diameter. It seems that the T-Tauri winds are likely to have commenced between 10^5 and 10^6 years after nebular contraction started, and to have lasted for up to 20 Ma. Thus, in terms of the time-scales in Section 5.1, planetary embryos would have formed from refractory and even moderately volatile compounds. But highly volatile com-

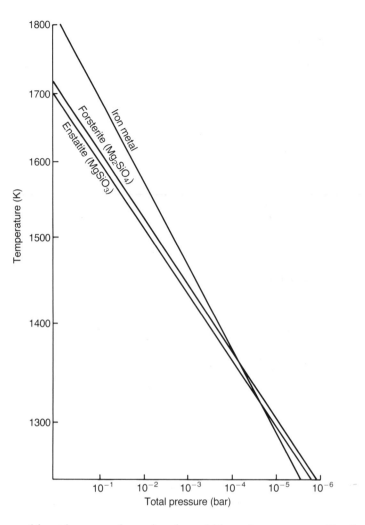

Figure 5.1 The effect of varying pressure, over the likely range for the Solar Nebula, on condensation temperatures for metallic iron and for the two most common silicate minerals likely to be produced from the nebula. (Source: Grossman, 1972.)

pounds would not have condensed and would have been removed by the intense solar winds.

Sadly, it is not yet possible to constrain the timing of events and predictions of chemical consequences in the condensing nebula with any greater precision, but the theory is able to account for gross planetary density differences. Beyond the asteroids are the low-density giant planets, believed to incorporate significant amounts of hydrogen, helium and volatile compounds. They also have many icy satellites, some as big as Mercury (see Rothery, 1992 for similarities between processes in the satellites of the outer planets and the terrestrial planets). The satellite densities are lower than those of the terrestrial planets, yet these bodies are probably layered with 'mantles' of water ice, and probably ammonia ice too, surrounding silicate 'cores'. This reflects their distance from the Sun. Water ice must have been able to condense in this part of the Solar System, requiring a temperature below 160 K, the pressure-adjusted condensation temperature for water at 4–5 AU. This leads to the concept of a **snow line** within the condensing nebula (Fig. 5.2(a)), the effect of which was to increase the abundance of solid particles at and just beyond the distance from the Sun where water condensed. Figure 5.2(b) illustrates a possible scenario for the **surface density** (total mass per unit area perpendicular to the nebular disc) for both gas and dust

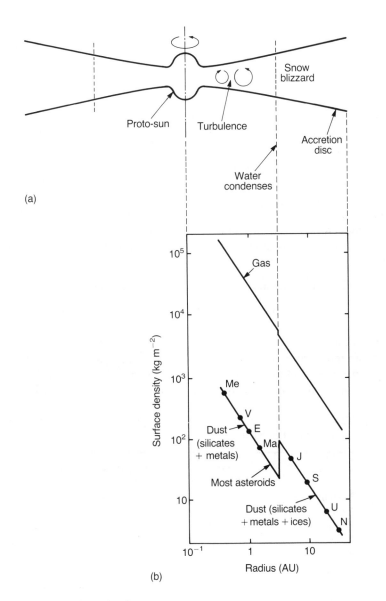

Figure 5.2 (a) Cross-section of the Solar Nebula during condensation, prior to the onset of the T-Tauri stage for the Sun. Grains of water ice had condensed at a distance around 4–5 AU from the Sun (1 AU = 1·5 × 10^8 km) giving rise to a snow line (dashed). (b) Equivalent distribution (mass per unit area) of gas and dust grains across the Solar System. Although the amount of dust decreased by several orders of magnitude away from the Sun, superimposed on this trend was a five-times increase at the snow line because of ice condensation. Even at this stage, there was about 100 times as much gas as dust. The planet positions are marked for easy reference, though none had been fully accreted. (After Hayashi *et al.*, 1985 and Stevenson and Lunine, 1988.)

components. It is this distribution of matter that was interrupted by the T-Tauri wind. Although a large part of the remaining volatile gases may have been accelerated right out of the Solar System at this stage, the presence of H and He in the giant planets suggests that embryonic planets beyond 5 AU were already large enough, gravitationally, to retain some of these gases against the strong solar wind (see Section 4.3). But by far the majority of the gas component in Figure 5.2(b), representing well over 90% of the Solar System mass outside the Sun, was lost at this stage.

Figure 5.2(b) makes the further interesting prediction that pre-planetary gas and dust density varied by a factor of 100 across the outer Solar System. A simple consequence of the very low densities in these outer reaches is that embryo formation would have been much slower. One estimate is that whereas Jupiter grew rapidly, perhaps developing into a substantial planetary embryo in 4 × 10^5 years, it may have taken Neptune as much as 30 Ma to reach the same state. This may explain why Uranus and Neptune are smaller giants than Saturn and Jupiter: the former two had attracted less H and He gas when the T-Tauri wind developed.

Taking stock, it seems that the chemical fate of the planets was, in many ways, sealed within less than the first million years of Solar System history. During this time, much of the initially cold, contracting nebula had vaporized. The re-condensation and particle coagulation processes for all refractory and some of the moderately volatile compounds were well advanced, certainly in the terrestrial planet zone. Dust had accreted into planetesimals, and runaway growth of most embryos had occurred. Beyond 5 AU, they incorporated water and other ices. The T-Tauri wind then swept most of the uncondensed matter right out of the Solar System. However, some of these gases had already been attracted gravitationally into the giant planets. Next we focus on terrestrial planet compositions and ask just how heterogeneous was the material from which they formed. It is time now to return to the Chondritic Earth Model.

5.2.2 Terrestrial planet and chondritic meteorite compositions

The objective here is to consider primary chemical differences between the dense bodies of the inner Solar System in which oxygen, silicon, iron, magnesium and possibly sulphur are the most important elements (Figs 4.7 and 4.12). As an initial constraint, planetary densities are used (Table 4.1), but these must be corrected to allow for the different pressures that occur inside bodies of different sizes. Table 5.2 reveals some interesting variations, with only Mars

Table 5.2 Densities of the terrestrial planets and chondritic meteorites (after Taylor, 1988).

	Density (Mg m^{-3})	*Estimated uncompressed density* (Mg m^{-3})
Mercury	5·435	5·3
Venus	5·245	4·0
Earth	5·514	4·0
Moon	3·344	3·3
Mars	3·934	3·7
Chondrites*	3·4–3·9	3·4–3·9

* This density range excludes carbonaceous chondritic meteorites that have large amounts of hydrated minerals, as these are unlikely to be typical of the bulk of the terrestrial planets.

falling inside the range for chondritic meteorites. The Moon is included here simply to note that its density is even lower than for an average chondrite. Now the only sufficiently abundant element able to determine such strong density differences is iron. Ignoring atomic *volume* for the moment, this is because iron has a relative atomic *mass* of 56, much greater than any of the next three most abundant elements (O = 16, Mg = 24, Si = 28). For example, the low iron (L and LL) chondrites, with an *overall* Fe/Si ratio of c. 0·5–0·6 (roughly the sum of the ratios on both axes of Fig. 4.10), give the lowest densities in the range quoted (Table 5.2), whereas the high-iron group (H) chondrites give the largest densities. The H-group chondrites have an average Fe/Si ratio of 0·8. Moreover, 60–80% of the iron in H chondrites occurs as metal, reflecting a paucity of oxygen, and thus high Fe/O ratios too. Thus we may conclude from Table 5.2 that, overall, Mercury, Venus and the Earth are likely to have an even higher Fe/Si ratio and perhaps less oxidized iron than any chondrites, whereas the Moon almost certainly has a lower Fe/Si ratio. For the Earth and Venus, this implies that slightly more of the iron occurs in a metallic state than the 60–80% in H-chondrites, and the same conclusions follow in a more extreme sense for Mercury. Similarly, the Moon must contain very little or no metallic iron. In turn, the amount of metallic iron available will be reflected in the size of the core inside each of these bodies (Chapter 6); leaving silicate (Si + O with Mg and any non-metallic Fe) in the mantle (Chapter 7).

A striking feature of Table 5.2 is the way that the uncompressed planetary densities and implied Fe/Si ratios decrease across the terrestrial planet zone and, although the chondritic meteorites are remarkably varied, their average fits this trend. This indicates that *iron/silicate fractionation* must have occurred during or after condensation but before accretion. The likely processes were discussed earlier, and this density evidence suggests that the dust grains from which the terrestrial planets ultimately formed were condensed, on average, at progressively decreasing pressures (Fig. 5.1), as well as temperatures, away from the Sun. The fact that the Moon appears to have too low an iron/silicate ratio, whereas the Earth's may be too high (see densities in Table 5.2), is probably connected with their early interaction at the embryo stage, as we shall see in Section 5.4. Perhaps the most surprising deduction is that the pressures varied markedly over the width of the asteroid belt, and this is confirmed by similar fractionation between H and L chondrites for other metallic elements (Table 5.3). However,

Table 5.3 Fractionation of metallic elements and sulphur between H- and L-group chondrites together with condensation temperatures.

Element	Abundance L/H	Condensation temperature (K) at 10^{-4} bar
Ir	0·57	1610
Ni	0·68	1354
Co	0·66	1351
Fe	0·74	1336
Pd	0·55	1334
Au	0.61	1225
Ga	0.93	918
S	1·07	648

the lowest-temperature condensates in this list, gallium and sulphur, are *not* strongly fractionated, implying that fractionation processes among these meteorites occurred at temperatures above 1000 K.

Another way of looking at variations in the principal elements is to plot atomic ratios, and this is done for chondrites in Figure 5.3, using Fe/Si and Mg/Si ratios. Taking the carbonaceous chondrite (CI – Section 4.5.2) element ratios as 'parental', then other groups can be accounted for in terms of the processes discussed above, involving changing the Fe/Si ratio by 'iron metal loss' and increasing the proportion of Fe_2SiO_4 (and/or decreasing Mg_2SiO_4) in the accreted material. Similar processes would account for the large range found in

Figure 5.3 Variation of
average Fe/Si and Mg/Si
ratios among the chondritic
meteorite groups (see Fig.
4.10) compared with an
estimate for the Earth which,
in simple terms, could have
CI chemistry with the
addition of iron metal.
Arrows indicate the effects of
iron gain/loss, which does not
change the Mg/Si ratio, and
of variations in Fe_2SiO_4 and
Mg_2SiO_4 constituents. If CI
chemistry is taken as parental,
the Fe–Mg chemistry of the
other chondrites may be
explained in terms of lower
amounts of iron and higher
amounts of Fe_2SiO_4 grains
involved in their accretion.
(The present amounts of
oxidized and reduced iron in
these meteorites (Fig. 4.10)
are a function of their oxygen
content, which is not
represented in this diagram.
(After Larimer and Wasson,
1988; see Ringwood, 1989, for
a different view.)

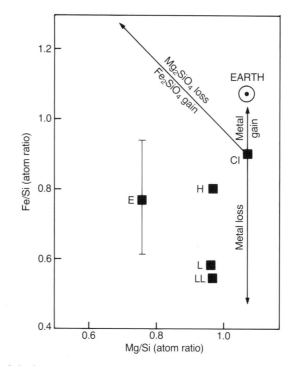

the terrestrial planets, though, as noted above, the Earth seems to have gained more metal than any of the chondrites.

Most of the compounds involved in the fractionations discussed so far are moderately refractory (Table 5.1). We also have evidence that volatile elements have been depleted in all the terrestrial planets. For example, we are fortunate to have data on the crustal ratio of potassium (moderately volatile) to uranium (refractory) for Venus, Mars, the Moon and, of course, for the Earth and chondrites. The reason why these elements have been chosen is because both are radioactive and emit gamma-rays, but of different energies. Thus orbiting space craft and landers were able to map the surfaces of Venus and the Moon using **gamma-ray spectrometers**. Lunar data are supplemented by surface samples and the Mars data come from SNC meteorites collected on the Earth (see Section 4.5.5). Although we measure the K/U ratio at planetary surfaces, it is likely to be the same at depth because the two elements tend to behave similarly in melting processes at the pressures of planetary interiors (see Section 5.3). They are both selectively concentrated in the silicate melts reaching planetary surfaces, such as mid-ocean ridge basalts (MORB) on the Earth (see Chapter 7 for details). In Figure 5.4 this means that the K/U ratio represents that incorporated during planetary accretion, whereas the K content, in surface material, is a product of post-accretion processes.

The highest K/U ratios come from chondritic meteorites (Fig. 5.4); observed values for the Earth's surface and estimates for **Bulk Earth** are lower by a factor of six. Based on the bulk composition of available material, this is almost entirely due to the Earth's lower volatile content (potassium concentrations are estimated at 180 ppm (parts per million) in the Earth, compared with 885 ppm in CIs). On available surface data (Fig. 5.4), the ratios of volatile to refractory elements for Venus and the Earth are similar. However, despite appearances in Figure 5.4, the K/U ratio for Mars is thought to be different, with the average for

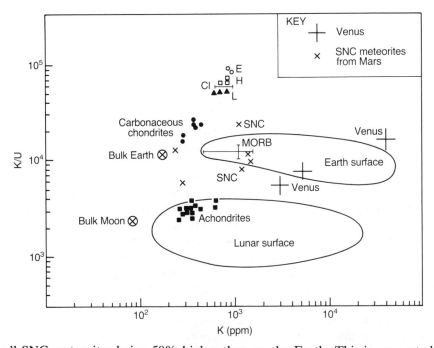

Figure 5.4 Comparison of volatile/refractory element ratios, represented by K/U, for achondrite and chondrite meteorites, terrestrial planets and the Moon plotted against potassium concentrations. The data points for Bulk Earth and Bulk Moon are estimates based on the back projection of surface rock data, for which the large areas indicated encompass the data. The three Venus points were derived by gamma-ray spectrometry during Venera soft landing missions (large crosses), and the small crosses are from SNC meteorites, almost certainly samples of the surface of Mars. The squares are values from differentiated basaltic achondrite meteorites which have lower K/U ratios than the chondrites. See text for further discussion. (After Taylor, 1988, 1992.)

all SNC meteorites being 50% higher than on the Earth. This is supported by isotopic data for SNCs which, in particular, require more volatile rubidium relative to refractory strontium in the mantle sources of Martian basalts than on Earth (see Taylor, 1988 for details). In contrast, the Moon clearly has the lowest K/U ratios. However, since estimated potassium concentrations in Bulk Earth (180 ppm) are only around twice those estimated by petrological models for Bulk Moon (83 ppm), the Moon must be enriched in refractory uranium relative to the Earth and, indeed, to chondrites (estimates for uranium are 0·014 ppm in CIs, 0·018 ppm in Bulk Earth and 0·033 ppm in Bulk Moon). Similar lunar enrichments seem to apply to other refractory elements, such as Ca, Al and Ti (Table 5.1). These data provide important constraints on the Moon's origin (Section 5.4). The only meteorites that have such low volatile/refractory element ratios, less than the Earth's and similar to the Moon's, are basaltic achondrites. These originate from parent meteorite bodies that had differentiated, and probably formed at a different distance from the Sun compared with the chondritic parents (Section 4.5.5). Significantly, they date at 4550 Ma, close to the beginning of the Solar System, so volatile element losses must have occurred early in the history of the Solar System.

So far as the terrestrial planets are concerned, significant volatile-element depletions occurred as compared with Solar Nebula values represented by CI meteorites. Thus Figure 5.4 provides further strong evidence for fundamental chemical heterogeneity between the planets and chondritic meteorites. These heterogeneities are believed to date from the period *before* planetary accretion since, if potassium were volatilized during accretion, the K/U ratio should decrease with increasing planetary size, which is not observed. Ignoring the special cases of the Moon and achondrites, there is reasonable evidence that potassium increased with distance from the Sun. Again, this may well reflect the temperature gradient across the inner parts of the nebula in the stages before planet formation. In the same way as argued earlier for water ice, it is likely that compounds containing moderately volatile elements such as potassium were incompletely condensed when planetesimals and planetary embryos were formed.

Similarly, good evidence of volatile element heterogeneities across the terrestrial planet and asteroid zone is provided by oxygen isotope differences between the Earth, Moon and meteorites and by noble-gas isotope ratios for the Earth's atmosphere, the Martian atmosphere and meteorites (Clayton *et al.*, 1985; Ott and Begemann, 1985). The details need not concern us here, so suffice it to conclude that the terrestrial planets are not chondritic, but they do have element abundances remarkably similar to those of the chondritic meteorites.

Taking stock of all these lines of evidence, the main cause of the decreasing densities of the terrestrial planets away from the Sun, after allowing for self-compression (Table 5.2), is decreasing iron metal/silicate ratios, in turn, probably resulting from decreasing condensation *pressures* in the early Solar Nebula. Other elements that are often metallic (eg. Ni, Co, Pd, Au) were probably fractionated in the same way. We have also identified differences in the ratios of moderately volatile to refractory elements (Fig. 5.4) which were probably caused by decreasing condensation *temperatures* again, away from the Sun. On this basis, meteorite compositions appear to reflect both local variations in temperature and pressure prior to their accretion. It is important to recognize that, so far, we have only been discussing processes that determined the composition of dust grains available locally within the Solar Nebula for subsequent incorporation into planets. This *still* represents a very early stage in Solar System history, well within the first million years, during which interval planetesimals formed and then began to coalesce into planetary embryos (see Section 4.3).

5.2.3 Chemical evolution during embryo formation

It is reasonable to expect that the chemical compositions of planetary embryos reflected the type of materials that had condensed in the particular region of the Solar Nebula where they formed. So it is hardly surprising that each planet reflects the distinctive planetary building blocks produced by the local pressure–temperature regime. This is generally a more successful approach than modelling the terrestrial planet compositions using mixtures of various meteorite components as was common in the 1970s. For example, the mixing of a metal-rich high pressure planetesimal component with a volatile-rich CI-like component to form the Earth might appear to be consistent with Figure 5.3. Such historical models are useful as a reminder that the planetesimals forming the embryos must have been internally heterogeneous at the centimetre scale, comprising a range of different original dust grain compositions, though each was homogeneous on a gross scale.

The theory introduced in Section 4.3 predicts that a planetary embryo was formed every 0·02 AU across the condensing nebula. Their sizes would have varied, of course, depending on how successful each embryo was in accreting planetesimals in competition with its nearest neighbours. So we have a picture of some 30–100 bodies of mass in the range between the Moon and Mars, and chemistry varying according to the prevailing pressure–temperature regime, spread across the inner Solar System ready to complete planetary accretion. However, in the asteroid belt, the accretion model suggests that the lower grain densities at 3–4 AU (Fig. 5.2(b)) would lead to slower embryo growth than in the adjacent high grain density region at 5 AU where Jupiter formed. It is believed that a strong depletion of material occurred as the asteroidal zone was scavenged by the development of Jupiter. This depletion would have prevented full embryo growth occurring in the outer parts of the asteroid belt, so that an intermediate size frequency distribution of planetesimals resulted (Fig. 4.5). However, it appears from the differentiated meteorites that the inner part of the asteroid belt, like the rest of the terrestrial planet zone, did pass through an embryo stage. The final step in planetary growth, collisions between embryos, would, in general, cause the smaller one to be fragmented. Although much of the mass accreted to the larger embryo, some was ejected from the collision at high velocity and escaped. This leads to several important consequences for the chemical compositions of the accreting planets.

Firstly, embryos with different compositions may have come together and homogenized internally, so we would not expect planetary bulk chemical compositions as deduced today simply to reflect a unique set of condensation conditions. Secondly, embryo collisions were giant impacts that released enormous amounts of energy as heat, which may have further modified the resulting chemistry by vaporizing some of the most volatile elements. For example, there is a suggestion that a large body impacting Mercury may have boiled away some of its silicate mantle, thus helping to account for its exceptionally high density (Table 5.2). An impact between the Earth and a Mars-sized object may have been responsible for the origin of the Moon (see Section 5.4), a process during which both underwent some chemical modification. Finally, although most of the small planetesimals and fragments left after embryo formation were probably swept up and accreted to the developing planets within 100 Ma, a large number remained and continued to impact on the planetary surfaces, but at a decreasing rate, for at least 600 Ma. The result was that extensive internal melting processes were initiated that led to post-accretional chemical differentiation – the subject of the next section – which commences with a review of chemical element affinities in fully formed planets.

5.3 POST-ACCRETIONAL CHEMICAL PROCESSES

Here we encounter a major change of emphasis, and instead of the low pressures of the Solar Nebula, where solids and vapours but no liquids occurred, we encounter much higher pressures at the surface and inside fully formed planets. A new set of chemical criteria apply (Section 5.3.1) which, in many ways, underpin the processes covered in much of the remainder of this book.

5.3.1 Element segregation: some geochemical rules

Many geological processes, such as weathering, sedimentation, metamorphism and partial melting, continually separate the chemical elements in what has been termed '**the rock cycle**'. For instance, segregation among sediments by physical and chemical processes leads to sandy beaches (SiO_2), limestone reefs ($CaCO_3$) and coal (dominantly carbon). The way in which elements behave both at the surface and within the planets is controlled largely by their electronic configurations and their affinities for different types of crystalline bonds. In differentiated planets and meteorites, the elements are sorted into clear groups based on silicate, sulphide and metal, and these are distinguished as follows:

1. **Lithophile** elements, which tend to occur with oxygen in oxides and silicates, e.g. K, Na, Ca, Mg, Al and Ti (Gr. *lithos*, stone). A subgroup includes those elements that tend to be gaseous at surface conditions on the Earth and so are **atmophile** (H, C, N, O and noble gases).
2. **Chalcophile** elements, which tend to concentrate as sulphides, e.g. Zn, Co, Pb, Cu (Gr. *khalkos*, copper).
3. **Siderophile** elements, which tend to be metallic, e.g. As, Pt, Ir and Au (Gr. *sideros*, iron).

Figure 5.5 Periodic Table of the elements, classifying their geochemical tendencies in natural systems, such as in the Earth. Electronegativity values appear below element symbols and a heavy line surrounds all elements that show lithophile (or atmophile) tendencies. Note 6 explains in more detail how electronegativity is related to electronic structure. In turn, the geochemical affinities of the elements are due to the nature of bonding as explained in the text. Notice that some elements near to the boundaries of a group overlap into different groups. Thus Fe, Cu, Ga and Ge may be lithophile, chalcophile or siderophile in the Earth.

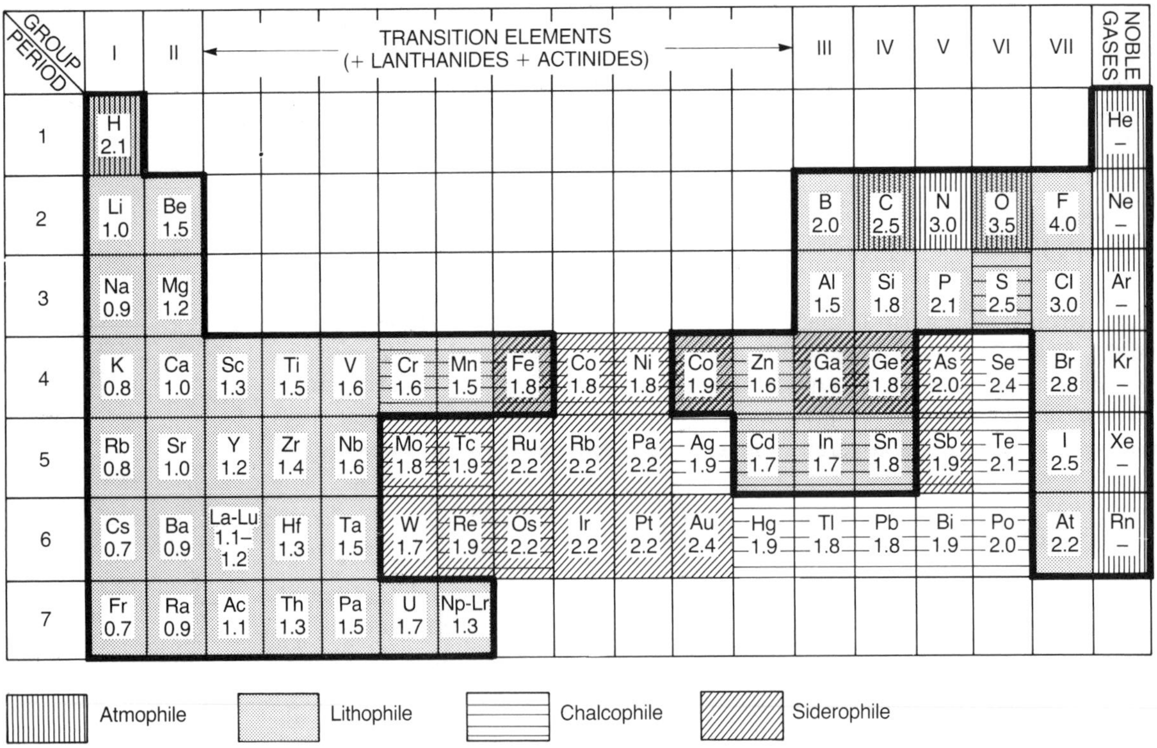

The bond-forming ability of elements depends on their positions in the Periodic Table (Fig. 5.5). Here we will simplify matters by using just one property of each element, its **electronegativity**, E, measured on a dimensionless scale from 0 to 4 devised by Pauling (1959). *This is the ability of an atom to attract electrons* (see Note 6) *and so to become a negatively charged anion*. Thus, the value is highest for fluorine and the other halogens, particularly those with small distances between their nucleus and the outer electron shell (F, 4·0). Among the other common elements, it is also high for oxygen (O, 3·5), and is moderate for sulphur (S, 2·5). But it is low for the more metallic elements which prefer to *lose* electrons to form positively charged *cations* (e.g. Mg^{2+}, Si^{4+}, Fe^{2+}; the uncharged atoms have E between 1·2 and 1·8). Apart from oxygen, sulphur and certain complex anions, all the elements in which we are interested form cations, and their electronegativities are given in Figure 5.5. Notice that it is the largest cations within a group (i.e. those with a high period number) that lose electrons

Figure 5.5

most easily because their outer electron shells are distant from the nucleus; they have the smallest electronegativity.

Elements combine by means of three types of bond, which depend largely upon their E values. If there is a large difference in the values of E for two elements, then a bond formed between them is likely to be **ionic**, characterized by strong electrostatic attractions (NaCl, for example). All the elements in Figure 5.5 that form cations with values of E less than 1·6 have an affinity for ionic bonding with oxygen and so exhibit lithophile behaviour in terrestrial planets. By and large, these are the alkali and alkaline-earth cations of groups I and II in the Periodic Table, notably Na^+, Mg^{2+}, K^+ and Ca^{2+} among common elements in the Earth, but also including Al^{3+} from group III. A few other elements with higher electronegativities (notably B^{3+}, C^{4+}, Si^{4+} and P^{5+}) are also lithophile because they form small ions with large charges, and this promotes the development of complex anions with oxygen to form borates, carbonates, silicates and phosphates.

Elements that have E values falling between 1·6 and 2·0 tend to be chalcophile, and this includes many of the transition elements, particularly those from period 4. This is because the small difference in electronegativity, such as that between these elements and sulphur ($E = 2·5$), favours the 'sharing' of electrons and hence **covalent** bonding. In fact, there is a continuous range of bond types from completely ionic to completely covalent, determined by variations in the electronegativities of combining elements, and explains why several elements have both lithophile and chalcophile behaviour in the Earth (see Fig. 5.5).

The third type of naturally occurring bond is that which occurs in **metals** which have closely packed regular structures, often containing just one type of atom. The positively charged metal ions are surrounded by a 'gas' of mobile electrons, which are free to move from atom to atom, giving rise to the high thermal and electrical conductivities of metals. In the Earth, elements with electronegativities lying between 2·0 and 2·4 belong to this group. These are the transition elements, particularly those with high period numbers; they are termed siderophile because they tend to occur along with metallic iron in the Earth. This raises an interesting paradox because iron, which gives its name to the siderophile group, has an electronegativity of 1·8 and so should be a chalcophile element according to these definitions. The explanation of this paradox is tied up with the overall relative abundances of different elements in the Earth (next section). Here is the summary of the theoretical relationship between electronegativity and geochemical affinities:

Lithophile: $E < 1·6$ but, in addition, small complex-forming anions of higher E value (e.g. among the common elements in the Earth: K, Na, Ca, Mg, Mn, Al, Ti in order of increasing E, plus silicates incorporating Si).

Chalcophile: $1·6 < E < 2·0$, except for those forming complexes (e.g. Zn, Fe, Co, Ni, Pb, Cu).

Siderophile: $2·0 < E < 2.4$ (e.g. As, Pt, Ir, Au).

5.3.2 Layering in the terrestrial planets: a simple chemical model

There are really two problems to consider in the rest of Section 5.3; firstly, the chemical reactions that sorted out which elements were destined to form the

89

cores and mantles of the terrestrial planets; and secondly, how and when the energy for the necessary melting processes was delivered. We will take a preliminary look at each problem in isolation, before combining them to give an understanding of planetary layering.

What can be deduced from our geochemical rules about the distribution of elements inside the terrestrial planets? This depends on the relative proportions of the major elements present, and the best estimate that we have for the Earth is shown in Figure 5.6 for the seven most common elements which account for *c.* 97% of the Earth's mass. To develop chemical compounds we use Figure 5.6(b), since it is the atoms of different elements which are combining. First, silicon combines with oxygen to form anion complexes such as $(SiO_4)^{4-}$. Then calcium, magnesium and aluminium, having the lowest E values among the principal elements, are rapidly used up to form, principally, Mg_2SiO_4, olivine (or $Mg_2Si_2O_6$, pyroxene). Thus three (or four) cations (mainly Mg^{2+} and Si^{4+}) combine with four (or six) anions (O^{2-}), so more atoms of oxygen are needed than the total of cation-forming atoms to produce electrically neutral mineral compounds. Oxygen is the most abundant atom in the Earth and so some oxygen

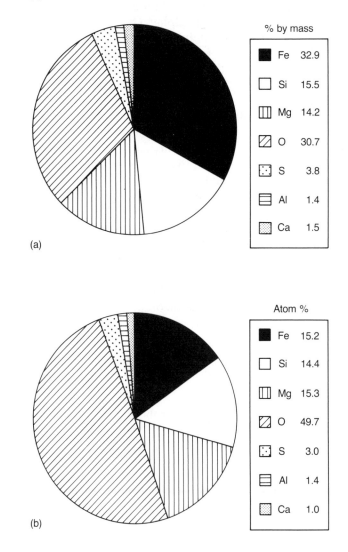

Figure 5.6 Relative proportions of the seven most common elements which account for *c.* 97% of the Earth's mass. (a) Shows proportions by weight, and (b) shows proportions by number of atoms, more relevant to chemical combinations in the Earth. (Note that there is some doubt about the sulphur value, and the abundance given here assumes that sulphur is the diluent in the outer core – Chapter 6.) These estimates were derived using chondrite abundances, adjusted for the higher Fe/Si ratio of the Earth, and were recalculated to 100% (see Fig. 5.3, also Morgan and Anders, 1980).

% by mass

■	Fe	32.9
□	Si	15.5
▦	Mg	14.2
▨	O	30.7
⋰	S	3.8
▤	Al	1.4
▨	Ca	1.5

(a)

Atom %

■	Fe	15.2
□	Si	14.4
▦	Mg	15.3
▨	O	49.7
⋰	S	3.0
▤	Al	1.4
▨	Ca	1.0

(b)

will remain when all the lithophile cations Si^{4+}, Mg^{2+}, Al^{3+} and Ca^{2+} (total atomic % c. 32) have been used up. This brings us to iron which, because of its electronegativity, prefers to be chalcophile and so it combines with sulphur. This uses up the available sulphur. But because it is the only abundant cation left, some of the remaining uncombined iron is forced to combine with the surplus oxygen and thus to be lithophile. However, because of the major abundance of iron (Fig. 5.8), it is inevitable that some iron will remain as metal after all the oxygen and sulphur have been used up, thus joining the siderophile elements in the terrestrial planets. A surplus is especially likely if oxygen is lost by degassing of volatiles such as H_2O, CO_2, SO_2, etc. during accretion, and so it is clear:

1. *that the oxygen content of a planet will determine the size of the lithophile silicate layer,*
2. *that the sulphur content will determine the size of the chalcophile layer; and*
3. *that the excess of cation-forming atoms over those required to combine with the available oxygen and sulphur will determine the size of the siderophile layer.*

The result of these combinations is to create three separate layers in the terrestrial planets. Provided that the temperature is high enough for melting (discussed later), these layers will form in order of increasing density with depth: Mg–Fe silicates, FeS, Fe metal. To anticipate later discussion (Chapters 6–9) these are the layers that we call the mantle, outer core and inner core of the Earth. But chemical separation is not an immediate, clean, 100% efficient process. So we need reactions (where M is any metallic element) such as:

$$\text{M sulphide} + \text{Fe silicate} \rightleftharpoons \text{M silicate} + \text{Fe sulphide} \qquad (5.1)$$

at the lithophile–chalcophile boundary, and

$$\text{M} + \text{Fe sulphide} \rightleftharpoons \text{M sulphide} + \text{Fe} \qquad (5.2)$$

at the chalcophile–siderophile boundary, to sort out the three categories of elements into their most compatible layers according to their electrochemical properties (Fig. 5.5). In the Earth, the efficiency of this process has been high, favoured by a history of high-temperature melting processes and internal activity, but separation is still not complete. For example, sulphur is quite abundant at the Earth's surface, and traces of gold and platinum are found in the crust, even though these elements must be strongly concentrated into planetary cores – apparently beyond the reaches of even the most ingenious entrepreneur! However, Figure 5.5 shows, for example, that elements such as aluminium, calcium and alkalis should be confined almost entirely to planetary silicate layers, whereas gold, platinum and nickel should enter their cores. So far as the Earth is concerned, there is one more caveat, and that concerns the overall sulphur abundance, which may be less than we have assumed (Fig. 5.6). This has consequences for the composition of the outer core, to which we shall return in Chapter 6.

5.3.3 Layering in the terrestrial planets: heat sources and melting

We now consider **heat sources in the early planets:** those that may have induced melting and hence allowed the different layers to separate according to their densities. It turns out that the planetary embryos themselves became hot enough to differentiate. Just how much heat energy is required? The specific heat capacity of terrestrial planet material is about 10^3 J kg^{-1} K^{-1} and the enthalpy of transition required to convert solid to melt is in the order of 3×10^5 J kg^{-1}. So to produce a *totally molten embryo* (4×10^{23} kg – 6% of the Earth's mass – Fig. 4.5) would require, to a first approximation, 8×10^{29} J (a melting point of 1700 K has been assumed). One obvious source of heat is through the accretion process itself, because impacts between planetesimals must have converted much of their kinetic energy into heat. At the early stages of embryo accretion, gravity was low and collision velocities small, so there was little heating. But as a planetary embryo grew, velocities increased and the temperature rose more rapidly.

To illustrate this, assume that all the kinetic energy is transformed into heat. Impact velocities leading to mass gain on a 10-km-diameter planetesimal are *c.* 10^2 m s^{-1}; rising to 5×10^3 m s^{-1} on our 5000-km-diameter planetary embryo (mass 4×10^{23} kg). Since kinetic energy increases with the square of the velocity, impacting material adds 5×10^3 J kg^{-1} to a planetesimal, $1 \cdot 25 \times 10^7$ J kg^{-1} to the embryo. So the heating effect escalates as bodies grow and if we assume an average of 10^7 J kg^{-1} for the last 20% of the mass added to the embryo (8×10^{22} kg) we obtain 8×10^{29} J, enough to melt the body totally! This calculation ignores heat losses, but it also ignores the heat energy gained during the first 80% of embryo accretion. We conclude that temperatures became high enough for planetary embryos to become molten as required for internal differentiation to occur. Note, however, that the highest temperatures probably occurred quite close to the surface, probably through a zone tens to hundreds of kilometres deep, where impact energy was carried downwards by intense seismic vibrations, but not at the immediate surface where efficient radiation of heat probably maintained a chilled crust.

What were the geochemical consequences of the high-temperature, near-surface layer formed during gravitational accretion? As subsurface temperatures rose, it is highly probable that a **magma ocean** developed throughout this region. Within the magma ocean (Fig. 5.7), iron-rich droplets formed either directly by

Figure 5.7 Schematic illustration of a magma ocean formed during planetary embryo accretion, where temperatures were highest in a zone extending from a few tens to several hundreds of kilometres beneath the surface. An outer chilled crust is assumed to be present because of radiation of surface heat. Within the molten layer, dense, iron-rich droplets accumulated into large blobs at the lower boundary, and then sank through the partially molten 'mantle' to accumulate as a central 'core'. (After Newsom and Sims, 1991.) See text for further discussion.

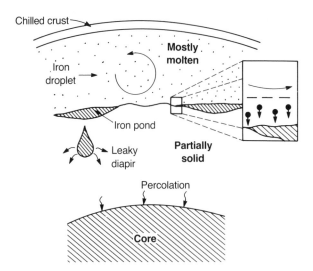

melting of elemental iron, or indirectly by reduction reactions initiated by the high temperatures, producing water, carbon dioxide and sulphur dioxide vapour, such as:

$$2FeO + C \rightleftharpoons 2\,Fe_{metal} + CO_{2\,gas} \tag{5.3}$$

Owing to their high density, the tiny droplets would accumulate at the bottom of the magma ocean, and there collect into large blobs, perhaps 10 to 100 km in radius. These would be heavy enough to percolate down through the partially molten mantle below the magma ocean to form the central core (Fig. 5.7). Mixtures of FeS with Fe are known to have lower melting temperatures than pure iron alone (Chapter 6), so chalcophile and siderophile elements would have been incorporated into the sinking blobs.

So far we have made the case that the fundamental process of core–mantle separation would occur in planetary embryos that were just a few per cent of the Earth's final size. Embryo accretion took about 50 000 years at the Earth's distance from the Sun (Fig. 4.5). Planetary growth then involved embryo–embryo collisions, giant impacts which continued for 10–100 Ma. Mutual disruption of two colliding embryos would have involved enormous amounts of energy; chaotic mixing of previously differentiated material would have been followed by a further period of segregation. The formation of dense, metal-rich blobs and their percolation towards the centre would have allowed the core to develop incrementally after each major impact on a growing planet.

The sinking of dense metal towards the centre of mass represents a release of gravitational potential energy which again appears as heat. Some $1 \cdot 3 \times 10^{30}$ J is estimated to have been liberated by core formation in the Earth by a succession of events following the impact of each planetary embryo incorporated into the Earth. This energy would be enough to raise the temperature by nearly 1000 K. In addition, self-compression increases towards the centre of the growing Earth also raised the temperature by *c.* 2000 K without heat energy being added (the adiabatic effect, discussed in Chapter 8). So we see that core formation tended to even out the radial temperature distribution in the Earth. Other potentially important heat sources in the early Solar System were discussed in Section 4.5, and include *short-lived radioactive isotopes*, the decay of which could have generated perhaps 5×10^{30} J of heat in the first 100 Ma of the Earth's history, and electromagnetic heating by the solar wind. With all this heat energy available, it is not surprising that modern estimates of the *post-accretion temperatures* at the Earth's centre are *c.* 7000 K, with a temperature of *c.* 5500 K at the core–mantle boundary. After a few hundred million years, the contemporary pattern of temperature decrease towards the surface would have been established, and the Earth's shallow magma ocean would be freezing, but not until its existence had left an indelible geochemical imprint on the chemistry of the upper mantle, as we shall see in later chapters.

An intriguing possibility is that, during parts of the accretion period, the magma ocean may have extended up to the Earth's surface. For this to have happened, the outgassed steam, CO_2, etc. (Eq. (5.3)) stayed in the gaseous state with a high vapour pressure at temperatures well above 373 K. A dense, hot atmosphere then developed, much as on the surface of Venus today, in which case, surface temperatures were so great that surface rocks were molten. Thus a *surface magma ocean* might have formed, and there may even have been times when the whole Earth was molten. The consensus is that after major accretionary impacts, the two large terrestrial planets, Venus and the Earth, were intermittently swathed in steam atmospheres with magma oceans extending from the surface to great depth (see Stevenson, 1988 for a review).

5.3.4 Further analysis of the Earth's geochemistry

Figure 5.6 showed how the total abundance of the main lithophile cation-forming atoms (Si^{4+}, Mg^{2+}, Al^{3+}, Ca^{2+}) in the Earth is less than required to combine with all the oxygen and sulphur. So some iron occurs in the lithophile silicate layer, while the remainder must occur in the core. Using Eqs (5.1) and (5.2) we predicted that all the other elements would tend to be segregated according to their electronegativity compared with that of iron. A simple test of these predictions is available, and in Figure 5.8 we compare relative element abundances in the accessible part of the Earth's lithophile layer, its crust, with analogous abundances in solar spectra (representing the Solar Nebula composition, cf. Fig. 4.12). This broadly confirms the predicted pattern of element segregation; most lithophile elements fall below the line of equal abundance in Figure 5.8 because they are enriched in the crust whereas, as we might expect, chalcophile and siderophile elements fall above the line and are strongly depleted. Note that iron plots well above the line, largely because most of the Earth's iron is in the core, with some in the lower mantle. Magnesium is the only lithophile element to be depleted in the Earth's crust because mantle partial melting processes tend to leave Mg in the residual (mantle) solids (Chapter 7). So the Bulk Earth (mantle plus crust) lithophile zone would be enriched in magnesium and depleted in iron relative to the Sun's atmosphere ratios. These comments are likely to apply equally to the principal mantle/core subdivisions in the other terrestrial planets; crust formation in the Earth is a product of the

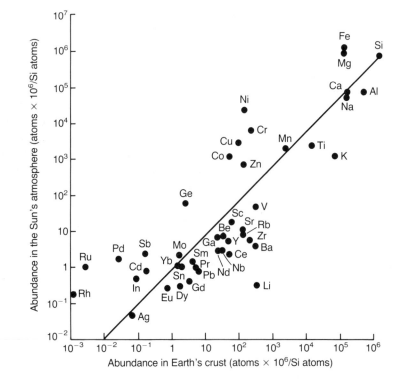

Figure 5.8 Comparison between element abundances in the solar atmosphere and in the Earth's continental crust, standardized as in Figure 4.12 by setting silicon at 10^6 atoms on each axis. Hydrogen He, C, O and N are not plotted as they are far more abundant in the solar atmosphere and were only partially accreted to the Earth. (After Wood, 1981 and references therein; with revisions from Anderson, 1989b.)

unique surface environments that occur here, and has been a continuous process occupying all of the Earth's history (Chapters 10 and 11).

Finally, what about some of the minor elements? We have some interesting data on siderophile elements from the Earth's main silicate layer, the mantle, obtained on nodules that are brought up from depth in lavas (Chapter 7). These show (Fig. 5.9) that the highly siderophile elements (Pt and Au) are strongly depleted in the mantle, as we might expect. But weakly siderophile elements (Mn, V, Cr, Ga) have almost chondritic abundances in the mantle, yet they should be in the core. So we need to explain these high concentrations. Weakly siderophile elements have an affinity for *silicate liquids*, whereas strongly siderophile elements prefer solids. So, if the Earth's mantle was largely or totally molten during core formation, the 'weak/strong' ratio would be increased, with the weakly siderophile elements being 'held back' in the silicate melt that eventually crystallized to form the mantle. Thus the abundance pattern in Figure 5.9 may provide further evidence favouring a molten mantle during the Earth's accretion and core formation. However, weakly siderophile elements are also the most easily oxidized. This has led to the alternative suggestion that if an oxidized body, perhaps from further out in the Solar System, impacted the growing Earth at a late stage of accretion, it may have enriched the mantle in weakly siderophile elements. Either or both these explanations of the data in Figure 5.9 may apply. The 'oxidized impactor' model reminds us that during violent collisions between embryos, it would not be surprising if material formed at different solar distances, and therefore with different chemistry, was involved in planetary accretion.

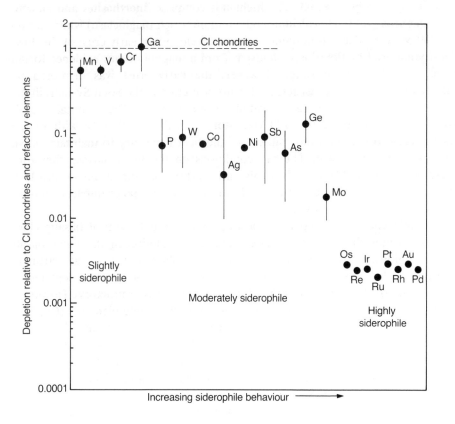

Figure 5.9 Depletion pattern for siderophile elements in the Earth's lithophile layer, plotted as the ratio of the abundance in the mantle to that in CI chondrites. (Mantle abundances are based on mantle-derived nodules; the chondritic abundances have been corrected for estimated pre-accretion volatile losses of slightly and moderately siderophile elements from the Earth-forming region of the Solar Nebula as compared with the chondrite-forming region. After Newsom and Sims, 1991.)

5.4 THE EARTH–MOON SYSTEM: A SPECIAL RELATIONSHIP

For centuries, the Moon has captured the imagination of scientists, but little information was available until space exploration resulted in a myriad of new and exciting discoveries. Many of these concern the history of lunar activity and the Moon's composition, which, in turn, reflects its origin. First, the basic facts: the Moon has a radius of 1738 km and a low mean density of 3.34 Mg m^{-3} (Table 5.2), giving it a mass only 1.2% that of the Earth, rather smaller than the typical terrestrial planetary embryos (Fig. 4.5). Nevertheless, it is one of the largest satellite bodies in the Solar System, which prompted the view that the Moon was once an independent body orbiting the Sun, subsequently captured into Earth orbit. Other hypotheses were that the Moon formed from a ring of small fragments orbiting the Earth that eventually accreted into a single body. There are now seen to be serious difficulties in producing the strongly volatile-depleted geochemistry (Fig. 5.4) from the low-energy impacts that would have occurred between small fragments. Owing to this objection, combined with the increased recognition that collisions between embryos were important during the planetary growth phase, acquisition of the Moon through a **giant impact** is now favoured (discussed below). This has the advantage that at least some of the material may have been vaporized leading to volatile losses.

The petrology of the lunar surface is visible in outline from the Earth – the lighter *highland* areas and the darker *mare* basins (so called because they were once thought to be oceans). The highlands comprise **anorthosites** and **anorthositic gabbros** (70–100% calcium- and aluminium-rich plagioclase) which are up to 4400 Ma old. The Moon seems to have developed an early crust of this low-density material by flotation of feldspar from a magma ocean: evidence for an initial partially molten state. However, the early crust was disrupted by cataclysmic impacts (the last stages of major accretion in the Solar System) in the period 4400–3800 Ma; these excavated the *mare* basins. These contain basalt lava flows dating from 3900–3100 Ma, after which this small body, with a large surface area to volume ratio, ran out of the heat necessary to maintain active volcanism at the surface. The dry conditions on the lunar surface therefore preserve a remarkable record of this early history of bombardment and lava effusions; the only subsequent events have been much smaller impacts continuing throughout the past 4 Ga.

A simple chemical comparison between lunar and terrestrial basalts (Fig. 5.10) illustrates the degree of volatile-element depletion in the lunar basalt source regions. Notice also that all the chalcophile and moderate–strongly siderophile elements plotted (see Fig. 5.9: Ag, Ni, Ge, Au, Ir) are depleted in the Moon, irrespective of whether they are volatile or refractory. However, weakly siderophile Cr and most of the refractory lithophile elements (Ca, Mg, Ti, Th, U, Ba) are enriched in the Moon. The abundances of Al and Ca in lunar basalts are lower than they might be otherwise, because of earlier differentiation of anorthite into the highlands suite. Volatile sulphur is high largely because troilite (FeS) has crystallized in lunar basalts, whereas in relatively oxygen-rich terrestrial environments SO_2 gas would have formed and been expelled. In summary, this type of analysis has led to the view that, relative to Bulk (silicate)

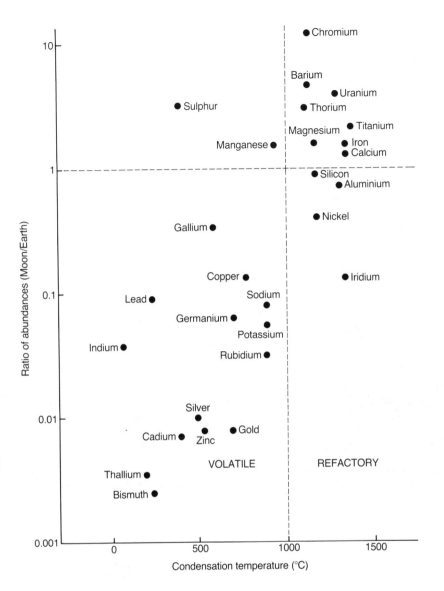

Figure 5.10 A comparison of element abundances in lunar and terrestrial basalts. The vertical dashed line at 1000 °C serves to separate elements into volatile and refractory groups and is roughly equivalent to the boundary between volatile and refractory mineral phases in Table 5.1. (See Taylor, 1982, for geochemical data.)

Earth, *Bulk Moon is*:

1. *depleted in volatile elements*;
2. *strongly depleted in siderophile elements*;
3. *enriched in refractory lithophile elements*.

To conclude this brief review, there is seismic and palaeomagnetic evidence for a small metallic core, at least 150 km in radius and, although this is only 0·1% of the lunar mass, other estimates range between 1 and 4%. Such a tiny core, together with the evidence of siderophile-element depletions in surface samples, makes it likely that the Moon *as a whole* is depleted in siderophiles. This is also consistent with the low density (Table 5.2), which indicates one of the lowest Fe/Si ratios within the terrestrial planet zone.

To produce this siderophile-poor geochemistry, seems to require an unusual separation of planetary embryo core and mantle material. The Moon seems to be made almost entirely of mantle silicates which have among the highest ratios of refractory to volatile elements in the terrestrial planets. Almost the only 97

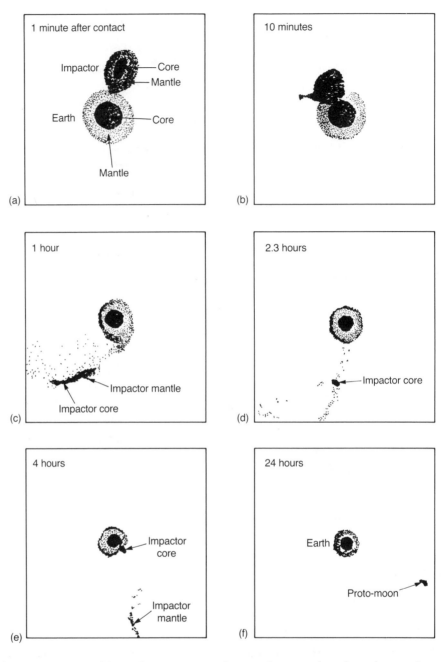

Figure 5.11 Computer simulation of the formation of the Moon by a giant impact, following an oblique collision between a Mars-sized impactor and the *c.* 90% accreted Earth. Prior to collision, both bodies had separated internally to form mantle and cores, but material was exchanged during the collision. In (c)–(f) impactor material is orbiting the Earth in an anticlockwise direction; the field of view is increased because the Earth stays roughly the same size throughout. (After Benz *et al.*, 1987; see also Newsom and Taylor, 1989 and Taylor, 1992 for further discussion.)

known way to achieve the necessary elemental separation, is to have a large, already differentiated embryo (i.e. with a silicate mantle and metal core) about the size of Mars come into a grazing impact with the Earth at a late stage during its accretion. The best dynamic fit in computer simulations (e.g. Benz *et al.*, 1987), that also satisfies the available geochemical data, involves a body of *c.* 0·14 Earth masses approaching an almost fully formed Earth at 5 km s^{-1}. The glancing collision raised large tides on the Earth and disrupted the impactor plus part of the Earth's mantle, causing vaporization of volatile elements (Fig. 5.11(a)–(c)). Some of the disrupted material was accelerated into orbit, some was absorbed by the Earth, and some was lost, but the impactor's core ploughed into the Earth where it decelerated and was absorbed in just a few hours (Fig. 5.11(d) and (e)). The remains of the impactor, mainly its mantle, and depleted of both siderophile and volatile elements, coalesced rapidly, forming a totally molten Moon, or accreted in stages, giving a partially molten Moon. There has been some debate about the respective contributions of the Earth's mantle and the impactor's mantle to the Moon. The arguments are complex and need not concern us here. However, the Earth probably gained a few per cent of both its core and mantle mass from the impactor. Indeed, this may be the late oxidized silicate component discussed earlier. Overall, however, the Fe/Si ratio of the Earth probably was raised and that of the Moon reduced as a consequence of this impact, as implied by their uncompressed densities in Table 5.2. In conclusion, the distinctive geochemistry of the Moon could have been produced in a single event, of the type quite plausible towards the end of the planetary growth period.

5.5 MARS, VENUS AND MERCURY: A POSTSCRIPT

It is helpful to our understanding of the Earth to examine briefly the three other planets produced by accretion and embryo collision in the inner Solar System. The geochemical models developed earlier predict that *overall* volatile element budgets would have increased and Fe/Si ratios decreased outwards in the Solar Nebula. For *Mars*, this is reflected in the lower uncompressed densities compared with the Earth (Table 5.2), in the higher K/U ratio (discussion of Fig. 5.4) and in a smaller core volume which follows from geochemical and moment of inertia considerations (Fig. 5.12). Thus Mars probably has the nearest to chondritic composition of the terrestrial planets.

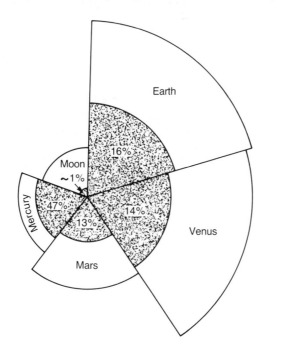

Figure 5.12 Comparative radial dimensions of the terrestrial planets and the Moon together with estimated sizes of their metallic cores. (Parameters from Spohn, 1991.)

99

Martian surface features vary from high-elevation, heavily cratered, ancient crust, resembling the lunar highlands, to young, relatively uncratered lowlands. The latter are basaltic plains with occasional immense volcanic edifices. It seems that the planet was volcanically active for much of its history, but this probably ceased by *c.* 1 Ga ago, simply because the lithosphere became too thick and rigid for magma to escape. Mars is much smaller than the Earth and so lost heat more rapidly. Today the Martian surface is extremely cold (212 K) and the atmosphere thin (6 millibars), but in the past Mars may have been warm and wet, with flowing water forming the impressive erosional channels seen in the Viking images. A denser atmosphere with both water vapour, now frozen into permafrost, and CO_2, now frozen into ice-caps, may have existed, with pressures around 1 bar. Atmospheric CO_2 is an important 'greenhouse' gas which may have contributed to surface warming when volcanic outgassing was at its peak (see Kasting, 1991, for a review).

Although *Venus* has long been regarded as the Earth's twin, recent space exploration has served to discover strong differences between the two planets. Surface conditions on Venus are extremely hostile, with temperatures of 730 K beneath a thick, dense atmosphere (surface pressure 90 bars), consisting mainly of CO_2 with corrosive sulphuric acid droplets. The lack of surface water not only prevents CO_2 and SO_2 being fixed into sedimentary rocks as in the Earth, hence their build-up in the atmosphere, but has had a profound effect on the tectonic evolution of Venus. For example, a 20–30-km-thick basaltic crust has developed by effusive *hot-spot volcanism* coupled with crustal thickening by intrusion from below. Crustal recycling which, on Earth, has produced the rich diversity of continental rocks, is therefore very limited or absent. The Magellan spacecraft, which reached Venus in 1990, discovered widespread volcanism and *broad* areas of both extensional and compressional deformation. The interpretation of these features is widely debated, but they are thought to reflect the high surface temperatures and a dry, stiff lithosphere. It is believed that much of the water in Venus may be trapped in a deep, volatile-rich partial melt zone in the mantle, which may also tend to absorb some of the stresses imposed by deeper mantle convection. On the Earth, with a cooler upper mantle, there is relatively little partial melt (Chapter 7), but oceanic rocks and volatiles are recycled into the mantle at *narrow* zones of high stress intensity where, unlike on Venus, the tectonic plates become sufficiently cool and thick to founder (Chapter 8). This has led to speculation that the young Earth was rather like Venus today but that, for various reasons, the formation of water oceans on the Earth subsequently led to divergent evolution. This is perhaps fortunate since, without those oceans, life itself would not have evolved!

Finally, we will consider *Mercury*, the smallest of the terrestrial planets and the closest to the Sun. Its remarkably high density demands that Mercury has a large metallic core occupying nearly 50% of its volume, and much of the silicate mantle was probably vaporized during a giant early impact. The surface is heavily cratered, probably old and inactive, and resembles the highland areas of the Moon. However, Mercury does have a magnetic field, though it has only about one-hundredth of the Earth's field strength. The favoured mechanism for magnetic field generation in the Earth (Chapter 6) is by crystallization in the core, a process that may also be occurring in Mercury. Venus may be too hot, internally, for this to happen yet. Figure 5.12 gives estimates of the relative sizes of the core and mantle in all the terrestrial planets.

SUMMARY

1. The main processes influencing planetary chemistry took place in the first 100 Ma following 4550 Ma ago, but most of the essential chemical ingredients were in place much earlier. Within the first 10^4–10^5 years, accumulation of the initially cold Solar Nebula caused strong heating and vaporization, particularly in the innermost, denser parts of the Solar System. Subsequent cooling allowed submicron-sized dust grains to form by recondensation, following an order of decreasing molecular volatility (Table 5.1).

2. Although individual pre-planetary grains were a heterogeneous mixture, average dust grain compositions varied radially across the nebula because of: (a) outwards decreasing pressures, due to grain density falling, which, for example, controlled the iron/silicate ratio (Figs 5.1 and 5.3), and (b) falling temperatures away from the hot nebular centre, which controlled volatile/refractory element ratios such as K/U (Fig. 5.4). A snow line existed at 4-5 AU, just inside orbit of Jupiter, beyond which low-density grains dominated the outer Solar System (Fig. 5.2). Grains accreted to form planetesimals, and chemical compositions then became fixed because strong T-Tauri winds removed over 90% of the potential planetary mass, still in the form of uncondensed gases.

3. Before the planets formed, a process taking *c.* 100 Ma for the Earth, planetesimals accreted gravitationally into planetary embryos over a much shorter period (*c.* 0·05 Ma). Each of these would have had a chemistry reflecting the pressure-temperature conditions in the part of the Solar System where it formed. Chance collisions between these embryos then produced terrestrial planets which have densities, determined principally by the iron/silicate ratio, decreasing away from the Sun (Table 5.2). The varied compositions of the meteorites reflect (a) local differences in pressure and temperature prior to accretion, and (b) the probability that some (chondrites from the outer parts of the asteroid belt) derive from small bodies that did not differentiate (see below).

4. Segregation of elements inside planets started when accretional heating raised temperatures high enough ($\leqslant 1700$ K) for melting to occur and for an internal magma ocean to develop, initially close to the surface where impact energy was most effectively converted into heat (Fig. 5.7). Core formation, with the separation of dense metallic components from silicates, released further gravitational energy as heat, and may have commenced when embryos had grown to a few per cent of the Earth's present mass.

5. At this point, siderophile and chalcophile elements in the Earth (Fig. 5.6) were separated into the core, dominated by iron. But iron was in excess and so joined Mg^{2+}, Al^{3+} and Ca^{2+} in the lithophile silicate mantle. All the other elements were sorted out according to their electronegativities (Fig. 5.5) via Eqs (5.1) and (5.2). At this stage, the number of cation-forming atoms in excess of those required to combine with the available oxygen and sulphur determined the relative sizes of embryo cores and mantles; volatile losses (Eq. (5.3)) may have caused further adjustments.

6. Giant impacts between embryos produced the planetary structures (Fig. 5.12) and compositions that we infer today. Such impacts may have vaporized part of the Mercurian mantle, may account for the slow retrograde axial rotation of Venus, and may explain the formation of the Moon. This involved an oblique impact with the accreting Earth of a relatively oxidized but internally

101

differentiated embryo (Fig. 5.11). Both the impactor and the Earth's shallow mantle were fragmented, partially volatilized in orbit and then coalesced to form the Moon. Some of the impactor's mantle and most of its core were accreted to the Earth, thus enlarging the Earth's core. The volatile- and siderophile-poor, refractory element-rich geochemistry of the Moon (e.g. Fig. 5.10) and the enrichment of slightly siderophile elements in the Earth's mantle (Fig. 5.9) may be a product of this early interaction.

7. The relative sizes of the terrestrial planets seems to have determined their history of surface volcanic and tectonic activity. Thus the largest bodies, Venus and the Earth, are the only active planets today. But far from being identical twins, they have evolved along markedly different paths, in which the strong contrast in surface temperatures has an important effect on internal as well as surface processes.

FURTHER READING

General books:
    Cattermole (1989): volcanic processes on all planets, including satellites.
    Cole (1984): review of the physical properties of planets.
    Rothery (1992): planetary satellites; surface and internal processes.
Advanced journals:
    Head *et al.* (1991) and Solomon *et al.* (1991): volcanism and tectonics on Venus.
    Kaula (1990): good account of contrasts between Earth and Venus.
    Liu (1988): water in the terrestrial planets and their atmospheres.
    Newsom and Sims (1991): account of core formation during accretion.
    Newsom and Taylor (1989): the giant impact model for Moon formation, and consequences for Earth-Moon geochemistry.
    Rama Murthy (1991): review of the chemical differentiation of the Earth.
    Russell *et al.* (1991): pre-solar grains in meteorites.
Advanced books:
    Heiken *et al.* (1991): useful reviews of various aspects of the Moon's geochemistry and formation.
    Kerridge and Matthews (1988): wide range of topics covering meteorites, planets, etc.
    Newsom and Jones (1990): compilation of modern papers covering most aspects of the Earth's accretion.
    Weaver and Danly (1989): review papers on early Solar Nebula processes and planetary embryo formation.

The Earth's core

6.1 PROBLEMS POSED BY THE CORE

In Chapters 2 and 3 we described the Earth in simple physical terms (seismic velocities and density), while in Chapters 4 and 5 we described the formation of the Earth and introduced the geochemical principles that determine how elements are partitioned. This is the first chapter that combines these physical and chemical types of information to examine a specific part of the Earth. The core has been chosen first, partly because it apparently lacks the complexity of the mantle and crust (though this apparent simplicity may be because of the difficulty of observing regions so deep in the Earth), and partly because an understanding of the core is needed more for a discussion of the mantle than vice versa.

We know from seismology (Chapter 2) that the core is divided into an outer liquid part and an inner solid part, with the outer core having a radius of 3485 km and the inner core a radius of 1225 km (or 0·547 and 0·191 of the Earth's radius), (Masters and Shearer, 1990). In Chapter 3 the density was shown to range from 9900 kg m^{-3} just below the core–mantle boundary to 13 100 kg m^{-3} at the Earth's centre, so that the core has 31% of the Earth's mass, of which only 1·7% is in the inner core (16 and 0·7% of the volume, respectively). From Chapters 4 and 5, it emerged that the only element of approximately the necessary density and of sufficient abundance is iron. What, then, are the problems to be solved?

The first problem is why the inner core is solid, despite its probable higher temperature than the outer core. A second is how the Earth's magnetic field is generated in the core. A third problem is that we must obtain a better match to the density of the outer core than is provided by pure iron: what other elements are likely to be there? These three problems are interrelated, and at present there is no general agreement on their solution.

6.2 THE EARTH'S MAGNETIC FIELD AND THE PROBLEM OF ENERGY

6.2.1 A description of the field

Most people regard the Earth's magnetic field merely as offering a useful means of navigation. Those who recollect experiments with iron filings might suggest that the Earth's field is very similar to that of a bar magnet. Indeed, to a good approximation, the Earth's field is like that of a short, powerful bar magnet (i.e. a **magnetic dipole**) situated near the centre (Fig. 6.1). Since the core is likely to be mainly iron it might be thought that it, or at least the solid inner core, forms a permanent magnet. However, there are compelling reasons why this cannot be so:

1. Above a certain temperature – the Curie temperature – any magnetic 103

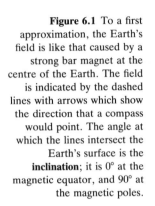

Figure 6.1 To a first approximation, the Earth's field is like that caused by a strong bar magnet at the centre of the Earth. The field is indicated by the dashed lines with arrows which show the direction that a compass would point. The angle at which the lines intersect the Earth's surface is the **inclination**; it is 0° at the magnetic equator, and 90° at the magnetic poles.

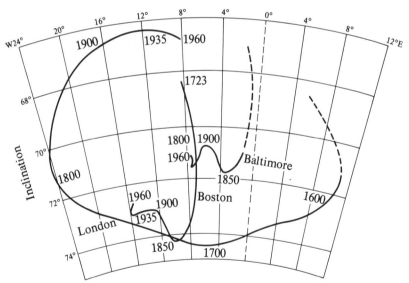

Figure 6.2 Variation of the direction of the Earth's magnetic field as recorded by various observatories over the past few centuries. **Declination** is the horizontal angle between magnetic and geographical north, while **inclination** is the angle below the horizontal (see Fig. 6.1). (Source: Jacobs, 1975. Reproduced with permission from Jacobs, J. A. (1975) *The Earth's Core* © Academic Press Inc., London., Ltd; latest edition is 1987.)

material loses its permanent magnetism. This temperature is well below the melting point. We know that the outer core is molten, and the inner core cannot be far below its melting point (see also Sections 6.4 and 6.6).

2. The magnetic field varies in time in both direction and strength. Figure 6.2 shows the variations in direction of both the horizontal component, or **declination**, and the vertical component, or **inclination**, at London and in North America since records began. This somewhat irregular **secular variation** of the magnetic field has required navigators to revise their maps periodically, relating magnetic to true (geographical) north. The records

show that the corresponding magnetic poles tend to move westwards, referred to as **westward drift**.

For these reasons, the magnetic field must be produced dynamically rather than statically, by some form of dynamo, as will be discussed in Section 6.2.2.

To extend records further back into the past, studies of palaeomagnetism (or 'fossil' magnetism) are used. This is possible because many rocks, particularly igneous ones, become magnetized in the direction of the magnetic field at the time they were formed; for instance, iron oxide minerals in basalts (similar to those used in magnetic recording tapes) pass through their Curie temperature as the rock cools, and often the magnetism is retained until the present. Orientated samples of the rock, when measured by sensitive magnetometers in the laboratory, reveal this ancient direction of magnetization. Samples from different localities and formed at different times allow the past field to be deduced. In principle, a knowledge of the Earth's field, past and present, should reveal much about the core which produces it, but currently our understanding is poor and therefore the deductions we can make are limited.

Spatially, the Earth's field can be split into a dipole component – like that of a

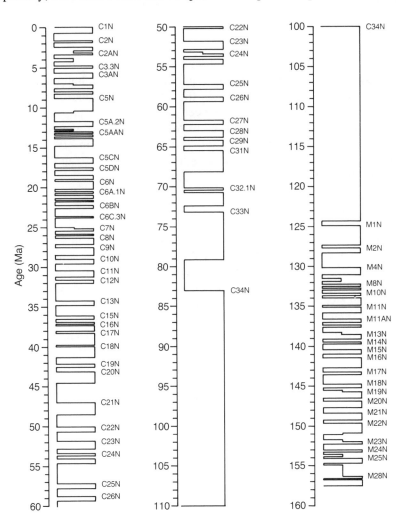

Figure 6.3 Polarity time-scale of the Earth's magnetic field. It shows how the field has alternated in polarity, being 'normal', i.e. in the same sense as the present field, when the vertical line is on the right-hand side of the column. The normal intervals – some multiple – are numbered at the 'top' of the interval (based on Harland *et al.*, 1990).

105

bar magnet – and non-dipole components, which are more complicated. Observatory and palaeomagnetic measurements reveal that these vary on a wide range of time-scales. The direction 'wobbles' about the geographical axis, as indicated by Figure 6.2, but over a period of a few thousand years the non-dipole components average out, and the average field is close to that of a dipole aligned along the rotation axis (this allows the palaeomagnetic direction to be used to determine the ancient geographical pole, and hence to measure continental drift, as will be discussed in Chapter 11). On a time-scale of millions of years, the field irregularly inverts or **reverses**, exchanging north and south poles, as Figure 6.3 shows for the past 160 Ma. Since the time taken for inversion is no more than 10 000 years and possibly much less, whereas the time spent with normal (i.e approximately parallel to the present field) or reversed polarity is measured in millions of years, few rocks are found to record the transition. However, enough examples are known to show that during an inversion the strength of the field is greatly reduced but not zero, and has poles far from the geographical poles. Moreover, these poles seem to recur in roughly the same locations for several successive inversions, indicating that the transitional field is 'locked' to some semi-permanent feature within the Earth (Laj *et al.*, 1991). Records show that the intensity of the field has varied, perhaps by a factor of two, but has probably persisted since 3500 Ma ago and perhaps even earlier, though records are sparse for the oldest times.

Figure 6.3 shows that the direction of the field was normal between about 124 and 83 Ma ago, known as the Cretaceous quiet zone, and that before and after it the rate of reversal was slow. This was the approximate time of a period of enhanced igneous activity, which has been attributed to a postulated super-plume, a great upwelling of hot material from the base of the mantle, discussed in Section 9.2.3. It has been suggested (Larson and Olson, 1991) that the removal of such a large amount of heat cooled part of the core and 'locked' the geodynamo into one polarity.

6.2.2 The geomagnetic dynamo

The dynamic nature of the Earth's magnetic field suggests a source in the fluid, mobile outer core, and it is agreed that the magnetic field is generated in the outer core by electric current loops, in turn powered by motions of molten material, forming a **geomagnetic dynamo** (Busse, 1975 and Bloxham and Gubbins, 1989 give technical summaries). However, details are not clear.

Figure 6.4 shows a simple disc dynamo analogue. In Figure 6.4(a) the disc, which is a conductor, rotates in a magnetic field, and an electromotive force (e.m.f.) is generated between the axle and the rim, but no current can flow until the circuit is completed by a wire joining the two (b). The current produces a secondary magnetic field, and in (c) the external magnetic field has been dispensed with; instead, the current is passed through a coil around the axle of the disc so that the secondary magnetic field that it produces is used for its own excitation. This is now a **self-exciting dynamo** which will continue to generate a field so long as it is rotated. It provides a useful analogy for the generation of the Earth's magnetic field by current loops in the core: once the dynamo system gets under way with a small initial magnetic field (perhaps that of the Sun), it requires only an energy source to maintain itself. If the current, but not the direction of rotation, is reversed, the field also reverses. The process of **self-reversal** of the Earth's field is undoubtedly more complex than this, but it has been successfully

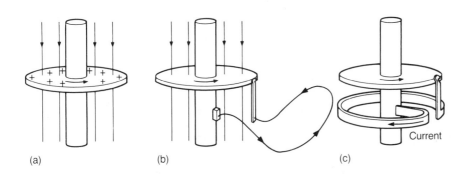

(a)　　　　　　　　　　(b)　　　　　　　　　　Current　　(c)

Figure 6.4 Illustration of a self-excited dynamo. In (a) the disc is made to rotate in a ·magnetic field and a potential difference is generated between rim and axle, but no current flows because there is no complete circuit. In (b) the external circuit allows a current to flow, and in (c) this current flows in a loop below the disc and so produces a vertical field, so that the original field can be removed. (After Bullard, 1971, from Gass, I. G., Smith, P. J. and Wilson, R. C. L. (eds.) *Understanding the Earth*, pp. 71–80, Artemis Press, Sussex.)

modelled using a pair of dynamos that generate fields for each other (see Jacobs, 1975, p. 153, for a discussion).

Computer modelling shows that the pattern of fluid flow is strongly affected by both the presence of the inner core – which suggests that an inner core has existed for as long as the magnetic field – and by the rotation of the Earth, which is why the field axis is aligned, on average, with the axis of rotation. However, not only does the liquid flow pattern determine the electric currents, and hence the magnetic field, but the field, in turn, exerts forces on the moving conducting liquid; it is because of these interactions that the problem is so difficult to solve.

A better understanding of the magnetic field would be obtained if we knew its form closer to where it is generated, in the core, rather than at the Earth's surface where much of the important detail is lost by the 'blurring' or fall-off caused by the distance. For instance, the field of a dipole decreases as the cube of the distance, while other components decrease even more steeply. Some detail has been resolved in the past few years by Bloxham and Gubbins (1985, 1989) who have overcome technical problems of extrapolating downward ('downward continuation') from the field at the Earth's surface. They have used detailed observatory records for the past few hundreds years and, particularly, satellite observations of the field. They have found that the greatest contribution to the dipole part of the field comes from four patches, situated 60° north and south of the equator, and 120° east and west. These seem to have stayed fairly constant over the past few hundred years and they may indicate the ends of two columns of liquid flow, parallel to the axis of rotation (Fig. 6.5). If so, molten iron would spiral up the columns, concentrating magnetic flux.

However, these columns do not account for the non-dipole parts of the field, and particularly their westward drift (Fig. 6.2). Further insight comes from deducing the flow patterns in the core. This can be partially achieved because the magnetic flux tends to be carried bodily along with the flow of molten iron, for the iron is such a good conductor – rather better under the conditions of the core than is copper under normal conditions – that flux tends to be 'frozen' within it (see Note 3). It is found that the surface flow is largely confined to the Atlantic half of the globe, in two huge swirls, a counter-clockwise one centred under Europe and a clockwise one under the tip of Africa, which combine to give a flow along the equator (there is very little flow in the Pacific half). This westward current is the most likely cause of the westward drift of the non-dipole field.

If the energy source were abruptly stopped – for instance, if all motions in the outer core ceased – the magnetic field would die away in less than 10 000 years,

107

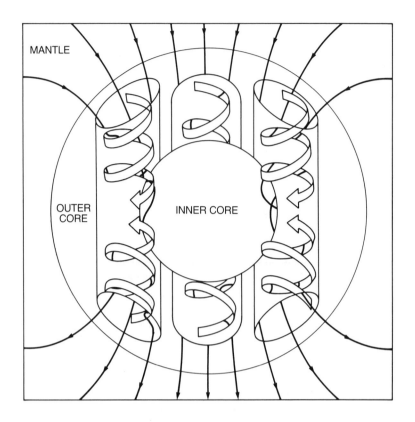

Figure 6.5 Possible liquid flows and magnetic field in the core. Liquid iron flowing in helical paths generates the field, which approximates to that of a dipole. (Bloxham and Gubbins, 1989.)

for the energy of the electric currents would be dissipated by the ohmic resistance of the core. This dissipation provides a lower limit on the energy needed to maintain the field, which is estimated to be about 5×10^9 watts (Melchior, 1986, p. 212). This is a lower limit because not all the energy leaving the core is converted into electrical currents that produce the magnetic field. How much energy is needed depends upon the mechanism that produces the motions. There are two possibilities, thermal convection due to a heat source, and compositional convection due to unmixing and separation of denser and lighter components.

6.3 ENERGY SOURCES FOR THE DYNAMO: (A) THERMAL CONVECTION

Thermal convection is the self-stirring that takes places in a liquid heated from within or below, such as the contents of a saucepan on a stove. As the fluid near the base becomes hotter, it expands, and the resulting lower density causes it to rise; at the top it cools, becomes denser and sinks, so completing the cycle. Thermal convection is considered in more detail in Section 8.4.1; suffice it here to note that thermal convection is a rather inefficient way of converting heat energy into mechanical motion, for several reasons. To produce convection in a compressible liquid – as is the core, because of the huge increase in pressure from top to base – it is not sufficient just to have the base hotter than the top, but the temperature gradient must exceed the adiabat, as explained in Section 3.5 in connection with the Adams–Williamson equation. Until the gradient exceeds the adiabat heat will simply conduct without inducing any convection. Even when the adiabat is exceeded, convection is resisted by the viscosity of the fluid. The adiabatic temperature gradient in the core is estimated to be about $0.5°C$ km^{-1} (Melchior, 1986, p. 571); the thermal conductivity of iron under core conditions has not been measured but it has been estimated from the related electrical conductivity to be about 40 W $°C^{-1}$ m^{-1}. Together, these values give a total heat flow down the adiabatic gradient and through the surface of the core of about 3×10^{12} watts. This is a significant fraction of the global heat flow, which is estimated to be about 4.2×10^{13} watts (Section 8.3) but, even so, 3×10^{12} watts must be an underestimate for the reasons given above. Clearly, if thermal convection is to drive the dynamo, giving such massive heat flow, a large heat source is needed. There are several possibilities.

A possible heat source is radioactive decay. One suggested element is potassium, of which today about 0.01% is the radioactive isotope ^{40}K, with a half-life of 1300 Ma. As discussed in Section 5.3, potassium is normally strongly lithophilic and so should be found outside the core and particularly in the continental crust, as seems to be the case. However, it has been suggested that, at core pressures, potassium may undergo a modification, with the single outer S electron being forced into the D shell, converting it into a siderophile element. In Chapter 5 it was argued, assuming the lithophilic nature of potassium, that the Earth retains only a third of the chondritic abundance of the element, the rest probably having been lost as a volatile phase during planetesimal formation. However, some of the missing potassium might be in the core. To account even for the heat flow down the adiabat, 3×10^{12} watts, would require about half the total that would be present in the whole Earth if it had a chondritic composition, so all the 'missing' potassium of a K-rich 'chondritic Earth' would hardly suffice to drive the dynamo. Neither does the case for potassium receive support from experiments at pressures of 1.5 GPa (admittedly less than half of the pressure at the core–mantle boundary): little or no potassium enters into solution, whereas uranium and thorium, our next candidates, dissolve in significant amounts (Murrell and Burnett, 1986).

When steel and uranium oxide, UO_2, are melted together at atmospheric pressure they form two immiscible liquids, one with 23% uranium oxide, the other with 99%. There is therefore a strong possibility that considerable amounts of radioactive uranium, and perhaps also thorium, could exist in the core (Feber *et al.* 1984). But it is less clear that the core contains sufficient to drive the dynamo, since the *total* heat production due to all the uranium of a 'chondritic Earth' at present is only about 8×10^{12} watts. Nevertheless, support for the presence of some U and Th in the core comes from studies showing that these elements are enriched in the first condensates, as confirmed by the inclusions in the Allende meteorite, and such condensates would form at about the same time as iron condensed.

If the amount of radioactive elements in the core is insufficient to power the dynamo, as seems likely, then the power may be provided by cooling of the Earth, which seems to be occurring (Section 8.3). However, it is unlikely that the heat released by cooling of the core alone is sufficient to power the dynamo and a more subtle mechanism probably operates. To appreciate how it works it is necessary to consider the compositions of inner and outer core.

6.4 COMPOSITIONS OF THE INNER AND OUTER CORES

One of the first explanations for the existence of the solid inner core within the liquid outer core assumed that they have the same composition but postulated that the inner core was solid because of the higher pressure there. In other words, the increase in melting point due to increasing pressure would be greater than any increase of temperature with depth (Fig. 6.6(a)). Later, it was calculated that the slope of the melting point curve for iron-rich compositions would be less than the adiabatic temperature gradient, so that this explanation for the solid core could only hold if the temperature gradient were so small that thermal convection would not occur. This turned attention to other explanations, though it is now known that the calculation was wrong because it did not allow for the effect of the extremely high core pressures (Melchior, 1986, p. 68).

It is more likely that the inner and outer cores have different compositions (Fig. 6.6(b)). Before considering the composition of inner and outer cores, we shall briefly review what we know about their physical properties, and, in particular, the inner core boundary. The existence of the inner core was discovered from arrivals in the core shadow zone (Section 2.3) and a solid inner core seemed to be likely. However, its solidity has never been demonstrated definitely from the arrival of seismic rays that have travelled as S-waves through the inner core and P-waves elsewhere, which is technically a very difficult task. More convincing evidence comes from long-period free oscillations, from which a shear wave velocity, V_s, of 3·45 km s^{-1} has been deduced (Masters and Shearer, 1990). And the observation that the velocity is anisotropic, being fastest parallel to the axis, is explicable for a solid which can have aligned crystals but not for a liquid (see Section 6.5). The gross variation of seismic velocity and density within the Earth are deduced by models such as PREM (Section 3.6), but these are unable to provide details. The observation that even seismic waves of quite short wavelength are reflected from the inner core shows that the boundary is no thicker than a few kilometres (it is a general rule that waves are reflected from a boundary only if the transition is no thicker than about a wavelength). But the observation that reflections from the inner core boundary are far weaker than reflections from the core–mantle boundary has been used to deduce that the inner core is only about 0·5 Mg m^{-3} denser than the outer core (Masters and Shearer, 1990; Smylie, 1992). These observations

Figure 6.6 Diagrams showing how melting points and temperature might vary with depth in the core and result in a liquid outer core between the solid mantle and inner core. In (a) the core has a uniform composition but its melting point increases more quickly with depth, because of the increasing pressure, than the temperature gradient, and the two curves intersect at the inner core boundary. In (b) the inner core has a composition different from that of the outer core, with a higher melting point. In both cases, the mantle is solid because its composition is different from that of the core, and has a higher melting point. (b) is believed to be the true explanation.

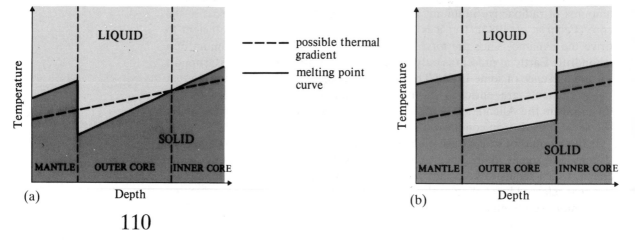

are consistent with a fairly small compositional difference between inner and outer core.

In the past few years it has become possible to produce, in specially constructed diamond anvil apparatuses, static pressures and temperatures equalling those of the top of the core. Higher pressures can be achieved very briefly by producing shock waves, using the impact of a projectile, but the properties that can be measured are more limited than in the static case. The core is almost certainly mostly iron, for iron has approximately the correct density and seismic velocities at core pressures; it is also a conductor (necessary to permit the geodynamo to function), and it is sufficiently abundant. But measurements show that though the outer core closely matches iron in seismic properties, its density is about 10% lower. There have been several suggestions for the outer core diluent.

To be acceptable, a diluent must:

1. be capable of alloying with iron under the present core conditions;
2. be capable of partitioning into the core in sufficient amounts;
3. be sufficiently abundant; and, of course
4. produce an alloy with iron that matches the seismic properties and density of the outer core.

Sulphur is one of two most favoured diluents, for several reasons. Iron readily combines with it to form iron sulphide, FeS, as explained in Section 5.3, and FeS and iron are miscible at high temperatures. Moreover, the Fe–FeS phase diagram (Fig. 6.7) contains a **eutectic**, that is, the melting point of the mixture is lower than that of either pure component, and this would assist the melting of iron and its separation to form the core during the early history of the Earth. Consider a cooling mass of fully mixed Fe–FeS liquid, with composition to the iron-rich side of the eutectic point, as is appropriate for the likely core composition. As a specific example, consider an alloy with 20 wt.% of FeS, at 1600°C, point A on Figure 6.7. Falling temperature would cause it to begin to solidify at B, with the formation of pure iron, the residual liquid being enriched in FeS. Further cooling would form additional solid iron, with the residual liquid being progressively enriched in FeS, B → C → D. Only when the liquid has become enriched to the eutectic composition, E, will complete solidification occur. Since iron is denser than FeS, it will settle throughout the cooling, so

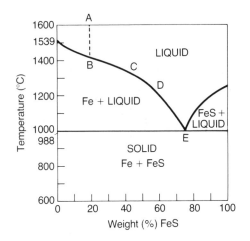

Figure 6.7 Phase diagram for Fe–FeS at atmospheric pressure. The two components form a eutectic, so that (unless the composition is at E) cooling results in pure Fe or pure FeS solidifying (depending on whether the composition is to the left or right of E), while the composition of the residual liquid moves towards E. The final mix to solidify has a composition E.

accreting to the inner core, according to this model. Although the phase diagram has not been determined for core pressures, it is unlikely to be so modified that the above description does not apply; the main modification is likely to be that all phase changes occur at higher temperatures because of the elevation of melting point by pressure.

The mantle is very markedly depleted in sulphur compared with chondritic meteorite abundances and, though probably much sulphur has been lost (since it is volatile), some may have partitioned into the core. Only an estimated 8–12% sulphur in the outer core is needed to match the outer core density (Ahrens, 1979), but this would represent 20–50% of the chondritic amount, which would mean that the Earth retained a higher proportion of sulphur than of most other similarly volatile elements. Nevertheless, support for sulphur as a diluent comes from iron meteorites which often contain considerable quantities of iron sulphide as the mineral **troilite**.

An alternative diluent, oxygen, has been advocated, particularly by Ringwood (Ringwood, 1977; Ohtani and Ringwood, 1984; Ohtani *et al.*, 1984). The argument is that at core pressures, oxygen, as FeO, is miscible with iron, though it is not at atmospheric pressure. Observations by Knittle and Jeanloz (1991) at pressures of up to 155 GPa, which exceeds that at the core–mantle boundary, and temperatures of 4200°C show that FeO becomes metallic in both liquid and solid forms. A phase diagram with FeO being fully miscible with iron was inferred. Since FeO is not miscible at low pressures it could not enter the core during the growth of the Earth (unlike FeS), but reactions between the mantle and core once the Earth had formed might have introduced it, though it is not known whether the reaction could have introduced enough in the time available (Jeanloz, 1990). Probably Fe–FeO does not form a eutectic; on the contrary, the addition of FeO would probably increase the melting temperature (see Section 6.6).

A case has been made for hydrogen. Measurements at pressures approaching those of the core show that iron hydride forms and has a much lower density than iron, so only a few per cent would be needed (Badding *et al.*, 1991). As with oxygen, there is the problem that it can be incorporated only at high temperature, and there is the additional problem that though hydrogen is extremely abundant in the Solar System, little uncombined hydrogen was likely to have been incorporated in the Earth because of its extreme volatility. However, it could have been produced by the reaction of iron and water:

$$Fe + H_2O \rightarrow FeO + H_2$$

The hydrogen then forms a hydride with unreacted iron (Fukai and Suzuki, 1986).

Also once popular as a diluent was silicon, mainly because it is found in some meteorites. Being somewhat lighter than sulphur (atomic weights respectively 28 and 32), slightly less silicon would be required for a given amount of iron. However, even several per cent by mass of silicon is much greater than the minute traces that have been found in the metal phase of enstatite meteorites. Another observation against silicon is that it does not form a eutectic mixture with iron, and so lacks the consequent lowering of the melting point that would assist core formation. Other elements have been discussed by Stevenson (1981). All were rejected as the sole diluent, though it is probable that some occur as minor constituents of the outer core.

It is also probable that the inner core is not pure iron, for it is the preferred location for siderophile elements to be concentrated (Section 5.3). However, as the abundances of most of these elements are very small, they will have little effect upon the density; the only exception is nickel, which has a meteoritic abundance about 5% that of iron. It is likely to partition into the inner core with iron, slightly increasing the density, but the density of the inner core is not known well enough to check this.

In summary, the outer core is mostly iron with roughly 10% of a lighter element, with sulphur the favoured diluent, though oxygen and hydrogen are other possibilities, while the inner core is solid iron probably with the addition of siderophile elements, notably nickel.

6.5 ENERGY SOURCES FOR THE DYNAMO: (B) COMPOSITIONAL CONVECTION

This second possible mechanism for generating a convective fluid flow in the outer core relies, like thermal convection, on light fluid rising, but the low density is due to a compositional difference, not a higher temperature. This is **compositional convection**. In geology an example of such a mechanism is a turbidity current, in which sediment-enriched water flows down-slope because of its higher density.

To operate in the core, the mechanism requires an inner core growing by material freezing on to its surface as a result of slow cooling. As the material at the base of the outer core cools it separates into two components. Considering Fe–FeS as an example: Fe freezes out on the inner core surface and therefore leaves immediately above it a liquid layer enriched in FeS. As this is lighter than the rest of the outer core, it tends to rise, in a similar way to the hot layer at the base of a tank heated from below.

But, unlike thermal convection, there is no limit – corresponding with the adiabat – that has to be exceeded before there is positive buoyancy, and this allows compositional convection to be more efficient than thermal convection. In fact, because of the low viscosity of the outer core, most of the gravitational energy released has to be dissipated by ohmic heating – resulting from the electric currents that produce the magnetic field – and this heating, being within the core – contributes to the buoyancy, tending to high efficiency. However, since the compositional convection requires cooling of the core, and latent heat is released as iron freezes on to the inner core, heat has to flow out of the core. Thus there is inevitably a thermal component to the convection, with an adiabatic temperature gradient and associated heat flow down it, and this reduces the efficiency (Gubbins and Masters, 1979; Roberts and Gubbins, 1987).

The growth of the inner core to its present size over the age of the Earth, would release heat at an average power of about 4×10^{11} watts, sufficient to drive an efficient compositional dynamo, but not an inefficient thermal one. Moreover, it is likely that the inner core began to form after the core itself, so the possible power is larger.

A critical test of compositional convection would be to establish that the density contrast across the inner core boundary matches that of the model. But there are too many uncertainties, in composition (remembering that there are probably several diluents in the outer core), temperature and the properties of materials under the conditions of the inner core boundary, for this to be possible

113

yet. Thus, at present, compositional convection is the preferred mechanism, but there is no evidence in direct support.

Convection also probably occurs in the inner core (though far less vigorously than in the outer core), because any material not far below its melting temperature is able to deform by solid-state creep, as is found in the mantle (Section 8.2). Convection is the likely cause of the seismic anisotropy reported in Section 6.4: the inner core probably has a hexagonal close-packed crystallographic form, which is anisotropic, and forms elongated crystals which would be aligned by flow.

6.6 TEMPERATURE IN THE CORE

The core temperature cannot be measured directly, but constraints are provided by the knowledge that the outer core is liquid, but the mantle and inner core are both solid. As their compositions and melting points are not well-established, only wide limits can be deduced. Consider first the inner core boundary. The highest temperature that the top of the inner core can have and be solid is if it were pure iron, whose melting point is 7300°C, but it might be reduced by a few hundred °C for the presence of nickel and other siderophiles. The lowest temperature that the base of the outer core could have and be liquid is that of a eutectic Fe–FeS alloy, 4400°C (Fig. 6.8).

We turn next to the top of the core. The temperature gradient in the outer core is likely to be close to an adiabat. (As explained in Section 3.5, thermal

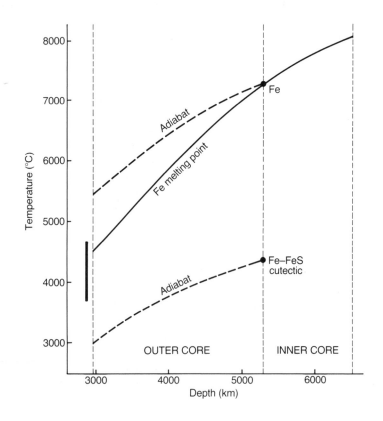

Figure 6.8 Temperature in the core. Melting points at the inner core boundary are given for the extreme compositions of Fe and Fe–FeS eutectic mix. From these, adiabats are drawn to the core–mantle boundary; and the melting point curve for iron is also shown. The bar at the left is the estimated melting temperature at the base of the mantle. All values are approximate. (After Jeanloz, 1990.)

convection cannot begin until the temperature gradient exceeds the adiabat, but once it has been exceeded, an increasing proportion of the heat transported is by convection rather than conduction. In a liquid with low viscosity only a small excess is needed for convection to be vigorous and to transport large amounts of heat. The viscosity of iron in the outer core is estimated to be no more than that of water at normal temperatures.) Therefore, a temperature of 4400°C at the base of the core for the Fe–FeS eutectic would correspond with about 3000°C at the top of the core, and gives its lower limit. The upper limit depends on the composition. An adiabat through 7300°C, the melting point of iron, would give about 5500°C for the top of the core; as the melting point curve of iron is steeper than the adiabat, this would allow the outer core to be pure iron yet liquid (this corresponds with Fig. 6.61(a) but, of course, it would not match the density).

However, the diluent may be oxygen, as FeO. This alone, added to iron, would raise the solidus temperature, and produce a contradiction at the base of the mantle, of having a liquid with a higher melting point next to the solid iron of the inner core. But if diluents which lower the solidus were also present, such as sulphur, several per cent of oxygen could be added and not exceed the melting point of iron.

We now need to add some other constraints. The base of the mantle is solid, which puts an upper limit on it of about 4200 ± 500°C. The sulphur content of the outer core must be far short of that needed to produce a eutectic mixture, so the temperature will be considerably above the eutectic, though addition of other elements may lower it somewhat. Together, these suggest that the outer core may contain sulphur, oxygen and other diluents in addition to iron, and the temperature of the core–mantle boundary may be 4000°C, or more. (This has implications for the temperature difference across the D″ layer, which will be discussed in Section 9.2.3.) This corresponds with a temperature at the inner core boundary of over 5500°C. The temperature in the inner core has an upper limit imposed by the melting point of pure iron, about 7300°C, rising to about 8000°C at the Earth's centre, but this is likely to be lowered by the presence of nickel and siderophiles.

SUMMARY

1. The core contains about 31% of the Earth's mass and 16% of its volume, whereas the inner core accounts for only 1·5% and 0·7%, respectively.
2. The core consists predominantly of iron, for only iron approximately matches the seismic velocities and density of the core, is a good electrical conductor, and is sufficiently abundant. To match the density of the outer core the iron must contain a light diluent.
3. The diluent must be capable of:

    (a) mixing with the outer core under present conditions;
    (b) partitioning into the core;
    (c) being sufficiently abundant; and
    (d) the mixture must have properties that match those of the outer core.

    Most likely diluents are sulphur (as FeS), oxygen (as FeO) and – less likely – hydrogen (as iron hydride). Sulphur is miscible at room temperature and pressure, and lowers the melting point of iron and could have been incorporated as the core grew, whereas oxygen raises the melting point and, being miscible only at high pressure, is more likely to have been added after the core had formed.

Hydrogen also is miscible only at high pressure, and probably could be incorporated only by chemical reactions between iron and water.

The outer core probably contains more than one diluent – perhaps sulphur and oxygen plus lesser amounts of other elements. The inner core probably contains minor amounts of other siderophile elements. Both outer and inner core probably contain nickel but this has little effect upon their properties.

4. The magnetic field at the Earth's surface approximates to that of a dipole and it has its origin in the core. Since it changes with time it must be generated, not by permanent magnetism in the solid inner core, but by electric currents in the outer core, which in turn arise from currents of the liquid conductor, by means of a complex dynamo whose details are not fully understood.

5. The dynamo needs a power of about 5×10^9 W just to make good the ohmic losses, but the power required to produce the liquid motions may be very much greater, depending upon the driving mechanism. The driving mechanism may be thermal or compositional convection.

6. For *thermal convection* to occur a temperature gradient exceeding the adiabat must exist in the outer core, producing lighter material which rises. To produce vigorous convection, the temperature gradient must significantly exceed the adiabat, which alone transports an estimated 3×10^{12} watts by conduction. Only radioactive decay would be able to supply this power on a continuing basis. Uranium and thorium may have partitioned into the core, but a large fraction of their abundances in a chondritic Earth would be needed. Potassium – in particular the isotope ^{40}K – is a less-likely possibility as it would require the unsubstantiated property of being siderophile at core pressures, and implausibly large amounts would be required to sustain an superadiabatic temperature gradient.

7. The favoured mechanism for powering the dynamo is *compositional convection*, driven by the buoyancy of the less-dense layer left when iron solidifies and accretes on to the surface of the inner core. This is a more efficient process than thermal convection, and probably can provide the required power. It has yet to be demonstrated that the actual density difference between inner and outer core is consistent with such a compositional difference.

8. The temperature of the inner core boundary lies between the melting point of iron, about 7300°C, and that of the Fe–FeS eutectic mixture, 4400°C. At the core–mantle boundary the limits are about 3000–4500°C. Additional considerations make it likely that actual values are nearer the upper limit. The temperature at the centre of the Earth is below 8000°C.

FURTHER READING

General journals:
Bloxham and Gubbins (1989): popular account of the Earth's magnetic field and, particularly, what happens in the core.
General books:
Merrill and McElhinny (1983): general review of the Earth's magnetic field, its history, origin and a review of planetary magnetism.
Verhoogen (1980): energy of the geodynamo.

Advanced journals:

Bloxham and Gubbins (1985): field and flows in the core.

Jeanloz (1990): general review of the core.

Jones and Drake (1986): discussion of the geochemical constraints on core formation and accretion models.

Merrill and McFadden (1990): discussion of models of the Earth's magnetic field and geodynamo.

Ohtani *et al.* (1984): a discussion of oxygen as the outer core diluent.

Advanced books:

Anderson (1989a): this book contains a technical review of studies on the composition of the core.

Melchior (1986): a rather mathematical treatment of the physical principles applicable to the core.

Stacey (1977): dynamo theory of the magnetic field.

7 The mantle and oceanic crust

7.1 INTRODUCTION

The mantle contains 84% of the volume and 68% of the mass of the Earth, but because it is separated from direct observation by the thin crust – only about 6 km thick beneath the oceans and an average of 35 km beneath the continental surface – there are many unsolved problems. Nevertheless, we know enough to give a broad picture of its structure and composition which is the principal aim of this chapter. We have few direct samples of the upper mantle, but rocks that derive from this region by partial melting provide important constraints on mantle compositions. Of these magmas, by far the largest quantity is that which gives rise to basalt at oceanic spreading ridges; hence this chapter also discusses the oceanic crust. Here we concentrate on the broad picture of what the mantle is like today, leaving the evidence for its fine structure, dynamics and historical evolution to the next two chapters.

What do we know so far? The mantle can be subdivided into three main seismic regions which are closely concentric with the Earth's surface: the upper mantle which includes the Transition Zone with jumps in velocity at *c.* 400 and 660 km depth, and the lower mantle. Values of density within these regions, shown in Figure 3.9, increase progressively with depth due to self-compression. Steep density gradients occur across the Transition Zone discontinuities, also at the core–mantle boundary, and there is a lower gradient around 100–200 km depth correlating with the low-velocity zone.

In Chapter 5 we used various lines of physical and chemical data to deduce the Earth's chemical composition (Fig. 5.6) and this has been refined to provide the analysis of the **Bulk Earth** (by convention, really the bulk silicate Earth) composition given in Table 7.1. This analysis is based on chondrite silicates with allowance for volatile losses during accretion. In terms of mineralogy, we expect the mantle to be composed predominantly of the same silicate minerals as found in meteorites: olivine (Mg_2SiO_4) and pyroxene ($Mg_2Si_2O_6$), where iron may substitute for magnesium. Calcium and aluminium silicates should also be present, and the other minor elements will either substitute for major elements in the same minerals, or form their own mineral species. These predictions are borne out by rocks derived from the upper mantle which include **nodules**, bits of solid mantle brought up in rising magmas, which are dominated by olivine and pyroxene. Other important materials are those from **kimberlite pipes**, the products of explosive gas–solid volcanic vents with sources at mantle depths, and **ophiolites**, which comprise both ocean crust and upper mantle rocks that have been upthrust in zones of plate convergence.

In Section 7.2 we describe these samples in more detail and use them to develop a model for the composition of the upper mantle. A link between this composition and the products of partial melting is then established in Section 7.3 using **experimental petrology**, which allows material to be subjected to mantle pressures and temperatures. The mechanism of magma production allows us to understand the processes operating at spreading ridges and other sites of basaltic volcanism (Section 7.4). Finally, in Sections 7.5 and 7.6 we consider the effects

Table 7.1 Bulk Earth cosmochemical model compared with compositional model for the undepleted upper mantle and typical analyses of peridotite, eclogite and ocean-ridge basalt.

	1	2	3	4	5
Oxide (per cent by mass)	Bulk Earth	Undepleted upper mantle	Peridotite	Eclogite	Ocean-ridge basalt
SiO_2	47·9	46·2	44·1	47·2	48·5
Al_2O_3	3·9	4·75	1·57	13·9	15·0
FeO	8·9	7·70	8.31	11·0	11·8
MgO	34·1	35·5	43·9	14·3	11·0
CaO	3.2	4.36	1·40	10·1	11·4
TiO_2	0·20	0·23	0·13	0·6	0·7
Cr_2O_3	0·90	0·43	0·34	—	—
Na_2O	0·25	0·40	0·15	1·6	1·5
K_2O	0·025	—	< 0·1	0·8	0·1
Mg/(Mg + Fe) atomic	0·87	0·89	0·91	0·70	0·63
Ca/Al (atomic)	0·75	0·84	0·81	0·67	0·69

Sources:
1 Morgan and Anders (1980).
2 Palme and Nickel (1985).
3–5 Anderson (1989a).

of increasing pressure and temperature on mantle silicates for understanding the mineralogy and composition of the Transition Zone and the lower mantle.

7.2 THE UPPER MANTLE: MINERALOGICAL AND CHEMICAL RELATIONSHIPS

7.2.1 Peridotite and eclogite

There is ample evidence that basalt magmas arise by partial melting in the upper mantle. Indeed, the top of the low-velocity zone (Chapter 2), 10–20 km beneath ocean ridges and 50–60 km deep in other areas of magma production, is commonly linked with the onset of partial melting. The question then is what source materials could melt to produce basalt. There are two main candidates: peridotite and eclogite:

1. **Peridotite** – a group name given to an extensive category of ultramafic rocks, typically composed of about 65% olivine $(Mg,Fe)_2SiO_4$ and 25% ortho-pyroxene $(Mg,Fe)_2 Si_2O_6$ with the other 10% made up usually of clinopyroxene $(Ca,Mg,Fe)_2Si_2O_6$ and garnet $(Mg,Fe)_3Al_2Si_3O_{12}$. Notice that the seven major elements in the Earth (Fig. 5.6) all occur in these minerals. Peridotites are found as thrust slices in uplifted zones resulting from continental convergence, as nodules in basalt lavas, particularly from oceanic islands, and in the diamond-bearing kimberlite pipes of ancient continental regions such as South Africa and Siberia (Fig. 7.1). It is the diamond content of some kimberlites that brought these pipes to fame; diamond is a high-pressure form of carbon which, from experimental studies, must originate at least 150 km down in the mantle.

119

Figure 7.1 A kimberlite pipe from which the garnet peridotite has been mined for diamonds. (From an original picture by P. G. Harris.)

2. **Eclogite** – a dense, high-pressure metamorphic rock with a bulk composition closely resembling basalt. Mineralogically, eclogites contain roughly equal parts of an aluminous pyroxene (omphacite, which is clinopyroxene with some CaMg replaced by NaAl) and garnet. Eclogites also occur in uplifted mountain belts such as the Alps and Himalayas, where they are clearly metamorphosed basalts, and they also form a variable component in kimberlite pipes.

The greater abundance of peridotite samples from the upper mantle, and the much better match between peridotite chemistry and that of Bulk Earth predicted from chondrite silicates (Table 7.1 and see later discussion), combine to favour peridotite rather than eclogite as a working model for the principal upper-mantle material. Nevertheless, there is ample evidence for some eclogite, as we shall find by taking a closer look at the available deep samples.

7.2.2 Kimberlites and ultramafic nodules

Most kimberlite pipes are dominated by garnet-bearing peridotite, though a few are made entirely of eclogite, which contains garnet as a main mineral. So garnet appears to be an important constituent at the depths (100–300 km) tapped by kimberlite, often reaching 15%, though olivine still predominates. One problem with the mineralogy of many kimberlite samples is that, as the gas–solid mixture rose through the mantle, it became disaggregated. So it is sometimes difficult to estimate the relative proportions of the different rock types present. Overall, around 10–15% of the material in these pipes is truly eclogite.

The diamond crystals in kimberlite pipes have proved informative, first because it is possible to distinguish between eclogitic and peridotitic diamonds on the basis of inclusions trapped in the growing crystals, and second by virtue of the carbon isotope analyses obtained from these crystals. These show that peridotitic diamond carbon has always been part of the mantle since the Earth formed, whereas many eclogitic diamonds contain carbon that has been recycled into the mantle from the Earth's surface. The probable mechanism for recycling is subduction of basaltic ocean crust, together with sediment; together these were metamorphosed to eclogite on entering the mantle. We return to this topic in later chapters.

Turning next to the nodules that occur in basalt lavas; most of these are 1–10 cm in size and consist of coarsely crystalline material (Fig. 7.2). They are usually from shallower depths than the basalt source region but, because basalts come from a range of depths, so too do these ultramafic nodules. Many of them are peridotites, more specifically dunites and harzburgites (Fig. 7.3), extremely rich in olivine, which has the highest melting temperature of a mantle mineral and is, therefore, the most refractory. The way in which these peridotites form is believed to be by partial melting of 'fertile' mantle (as discussed below) which extracts basalt and leaves a residue rich in olivine. Olivine is also the upper mantle mineral that has the lowest SiO_2 and highest $MgO + FeO$ content, so olivine-rich refractory residues are the most ultramafic materials from the mantle. Interestingly, the sections of uplifted upper mantle beneath the ocean crust seen in ophiolites contain among the most depleted harzburgites and dunites, which is not surprising as the topmost mantle has been drained of its fusible components in ocean crust formation – more details on this are given in Section 7.4. In summary, olivine-rich peridotite nodules are ultramafic, lying on the base of the tetrahedron in Figure 7.3. They cannot represent fertile **undepleted mantle** (i.e. mantle from which little or no melt fraction has been extracted).

Other highly significant nodule compositions also occur commonly in basalt.

7.2   The upper mantle: mineralogical and chemical relationships

Figure 7.2 Olivine nodule in basalt lava. The outer part of this sample is dark-grey, fine-grained basalt; the nodule is the enclosed coarse aggregate of green olivine crystals.

121

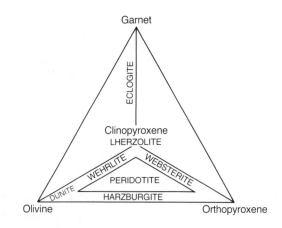

Figure 7.3 Nomenclature of
ultramafic (base of
tetrahedron) and mafic (apex
of tetrahedron) upper-mantle
rocks expressed in terms of
the four principal mineral
components. All the
ultramafic rocks represented
here go under the general
name peridotite. (Note that
clinopyroxene is calcium-
bearing, usually (Ca, Mg,
Fe)$_2$Si$_2$O$_6$, whereas
orthopyroxene is
(Mg,Fe)$_2$Si$_2$O$_6$.)

These are the **lherzolites**, which contain variable proportions of olivine and the
two pyroxenes, but which also lie within the volume of the tetrahedron (Fig. 7.3)
and include spinel or garnet (Table 7.2). Why do we see either spinel (MgAl$_2$O$_4$)

Table 7.2 Mineralogy of spinel and garnet lherzolites.

	Spinel lherzolite		*Garnet lherzolite*	
	Average	*Range*	*Average*	*Range*
Mineral	(vol. %)	(vol. %)	(vol. %)	(vol. %)
Olivine	66·7	55–90	62·6	55–80
Orthopyroxene	23·7	5–35	25	20–40
Clinopyroxene	7·8	3–14	2	0–10
Spinel	1·7	0·2–3	—	—
Garnet	—	—	10	3–15
Phlogopite (mica)	—	—	0·4	0–0·5

Adapted from Maaloe and Aoki (1977); see also Palme and Nickel (1985).

or garnet (Mg$_3$Al$_2$Si$_3$O$_{12}$), but not both, in lherzolites? The answer is that an
aluminium-bearing mineral occurs in many upper mantle rocks, and this tends to
be spinel at low pressures (depth less than about 50 km) but garnet at higher
pressures – further details are given in Section 7.3. Significantly, lherzolites are
less depleted than the peridotites considered above, as we can now demonstrate.

7.2.3   Relationships between mantle samples

Let us consolidate our understanding of the relationships between residual and
relatively fertile, undepleted mantle material with the aid of chemical analyses.
The data in Table 7.3 are averages which form the basis of a rough comparison
between the groups. Of the three mantle compositions given, lherzolite is closest
to the composition of basalt, and therefore must be *least* depleted. For example,
compounds that show the greatest variation across the table: TiO$_2$, Al$_2$O$_3$, CaO,
Na$_2$O, all enter the melt phase during partial melting in preference to the
residuum.

Figure 7.4 focuses on lherzolite as a possible undepleted mantle source from
which basalts could be produced by partial melting, following which the residual
composition would be driven back in the opposite direction. The concentrations

Table 7.3 Typical chemical compositions of principal upper-mantle rock types and oceanic basalts.

	1	*2*	*3*	*4*
Oxide	Olivine	Ophiolitic	Garnet	Ocean-ridge
(per cent by mass)	nodules	harzburgite	lherzolite	basalt
SiO$_2$	42·5	44·5	45·3	48·5
Al$_2$O$_3$	0·5	1·7	3·6	15·0
FeO	7·1	9·6	7·3	11·8
MgO	49·5	42·6	40·3	11·0
CaO	0·3	1·4	3·0	11·4
TiO$_2$	0·01	0·04	0·16	0·7
Na$_2$O	0·05	0·1	0·3	1·5

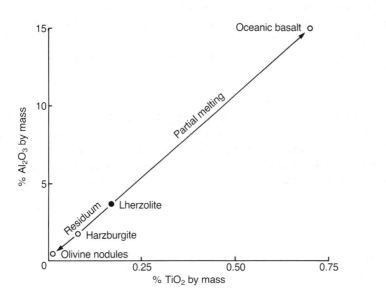

Figure 7.4 Plot of Al$_2$O$_3$ vs. TiO$_2$ for mantle rock types and oceanic basalt in Table 7.3, also showing the directions in which partial melt and residuum progress from a lherzolite parent.

of the two compounds plotted are almost negligible in pure olivine nodules, and so we might regard this as the most depleted residual material sampled, more so than the harzburgite nodules considered here, which are only partially depleted. If all of the TiO$_2$ and Al$_2$O$_3$ of garnet lherzolite were concentrated into the basalt melt, leaving none in the residuum, then the concentration factor from melt to undepleted source is just over four. (For example, for Al$_2$O$_3$ this is $15\cdot0/3\cdot6 = 4\cdot17$). This places a *maximum* limit of 20–25% on the degree of partial melting involved to produce the basalt in Table 7.3. Of course, if a lesser volume of melt is produced, then the available TiO$_2$ and Al$_2$O$_3$ in garnet lherzolite would not be used up, leaving a source only partially depleted in these compounds. However, more than 25% melting of lherzolite could not yield a product melt so rich in them as oceanic basalt.

This does not mean that all the upper mantle has melted to anything like this degree, for various reasons. Firstly, there are various magmas, notably **alkali basalts** of oceanic islands, which contain more TiO$_2$, Al$_2$O$_3$, etc. than do ocean-ridge **tholeiites**. Secondly, materials as depleted in these elements as the olivine nodules analysis in Table 7.3 are rare, and it is more normal to find peridotites containing small amounts of Al–Ti-bearing minerals just above the base of the

123

tetrahedron in Figure 7.3. Thirdly, fluid dynamic considerations demonstrate that basalt melts are of such low viscosity that they tend to separate from their source regions at very small melt fractions, $c.1\%$ (further discussion in Section 7.3). With such small percentages of melting, repeated cycles of melt extraction are necessary to produce the depleted upper mantle samples.

Taking stock, the least-depleted mantle samples that we have are spinel lherzolites, giving way downwards to garnet lherzolites with the same chemical composition: see Section 7.3 for more details of this transition. In simple terms, on melting, the melt is dominated by this aluminous phase (i.e. garnet or spinel melts) plus clinopyroxene, leaving a residue dominated by refractory olivine, usually with some orthopyroxene. They fall closer to the base, in fact also nearer to the front of Figure 7.3, than lherzolite, and all of them have been called peridotite.

This leaves the question of just how much depleted material there is in the upper mantle. Basalt generation clearly has been depleting this region throughout the Earth's history and it is likely that the whole of the upper mantle is depleted but, as we have seen, to a varying degree. Perhaps even lherzolites are slightly depleted compared with the *original* upper mantle. To anticipate the discussion in Chapter 9, we believe this is the case, and hence the model chemistry of undepleted upper mantle given in Table 7.1 contains more of the critical compounds that contribute to partial melts (namely, Al_2O_3, CaO, TiO_2, Na_2O) than does the garnet lherzolite analysis in Table 7.3. Moreover, in Table 7.1 the estimate given for undepleted mantle chemistry differs slightly from Bulk Earth in ways that are important for the early chemical evolution of the mantle (Chapter 9). Returning to the mantle as it is today, next we examine in more detail where and what magmas are produced when lherzolite is subjected to melting conditions in the upper mantle.

7.3 EXPERIMENTAL PETROLOGY AND THE UPPER MANTLE

It is a property of the complex mixtures of silicate minerals in natural rocks such as mantle lherzolite, that they melt over a wide temperature range. This is because different minerals melt at different temperatures, and some individual minerals characterized by **solid solutions** (such as between Fe- and Mg-rich olivines) melt over a temperature range. Indeed, it is because Fe-rich liquids are produced first from such minerals that mantle-derived magmas have lower $Mg/(Mg + Fe)$ ratios than their sources.

7.3.1 A phase diagram for the upper mantle

The experimental technique used to find the conditions of melting consists of taking a series of identical small rock charges and subjecting each to a different temperature (T) all at the same pressure (P). Examination of the products reveals the temperature at which the first sign of melting occurs, defined as the **solidus** temperature, and by repeating the process at different pressures, a solidus curve can be drawn in $P–T$ space – see Figure 7.5 for lherzolite solidus. Similarly, the **liquidus** is defined as where the rock becomes totally molten; in the interval between the liquidus and solidus, a mixture of crystals and liquid

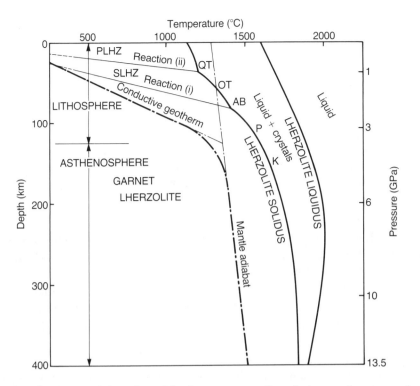

Figure 7.5 Lherzolite compositions, melting relations and products, and temperature conditions beneath a normal (*c.* 60 Ma old) oceanic area. Heavy solid lines indicate dry solidus and liquidus curves for lherzolite in the absence of volatiles. Compositions of magmas resulting from partial melting at the solidus are QT = quartz tholeiite, OT = olivine tholeiite, AB = alkali basalt, P = picrite, K = komatiite. Faint solid lines indicate boundaries between different forms of lherzolite in the solid state; PLHZ = plagioclase lherzolite, SLHZ = spinel lherzolite. The heavy dot–dash curve indicates change of temperature with depth (geotherm) and is characterized by an upper part where heat transfer is by conduction, leading to a rapid variation of temperature with depth, and a lower part that follows the convective mantle adiabat. The 125 km thickness of the lithosphere beneath this 60 Ma oceanic area is defined by the intersection between the projected conductive and convective portions of the geotherm. (Sources: Scarfe and Takahashi, 1986; McKenzie and Bickle, 1988; Wyllie, 1992, with adaptations.)

occurs, i.e. a partial melt, with the amount of melt increasing towards the liquidus. For lherzolite this interval is about 500°C wide near the surface, but it diminishes to just 50°C at the base of the upper mantle.

The negative *P–T* slope on the lherzolite liquidus at pressures above 7 GPa is particularly interesting. It tells us that the density of the melt is greater than that of the residual crystals (mainly olivine) with which it is in equilibrium, probably because the melt is more compressible than the associated crystals. This means that any partial melts produced at depths greater than about 200 km are unlikely to rise towards the surface. So we concentrate our discussion of melting on the shallower mantle. Here, in the subsolidus region of the phase diagram, two boundaries are shown between different forms of the same lherzolite chemistry: these are **phase changes**. The almost horizontal slope of these boundaries implies that the reactions across them are highly pressure sensitive; small increases in pressure produce increases in density. We have already discussed the higher-pressure spinel–garnet transition; a reaction that may be written:

$$2 (Ca,Mg,Fe)_2Si_2O_6 + MgAl_2O_4 \rightleftharpoons Ca\,Mg_2Al_2Si_3O_{12} + (Mg,Fe)_2SiO_4$$

$$\underbrace{clinopyroxene + spinel}_{low\ pressure} \qquad \underbrace{garnet + olivine}_{high\ pressure}$$

At even lower pressures, closer to the Earth's surface, the natural form is plagioclase lherzolite as follows:

$$2(Mg,Fe)_2SiO_4 + CaAl_2Si_2O_8 \rightleftharpoons MgAl_2O_4 + (Ca,Mg,Fe)_2Si_2O_6 + (Mg,Fe)_2Si_2O_6$$

$$\underbrace{olivine + feldspar}_{low\ pressure} \qquad \underbrace{spinel + clinopyroxene + orthopyroxene}_{high\ pressure}$$

Note that these phase changes affect only a small proportion of the lherzolite **125** mineralogy (cf. Table 7.2); the bulk of the olivine and orthopyroxene remain unaffected. The slight density differences between the three lherzolite assemblages cause changes in the shape of the solidus curve where it is intersected by the phase boundaries, hence the cusps on the solidus in Figure 7.5. So, as indicated by the phase diagram, the top *c.* 25 km of relatively undepleted mantle will be plagioclase lherzolite; in the interval *c.* 25–75 km depth this becomes spinel lherzolite, giving way downwards at *c.* 75 km depth to garnet lherzolite.

Next, we need to know where the pressure–temperature conditions in the mantle are such that melting will occur at or above the solidus. *Pressure* in the mantle is lithostatic, varying according to the weight of overlying rocks, which, in turn, is a function of their thickness (h) and density (ρ):

$$P = \rho g h$$

where g is the acceleration due to gravity (as in Eq. (3.2), which gives $g = GM_E/r^2$). In the crust and uppermost mantle where the average density is 3000 kg m^{-3}, and if h is in km, we have:

$$P = 3 \times 10^7 h \text{ N m}^{-2} \text{ (Pa)} = 30h \text{ MPa (useful for crustal depths)}$$
$$= 0.03h \text{ GPa (useful in the mantle).}$$

Of course, density increases with depth (Fig. 3.9) and that is why the pressures in Figure 7.5 are a non-linear function of depth; this effect becomes even more marked deeper in the mantle (Sections 7.5 and 7.6).

Temperature increases with depth, otherwise known as geothermal gradients or **geotherms**, depend on several factors such as the heat productivities of the rocks, their thermal conductivities if heat transport is by conduction, and the type of heat transport: convection or conduction. Near-surface rocks are cold relative to the melting point, and therefore they are mechanically rigid on geological time-scales and thus form a **thermal boundary layer** insulating the heat content of the convecting interior. The concept of a rigid outer boundary layer around the Earth is important as a simple explanation of the *lithosphere* (see Chapter 8). Heat is transported by conduction and so the geotherm is steep. The geothermal gradient in 60-Ma-old oceanic lithosphere (Fig. 7.5) is a fairly uniform 11°C km^{-1} and, if extrapolated to 150 km depth, this gradient would encounter the solidus at 1650°C, resulting in partial melting. However, the mantle becomes hot enough to convect at around 1350°C and, of course, convection tends to prevent the temperature gradient exceeding the adiabat (see Section 8.5). It is no coincidence that the adiabat runs parallel to the solidus, for mantle rocks lose their rigidity as they approach melting temperatures. Our oceanic geotherm therefore curves around to meet the mantle adiabat, and the lithosphere–asthenosphere boundary, at 125 km in this case, occurs where the two projected *P–T* gradients intersect.

The geotherm we have described is by no means unique and the variation of temperature with depth is different depending on location. For example, if the conductive temperature gradient were greater, the lithosphere would be thinner. In the limiting case with no thermal boundary layer, convecting material would reach the surface at 1280°C, known as the **potential temperature**, where the mantle adiabat intersects zero pressure. We return to this in Section 7.4.

7.3.2 Causes of mantle melting

Since the typical mantle geotherm in Figure 7.5 and the lherzolite solidus do not intersect, we need to consider whether either or both can be deflected to promote partial melting. This could be achieved in three ways, firstly by volatiles depressing the solidus temperature, secondly by rising hot material raising the temperature, and thirdly by reducing the pressure.

First the solidus; the one we have been considering is in the absence of volatile components such as carbon dioxide and water, small amounts of which occur combined in minor mantle minerals such as magnesite, $MgCO_3$, amphiboles (e.g. $Ca_2(Mg,Fe)_5(Si_4O_{11})(OH)_2$) and mica (e.g. $KMg_3(AlSi_3O_{10})(OH)_2$). Notice from Figure 7.6(a) that beneath about 300 km along the typical mantle geotherm, CO_2 and H_2O are likely to remain fixed in minerals, so it is only above this depth that they may be available to influence melting. Figure 7.6(a) shows the lherzolite solidus in the presence of as much CO_2 and H_2O as the melt can absorb; this occurs at a temperature some 300–500°C below the dry solidus. But in the real mantle, volatiles are rapidly used up and their abundances are usually inadequate to promote more than a few per cent of 'wet' melting at normal mantle temperatures. Figure 7.6(a) shows that this partial melt is most likely to be present in the mantle between 100 and 300 km depth, particularly towards the top of this region. Significantly, some of the lowest seismic velocities recorded in the mantle occur just below 100 km depth in the low-velocity zone beneath tectonically active areas (Section 2.3).

Turning next to the way in which *thermal* perturbations may induce mantle melting, Figure 7.6(b) illustrates the effect of rising hot, low-density mantle 'burning' into the base of the lithosphere until it induces partial melting by intersecting the dry solidus at 100 km depth. With time, a large shallow melt zone (shaded in Fig. 7.6(b)), extending almost to the surface, could develop; this type of upwelling may occur beneath some 'hot spots' and ridge zones, such as in Iceland (discussed further in Chapters 8 and 9). Finally, there are circumstances (Fig. 7.6(c)) in which mantle at normal temperatures can be caused to melt; this requires that the thermal boundary layer is thinned dramatically, bringing about the necessary decompression. This will occur where the lithosphere is under strong *extensional stress* – for example at ocean ridges. This model of **decompression melting** is developed further in Section 7.4.

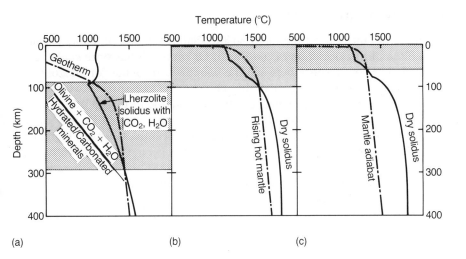

(a) (b) (c)

Figure 7.6 Development of a zone of possible melt production (shaded) under three different scenarios. (a) Lherzolite solidus is depressed by the presence of volatile species in the mantle, principally CO_2 and H_2O produced by the breakdown of minerals such as magnesite and amphibole above the feint solid line; intersection with normal geotherm (from Fig. 7.5) dictates the depth limits of the possible melt zone. (b) Rising high-temperature material in the mantle with a potential temperature of 1500°C thins the lithosphere from beneath and causes melting at the dry solidus (from Fig. 7.5) in the upper mantle. (c) Similar to (b) but with 1280°C potential temperature (the normal mantle adiabat) brought closer to the surface due to decompression (depressuring) of the shallow mantle. (Sources: Mckenzie and Bickle, 1988; Wyllie, 1992, with adaptations.)

127

7.3.3 Melt compositions and their extraction rates

Experimental data on melt compositions produced at and above the lherzolite solidus show a trend with increasing source depth (pressure) towards decreasing SiO_2 and increasing MgO: specifically, from quartz tholeiite to alkali basalt, with high-magnesian picrite and komatiite from the greatest depths (Fig. 7.5). Thus melts produced immediately beneath ocean ridges (cf. Fig. 7.6(b) and (c)) are tholeiitic, whereas deeper melt zones around 50–70 km depth produce alkali basalt magma. As the source depth increases, so the amount of each type of magma that has reached the surface decreases; picrites and komatiites are therefore comparatively rare. Komatiites are almost exclusive to early Precambrian terranes, where they may testify to the greater temperatures and/or thermal turbulence in the early upper mantle (Chapter 11).

Comparison of the thickness (*c.* 50 km) of the melt production zone in Figure 7.6(c) with the resulting thickness of normal ocean crust (6–7 km) indicates that some 10–15% of the shallow mantle must have melted (i.e. well within the upper limit deduced earlier from Fig. 7.4). These are the tholeiitic magmas which occur in large volumes, but, as noted in Section 7.2.3, basaltic melts have such low viscosities that melt fractions as low as 1% should segregate and rise (see McKenzie, 1985 for calculations). They travel upwards at speeds of 5–10 cm per year, so that it will take 10^5–10^6 years for them to arrive from sources 10–50 km deep. It follows that even zones of large volume melt production (10–15%) in the shallow mantle rarely contain more than a few per cent of melt at any time, for as the melt forms it is rapidly extracted, and is then followed by further melting and extraction. Since the speeds of sea-floor spreading and basalt magma ascent are similar, this implies that a large-volume zone with small melt fractions exists in the upper mantle, feeding surface activity focused at the ocean ridges (see Section 7.4). The more mafic magmas produced at greater depths have even lower viscosities and so separate at very small melt fractions (0·1%). Moreover, only relatively small volumes of melt are produced that take 10^6–10^7 years to reach the surface.

7.3.4 The early Earth magma ocean

In Chapter 5 we introduced the fact that the shallow mantle probably became totally molten during accretion of the Earth. Of course, it has since cooled and crystallized. The convergence between the liquidus and solidus for lherzolite at *c.* 13 GPa (Fig. 7.5) indicates that this may be the pressure at which final crystallization occurred. This is because the last liquids to crystallize in multi-component systems lie close to **eutectic** mixtures, characterized by the coincidence of the liquidus and solidus in *P–T* space. Significantly, however, lherzolite has a small temperature interval between these curves, and it is known that at high pressures the first mineral to crystallize is a form of garnet. Separation of these crystals from the melt by their downwards displacement (fractionation) deeper into the Transition Zone or lower mantle may account for the small differences in the geochemistry of Bulk Earth and the original undepleted upper mantle discussed earlier (Table 7.1 and see also Chapter 9).

The experimental petrology discussed here has guided our path through some of the intricacies of upper mantle phase changes, including the conditions of melting, the compositions of the melts produced and their extraction. Similarly, it has a major role to play in our understanding of the Transition Zone (Section

7.5). But first we examine the structure and composition of oceanic crust, to complete our discussion of processes at spreading centres.

7.4 STRUCTURE AND EVOLUTION OF SPREADING RIDGES

Volcanic activity on the Earth is dominated by the ocean ridges which produce about 20 km^3 per year of basalt. This activity is just a surface expression of the processes that create oceanic plates, which we now examine from a petrological standpoint. Oceanic plates are recognized as transient features, deriving from the mantle and returning to it in a time of the order of 100 Ma. This has many implications; for example, sometimes an oceanic plate escapes destruction at subduction zones in the final stages of ocean closure when part of the oceanic plate is carried or **obducted** on to one of the forelands of two converging continents, to give rise to a characteristic 'ophiolite suite'. The origin of the term comes from the Greek *ophis* which means snake or serpent, and it was originally used to refer to the characteristic serpentinized (metamorphosed) peridotites. The term has now come to embrace a whole suite of genetically related rocks with a special significance for ocean lithosphere formation. Many ophiolite sequences have been recognized, and perhaps one of the best known occurs in the Troodos Massif of Cyprus. The obduction of Troodos occurred during the closure of the Tethyan (early Mediterranean) Ocean during Tertiary times, and the actual underthrusting of Cyprus by the African continental foreland was a Miocene event.

7.4.1 Ophiolite petrology

The significance of ophiolite studies is that the petrology of ocean crust and upper mantle layers can be related to the seismic properties of these layers which characterize all oceanic regions (Fig. 7.7). Of course, the occurrence of ophiolite sequences in continents means that they may not be typical samples of the

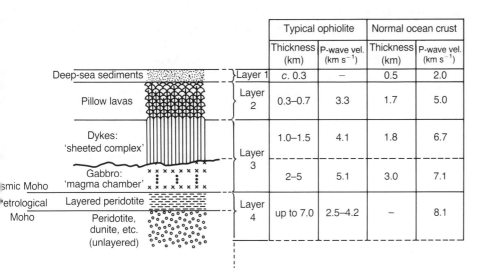

		Typical ophiolite		Normal ocean crust	
		Thickness (km)	P-wave vel. (km s^{-1})	Thickness (km)	P-wave vel. (km s^{-1})
Deep-sea sediments	Layer 1	*c.* 0.3	–	0.5	2.0
Pillow lavas	Layer 2	0.3–0.7	3.3	1.7	5.0
Dykes: 'sheeted complex'	Layer 3	1.0–1.5	4.1	1.8	6.7
Gabbro: 'magma chamber'		2–5	5.1	3.0	7.1
Layered peridotite / Peridotite, dunite, etc. (unlayered)	Layer 4	up to 7.0	2.5–4.2	–	8.1

Seismic Moho / Petrological Moho

Figure 7.7 Petrological, seismic and thickness data for a typical ophiolite sequence compared with seismic layers for the ocean crust. (Note that the layers are not drawn to scale, and the gabbro layer, in particular, is usually much thicker than is illustrated here.) (Data from Gass, 1982; Lewis, 1983; Spray, 1991.)

129

regions they are thought to represent, and four particular features emphasize this uncertainty:

1. Some ophiolite sequences are rather thin, containing *c.* 4 km of ocean crust materials overlying mantle, compared with 6–7 km of crust in most parts of the oceans; the pillow lava layer ranges from 2 to 8·5 km in thickness and sometimes the gabbro layer is completely missing (see Lewis, 1983, for data).
2. The magnetic anomalies that characterize ocean crust elsewhere have not yet been discovered from ophiolites.
3. Seismic velocities are generally much lower in ophiolites than in normal ocean crust.
4. Geochemically, many ophiolite sequences have more in common with **back-arc marginal basins** (i.e. on the continental side of subduction zones) than they do with ocean ridge magmas.

There are various explanations of these differences. For example, ophiolites are often 'dismembered' by the thrusting process. Magnetic anomalies may be lost through extensive hydrothermal metamorphism, usually observed in ophiolites; moreover, most of the ophiolites so far studied comprise ocean lithosphere formed during a long period of normal polarity during late Mesozoic times. Hydrothermal metamorphism may also account for the lower seismic velocities, due to the extensive mineralogical changes that occur. For example, expanded clay mineral and serpentine lattices replace many of the primary silicate minerals, particularly olivine.

Accepting that ophiolites really do represent slices through the top 10 km or so of ocean lithosphere, albeit perhaps often from marginal basins, what information do they provide about the materials present and their evolution? The upper portion of the sequence comprises fine-grained Fe–Mn-rich mudstones, cherts, shales and limestones which match up with deep-sea sediment cores. Both these sediments and the underlying tholeiitic lava sequence sometimes contain base-metal (e.g. Cu–Zn) sulphide deposits. The key to their origin comes from the spectacular discovery in the 1970s of so-called **'black-smokers'** where hot water gushes from sea-floor vents at 350°C or more forming a dense plume of black 'smoke' made of tiny metal sulphide particles. The hot water is sea-water that has penetrated several kilometres into the crust at spreading ridges; it becomes highly acidic due to mixing with discharging volcanic gases and is capable of cooling, cracking and leaching base metals from the crustal rocks. They are deposited as concentrated sulphides at or below the surface where the water is cooled and neutralized. As the same time the rocks of the ocean crust may become heavily metamorphosed.

Pillow lavas (Fig. 7.8) are usually prominent within the basaltic layer. They are typical products of subaqueous eruptions where the lava, chilled on contact with water, rapidly forms a rind, which expands to a bulbous or tubular shape as pressure increases within the rind. These surface lava flows apparently were fed by a series of multiple dykes – the so called **sheeted complex** – from a mafic magma chamber, now in the form of coarse-grained gabbros (Fig. 7.7). Figure 7.8 illustrates the lower part of layer 2, with pillow lavas being cut by later dykes. Continuing downwards, the dykes give way to layered plutonic gabbros at the base of seismic layer 3. These intrusive rocks are notable for their igneous lamination, which takes the form of lighter, felsic and darker, more mafic banding produced by crystal fractionation processes. Early crystallizing minerals were mainly olivine with some pyroxene. As both are relatively dense compared

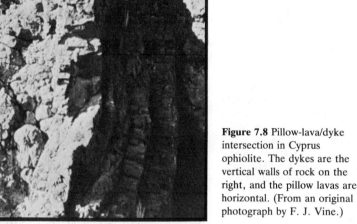

Figure 7.8 Pillow-lava/dyke intersection in Cyprus ophiolite. The dykes are the vertical walls of rock on the right, and the pillow lavas are horizontal. (From an original photograph by F. J. Vine.)

with mafic magma, they tended to accumulate in high concentrations towards the base of the chamber, where they formed **crystal cumulates** – the layered peridotites – at the top of seismic layer 4 (Fig 7.7). Small volumes of low-alkali silicic rocks (plagiogranites) sometimes formed from the final **residual liquids** which crystallized near the top of the mafic magma chambers.

The seismic distinction between ocean crust and upper mantle is thought to occur where gabbros grade downwards into layered peridotites at the layer 3/ layer 4 junction: this is the **seimic Moho**. But the petrological boundary – the **petrological Moho** – occurs at the boundary where layered peridotites, produced by gravity settling in *crustal* magma chambers, are separated from the massive peridotites that have always been part of the upper mantle. Not surprisingly, the mantle section is composed mainly of depleted harzburgite and dunite (Fig. 7.3), together with localized patches of lherzolite.

7.4.2 A spreading-centre model

Here we combine the petrological evidence from ophiolites with the information on mantle melting in Section 7.3, to give a model for spreading centre processes.

Beneath the ocean ridge we will assume a normal mantle geotherm (Fig. 7.6(c) and Fig. 7.9 in cross-section). In these circumstances, extensional processes cause decompression melting. Isotherms in the mantle are fairly horizontal and are displaced upwards only where the lithosphere is thinned and deeper mantle material rises, following the mantle adiabat, until it intersects the lherzolite solidus. Tholeiitic basalt magmas are produced as small melt fractions within the zone of extensive silicate melting, down to 50 km, and are rapidly extracted through the ridge zone. Repeated cycles of partial melt formation and extraction may continue for perhaps 5 Ma as the newly formed lithosphere spreads away to either side of the ridge.

Smaller amounts of melt (down to *c*. 0·1% melting) from the region below

131

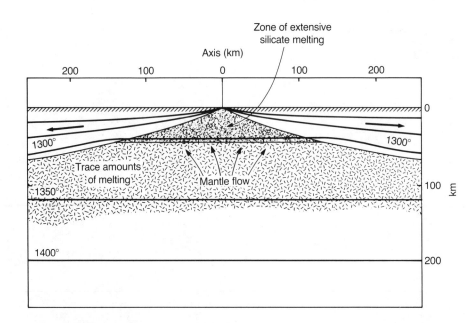

Figure 7.9 Summary of the thermal structure beneath a passive spreading ridge with a mantle potential temperature of 1280°C. Asthenospheric material rises (feint arrows) to form an extensive melt zone beneath the point of plate separation, due to strong extensional stresses (heavy arrows) that cause the upper mantle to become compressed. (After McKenzie and Bickle, 1988.)

may also be fed upwards; these would be produced when upwards flow of shallow mantle material (Fig. 7.9) in turn decompresses the deeper mantle, perhaps promoting dehydration reactions and, therefore, partial melting (cf. Fig. 7.6(a)). Note that the illustration in Figure 7.9 is a description that applies to any zone of high extensional lithospheric stress, oceanic or continental, where melting is induced by decompression. These are so-called **passive rifts**, as opposed to **active rifts**, which are driven by upwelling convection within the mantle (Chapter 8). Both types of rifting occur in the ocean basins, but the more usual form of ocean ridge is believed to be the passive variety, as this allows the sites of sea-floor spreading to migrate (Section 8.4.2).

The evidence from ophiolites suggests that ascending batches of melt become trapped below a thin thermal boundary layer in a crustal magma chamber beneath the ridge axis (Fig. 7.10). Here magma is stored before its eruption through narrow dykes to form the submarine volcanic lavas of the ocean ridges. As the two sides spread apart, most of the magma cools and crystallizes within the newly created ocean crust. The magma chamber peels off solid layers at its edges while growing from its centre, and it was at first thought to be linear and continuous beneath the ocean ridges, and therefore this became known as the 'infinite onion' model (Cann, 1970). More recently, seismic studies have shown that the molten part of the chamber is confined to a narrow central conduit, and that this is surrounded by a 6–8-km-wide zone of hot rock with just a few per cent of partial melt. A small, mushroom-shaped chamber, perhaps 4 km wide and a few hundred metres deep, occupies the axial zone and buoyantly supports the overlying ridge.

In summary, within the crustal magma feeder system, crystallization of high-temperature minerals causes the layered peridotites to be precipitated, while occasional ejection of magma to the surface builds up seismic layer 2 and the upper part of layer 3 (Fig. 7.7). The remaining gabbroic crystal–liquid mush also develops layering as residual liquids are squeezed out of the solid to form more silicic laminae alternating with mafic bands of earlier-formed crystals. Ultimately, the whole mass is crystalline; it then cools and contracts as sea-floor spreading takes it away from the ridge zone (Section 8.4.2). Thus Figures 7.9

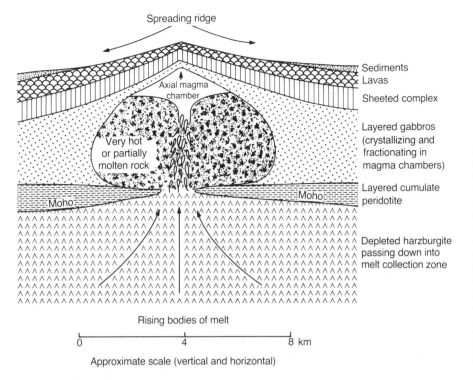

Figure 7.10 Schematic cross-section through the top 10 km of an ocean ridge to illustrate the evolution of the ocean crust and upper mantle structure and composition (see Gass, 1982). Note that only the central and upper parts of the mushroom-shaped magma chamber are more than 50% molten; this region is surrounded by a wider reservoir of extremely hot rock but probably with only a few per cent partial melt (Detrick *et al.*, 1987; Burnett *et al.*, 1989).

and 7.10 provide a summary at two scales of the melting regime and the petrological development of oceanic crust at spreading centres.

7.5 THE TRANSITION ZONE

This relatively inaccessible region, usually included as part of the upper mantle, is most simply defined by two major steps in seismic velocity–depth profiles at its upper and lower bounds: 400 and 660 km (Fig. 7.11). These steps correlate with jumps in the density–depth profile (Fig. 3.9) and, according to the constrained reference mantle model (modified PREM), a total density increase from 3·58 to 4·37 Mg m^{-3} occurs across the zone. In the 1930s, F. Birch and J. D. Bernal first proposed that the Transition Zone represents a region of *high-pressure phase changes* for silicate minerals. The new minerals, appropriate to the conditions of pressure and temperature in the lower mantle, would be denser and have more closely packed internal atomic structures. More recently, appropriate phase changes have been identified that account for the observed seismic velocity and most of the inferred density changes.

Attempts to resolve the precise depth of the mantle discontinuities have been numerous, and an approach of particular merit involves the stacking of long-period reflections in various seismic phases. Reflections that stack well, even if weak, provide good evidence for velocity steps in the mantle. Several sets of results from different laboratories confirm the existence of the two main discontinuities, with revised depths of 414–415 and 659–660 km in normal mantle. However, a third, previously unresolved discontinuity at 519–520 km depth has also been identified with a much smaller velocity change than the

133

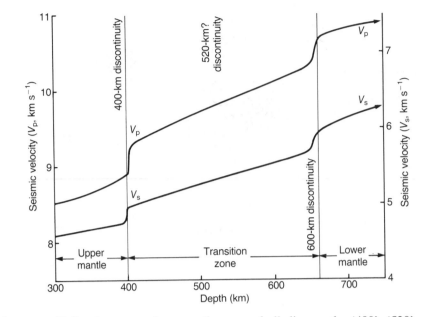

Figure 7.11 Seismic P-and S-wave velocity variations across the Transition Zone in the Earth's mantle showing the two principal discontinuities at *c.* 400 and *c.* 660 km depth. (Data are from the constrained reference mantle model, see Chapter 3.) The evidence for the 520 km discontinuity is discussed in the text.

other two. Following normal conventions we shall discuss the '400', '520' and '660' km discontinuities below.

7.5.1 The mineral physics of the Transition Zone

Experimental petrology has provided the key to understanding phase transformations in mantle minerals at pressures in the range 13·5–23·5 GPa (400–660 km depth). Until the 1960s it was impossible to attain the necessary pressures, but a good understanding was obtained using **germanate analogue minerals** (based on GeO_4^{4-}) which have identical mineral structures to silicates (SiO_4^{4-} in olivine), but their phase transformations occur at much lower pressures. For example, it was discovered that magnesium germanate (Mg_2GeO_4) transforms from a relatively open orthorhombic structure to a more closely packed, cubic, spinel-like structure, with an 8% density increase, at about 4 GPa pressure. This implies that natural olivine (Mg_2SiO_4) should undergo a similar transformation to a **spinel structure** in the mantle.

Detailed study of Transition Zone phase change reactions has shown most of them to be **reconstructive transformations,** in which the high-and low-pressure polymorphs are structurally very different. A familiar example is carbon, a minor mantle constituent which, at about 4 GPa (150 km depth), transforms from graphitic carbon with an open structure to the more tightly packed diamond structure with a density increase from 2 to 3·5 Mg m^{-3}. A large energy barrier must be overcome to break bonds and reorganize the constituent atoms. Like diamonds, the high-pressure silicate polymorphs produced in the Transition Zone do not readily revert to their low-pressure forms immediately after synthesis experiments. So their densities at 'zero-pressure' (i.e. atmospheric pressure) may be measured directly (see Table 7.4).

The simplest reconstructive transformation that occurs in silicates is that of pure SiO_2. A high-pressure silica polymorph known as **stishovite** (after S. M. Stishov) was first synthesized at about 10 GPa and was then recognized from meteorite impact craters. Stishovite has a zero-pressure density of 4·29 Mg m^{-3}

134

Table 7.4 Formulae, co-ordination, density and compressional wave velocity for some mantle minerals simplified to Mg-rich end members; in the Earth's mantle, all these minerals (except stishovite and corundum) will contain $c.10–15\%$ Fe substituting for Mg.

Mineral	Formula*	Zero-pressure density‡ (Mg m^{-3})	V_p estimated at depth‡ (km s^{-1})	Depth
Orthopyroxene	$^{VI}Mg_2{}^{IV}Si_2O_6$	3·31	7·8	
Clinopyroxene	$^{VIII}Ca^{VI}Mg^{IV}Si_2O_6$	3·32	7·7	200 km
Olivine	$^{VI}Mg_2{}^{IV}SiO_4$	3·37	8·3	
Garnet	$^{VIII}Mg_3[^{VI}Al]_2{}^{IV}Si_3O_{12}$	3·68	9·0	400 km
β-spinel	$^{VI}Mg_2{}^{IV}SiO_4$	3·63	9·4	
γ-spinel	$^{VI}Mg_2{}^{IV}SiO_4$	3·72	10·3	
Majorite	$^{VIII}Mg_3[^{VI}Mg, {}^{VI}Si]^{IV}Si_3O_{12}$	3·59	9·7†	
'Perovskite'	$^{VIII}Mg^{VI}SiO_3$	4·15	10·6†	600 km
Magnesiowüstite	^{VI}MgO	4·10	9·1	
Stishovite	$^{VI}SiO_2$	4·29	11·9	
Corundum	$^{VI}Al_2O_3$	3·99	11·5	

* The roman numerals ^{IV}, ^{VI} and ^{VIII} refer to the number of oxygen atoms that can be packed around the relevant Mg, Si, Ca or Al atom in the mineral structure, i.e. the co-ordination number. The ^{IV} refers to tetrahedral co-ordination, ^{VI} to octahedral and ^{VIII} to cubic co-ordination.
† Estimated values.
‡ Data from Duffy and Anderson (1989); mineral velocity data are given for a particular depth as indicated to the right of the table.

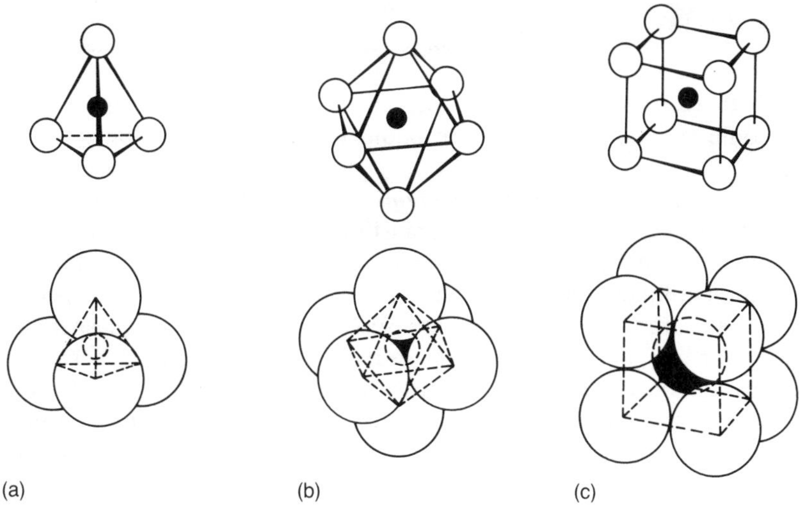

(a)   (b)   (c)

Figure 7.12 Illustration of co-ordination number for atomic species with different ionic sizes in crystal structures: (a) tetrahedral (fourfold co-ordination); (b) octahedral (sixfold co-ordination) and (c) cubic (eightfold co-ordination). In each case, the upper diagram shows the geometry of the co-ordination polyhedron and the lower diagram shows the atoms in their natural close-fitting co-ordination positions. In mantle silicates, oxygen is the largest abundant ion, and so co-ordination number reflects the way in which smaller magnesium, iron, silicon, etc. ions fit into mineral structures.

and is characterized by **octahedral co-ordination** of silicon and oxygen (i.e. six oxygens around each small silicon atom), whereas normal low-pressure silica, or quartz, has a density of $2·65$ Mg m^{-3} and **tetrahedral co-ordination** (four oxygens around each silicon): see Figure 7.12 for co-ordination details.

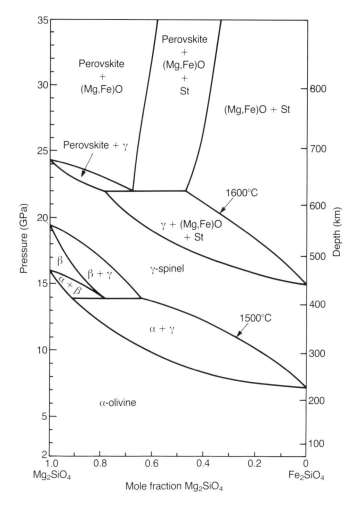

Figure 7.13 Phase diagram for Mg$_2$SiO$_4$–Fe$_2$SiO$_4$ in the pressure range 2–30 GPa approximately following the mantle adiabat for a 1280°C potential temperature. The lowermost set of curves was determined at 1500°C and the higher set at 1600°C. Notice that the phase changes occur at greater depths in more magnesium-rich rock systems. The β and γ are both forms of spinel; St = stishovite; (Mg, Fe)O = Magnesiowüstite; for further details of mineral phases see Table 7.4. (After Anderson, 1989a.)

Although there is very little free silica in the mantle assemblages described earlier, some may be produced by reactions in the Transition Zone.

Olivine–spinel–perovskite transformations. Although an increase of co-ordination number is common in the Transition Zone, this is not the explanation for the olivine–spinel phase change. Table 7.4 shows that in both minerals the co-ordination numbers for magnesium and silicon are six and four, respectively. But, because large oxygen atoms are susceptible to being squashed under pressure, some of the bond lengths shorten. Typical distances between the centres of Mg and O ions are 211 pm (picometres, m × 10^{-12}) in olivine and 208 pm in spinel while the volume occupied by each oxygen ion reduces from 18·1 to 16·8 × 10^6 pm^3; overall, the density increases by 7·7%. Note that this form of spinel is Mg$_2$SiO$_4$, not MgAl$_2$O$_4$ or MgCr$_2$O$_4$ which are present at shallower levels (in spinel lherzolite) in the mantle.

Figure 7.13 shows the pressure conditions under which natural olivines (the series Mg$_2$SiO$_4$–Fe$_2$SiO$_4$) transform to the spinel and other structures when the temperature follows a mantle geotherm with a 1280°C potential temperature (1500°C at 400 km deep – see Fig. 7.5). In upper-mantle olivines (α in Fig. 7.13) the typical mole fraction of Mg$_2$SiO$_4$ is 0·90–0·91, so β-spinel will start to be produced just short of 14 GPa, that is, just below 400 km depth. Since olivine is the dominant upper-mantle mineral phase, accounting for 60–70% of lherzolite (Table 7.2), it is not surprising that the olivine–spinel transformation has become firmly linked with the 400-km discontinuity. A further point about this phase change is that its depth will change if the mantle temperature is significantly different. For example, if the temperature is 200°C *less*, as may occur in deeply subducting cold oceanic lithosphere, the phase change will occur about 10 km *shallower* closer to the surface. (This is significant for plate driving forces, as we shall see in Section 8.4.)

Figure 7.13 shows that β-spinel is stable only through a narrow pressure interval; indeed it only exists in magnesium-rich mantle. Across a broad pressure interval around 17–18 GPa β-spinel gives way to a still more compact variant known as γ-spinel which has a 2% higher zero-pressure density (Table 7.4). So the β–γ spinel phase change may account for the small discontinuity at 520 km depth. Gamma-spinel remains a stable mineral across most of the Transition Zone but, in turn, it is replaced by $MgSiO_3$ in the **perovskite structure** at around 23·5 GPa. As the most important constituent of the lower mantle (itself 55% of the Earth's volume), Mg-perovskite is almost certainly the most abundant mineral in the Earth. The type mineral perovskite, $CaTiO_3$, is fairly uncommon at the Earth's surface. The sizes of the calcium and titanium ions (ionic radii: 99 pm and 69 pm) are such that they fit simply into cubic (large Ca^{2+} ions) and octahedral (smaller Ti^{4+} ions) co-ordination with oxygen (Fig. 7.12). But at 23·5 GPa the oxygen atoms are compressed down to $13·5 \times 10^6$ pm^3, about 75% of their surface volume, so each type of co-ordination site is also now smaller. Magnesium (Mg^{2+}, 66 pm radius) takes the cubic sites and silicon (Si^{4+}, radius 42 pm) the octahedral sites, so forming perovskite-structured $MgSiO_3$, which is 11.5% more dense than γ-spinel and 23% more dense than olivine at zero pressure. (The perovskite structure can be imagined in terms of a cube with oxygens at each of the six face centres, each shared equally with an adjacent cube. The single octahedral Si^{4+} site is at the cube centre, and Mg^{2+} sites exist at each of the eight corners but are shared with seven adjacent cubes. So, for any one cube we have 8/8 Mg, 1/1 Si, 6/2 O: that is, $MgSiO_3$.)

Now, the balance of component atoms has changed in going from γ-spinel to perovskite, so we also have some MgO (and FeO), derived indirectly from olivine:

$$(Mg_{0.9}Fe_{0.1})_2SiO_4 \rightarrow (Mg_{0.9}Fe_{0.1}) SiO_3 + (Mg_{0.9}Fe_{0.1})O.$$

spinel structure   perovskite structure   magnesiowüstite

So the phase transition at 23·5 GPa (Fig. 7.13), which tidily correlates with a depth of about 660 km in the Earth's mantle, involves the production of two new minerals in approximate **volumetric** proportions of three parts perovskite to one part **magnesiowüstite** (the volume ratio is due to the number of large oxygen atoms in each formula). Note that pure MgO is often called periclase, and FeO, wüstite. As with the olivine to spinel phase change at 400 km depth, the depth of the spinel to perovskite plus magnesiowüstite transition varies with temperature, but in the opposite direction. This time it occurs at a *greater depth* if the mantle is locally *cooler*, and vice versa. Again, the significance is discussed in Section 8.4.

Pyroxene and garnet transformations. Phase changes in the other important minerals of upper-mantle lherzolite – orthopyroxene, clinopyroxene and garnet (Table 7.2) – follow similar principles to those for olivine. Orthopyroxene, the

137

second most abundant mineral above 400 km, transforms at *c.* 13·5 GPa into spinel and stishovite:

$$Mg_2Si_2O_6 \; \leftrightharpoons \; Mg_2SiO_4 \; + \; SiO_2$$

orthopyroxene   spinel structure   stishovite

The density increase is nearly 15%, much higher than for olivine at this pressure, because of the lower starting density of orthopyroxene and the presence of high-density stishovite in the products (Table 7.4). Alternatively, in the presence of more than 20 mole per cent garnet, orthopyroxene transforms with an 8·5% density increase at 13·5 GPa as follows:

$$2Mg_2Si_2O_6 \; \leftrightharpoons \; Mg_3[Mg,Si] \, Si_3O_{12}.$$

orthopyroxene   majorite

Majorite was named after the Australian mineralogist who first synthesized it (Ringwood and Major, 1971). We have written its formula in this way to emphasize its similarity to garnet ($Mg_3Al_2Si_3O_{12}$ where aluminium is in octahedral co-ordination; Table 7.2). The substitution in the octahedral sites of [Mg,Si] and [Fe,Si] for [Al_2] groups in garnet means that the two minerals form a *solid-solution series* between the following end members (including, now, the Fe^{2+} and Ca^{2+} cations):

$$(Mg,Fe)_3[(Mg,Fe),Si] \, Si_3O_{12} \leftrightharpoons (Ca, Mg,Fe)_3[Al_2]Si_3O_{12}$$

majorite   garnet

Garnet itself does not undergo a phase change at 400 km depth, and so the majorite–garnet solid solution probably is stable across the Transition Zone. However, it breaks down at around 660 km depth to give principally perovskite-structured minerals:

$$(Mg,Fe)_3[(Mg,Fe),Si]Si_3O_{12} \leftrightharpoons 4(Mg,Fe)SiO_3$$

majorite   perovskite

$$(Ca,Mg,Fe)_3[Al_2]Si_3O_{12} \leftrightharpoons 3(Ca,Mg,Fe)SiO_3 \; + \; Al_2O_3$$

garnet   perovskite   corundum

So we see that garnet, the main mineral carrying Ca^{2+} and Al^{3+} in the upper mantle, finally undergoes a phase change at *c.* 660 km depth, giving rise to Ca-structured perovskite, together with another high-density mineral, corundum. In fact, the pressure of this phase change varies between 22 and 26 GPa (630–720 km depth), decreasing as the proportion of majorite increases.

Finally, what about clinopyroxene, which is really a rather minor constituent of lherzolite? Quite simply, clinopyroxene is stable on its own across the range of Transition Zone pressures, producing two perovskites (the Ca and the (Mg,Fe) varieties) at *c.* 660 km depth. A majorite form is known: $(Ca,Mg,Fe)_3[(Mg,Fe),Si]Si_3O_{12}$, but tends to be confined to garnet-rich mixtures of the two minerals.

Although, in this section, we have confined attention to the minimum number of most likely mineral phase changes required to obtain a good understanding of the Transition Zone, there are various other possibilities, of which two deserve mention. Firstly, even though the system CaO–FeO–MgO–Al_2O_3–SiO_2 (sometimes called the **CFMAS system**) accounts for over 97% of the chemistry, there are other minor constituents and therefore other minor minerals. Secondly, anomalously low temperatures in the Transition Zone may bring other dense

minerals into the picture. In particular, the high-velocity mineral $^{VI}Mg^{VI}SiO_3$
(i.e. with octahedral Si^{4+}) in the **ilmenite structure** ($FeTiO_3$) may be more
important than majorite or garnet in cooler areas.

7.5.2 Petrological model for the Transition Zone

On the basis of the reactions discussed above, Figure 7.14 provides a summary of
the probable changes of mineralogy with depth in the Transition Zone.
Remember that zero-pressure densities are quoted because experimentally
synthesized materials are measured at zero pressure. At the top of the Transition
Zone, zero-pressure and actual densities are quite similar, but as pressure
increases at deeper levels, this is less true because of self-compression effects.

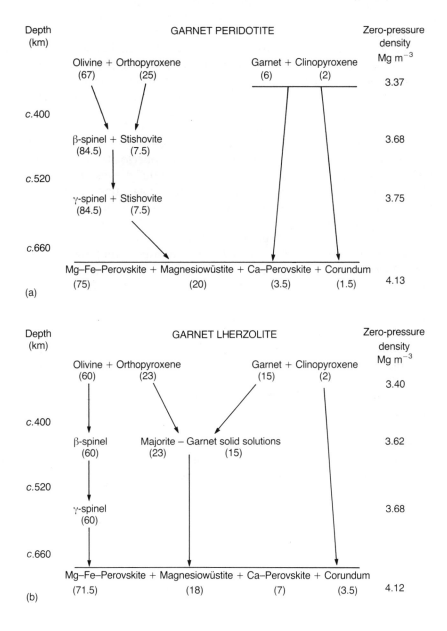

Figure 7.14 The variation of
mineralogy with depth in the
Transition Zone for two
different starting
compositions: (a) garnet
peridotite (i.e. depleted
garnet lherzolite, cf. analysis
3, Table 7.1, though this rock
is richer in silica); (b) 'fertile'
garnet lherzolite (cf. analysis
3, Table 7.3). The
approximate depths of phase
changes along the normal
mantle adiabat are given
(left), together with the mass
proportions of different
minerals and structures in
brackets, and the zero-
pressure density for each
assemblage (right). (See text
for discussion and note that
Mg–Fe-perovskite is
considered more likely to
occur below 660 km depth
than a combination of more
magnesiowüstite plus
stishovite.) (This is an
interpretation of data
presented by various
researchers, notably
Anderson, 1989a; Price et al.,
1989; Ito et al., 1990
(Shearer, 1991.)

139

This goes some way towards explaining the steeper overall density increase in modified PREM (3·58–4·37 Mg m^{-3}) than in the Figure 7.14 scenarios.

Figures 7.14 (a) and (b) show two compositions lying within the normal range for garnet lherzolite (Table 7.2) the first of which is relatively depleted (cf. Section 7.2), so we have called it garnet peridotite. They were also chosen to illustrate the different effects in the Transition Zone of high and low orthopyroxene/garnet ratios. Thus, at the 400-km discontinuity, a low-garnet mantle produces a much larger amount of β-spinel, together with stishovite, while the garnet and clinopyroxene components remain unchanged. In contrast, high-garnet mantle (Fig. 7.14(b)) follows the majorite–garnet solid solutions route with β-spinel being produced only from olivine. The overall density change across the 400-km discontinuity is higher in the first (9·2%) than in the second (6·5%) model, largely because stishovite is a product. Thus we would expect to see a stronger velocity and density contrast at 400 km where low-garnet mantle containing orthopyroxene is involved.

There is a continuing debate about the nature of the 520-km discontinuity. In Figure 7.14 we have given the simplest explanation, that β-spinel converts to γ-spinel at about this depth. The problem is that available experimental data indicate that the phase transition should occur at a shallower depth (*c.* 500 km) than observed. Moreover, even our first model, with 84·5% spinel, would produce only a 1·9% density increase, rather on the low side according to reflection seismology data. A strong case has been put by Anderson (1989a, 1992) that old subducted oceanic plates, in the form of eclogite, are buried in the Transition Zone. If so, the ensuing compositional differences, when added to the effects of phase changes, may provide the explanation. Currently, this problem is not resolved, but we return to the subject of mantle chemical heterogeneities in Chapter 9.

Accepting then that the lower parts of the Transition Zone probably are dominated mineralogically by γ-spinel and majorite–garnet solid solutions, Figure 7.14(b) illustrates the conversions that these minerals would experience at 660 km. The Mg_2SiO_4 spinel splits into Mg–Fe-perovskite (*c.*70%) and magnesiowüstite (*c.* 30%); Mg–Fe-majorite produces Mg–Fe-perovskite; garnet produces Mg–Fe- and Ca-perovskite (*c.* 35% and 41%, but depending on the starting composition) plus corundum (*c.* 24%); finally, clinopyroxene splits into roughly equal proportions of the two perovskites. The overall density increase is 12%, larger than at the 400-km discontinuity. Since this accompanies the most dramatic change in co-ordination number for Mg^{2+} and Si^{4+} cations, moving from the spinel to perovskite structure, many of the new minerals will be much more incompressible and so have dramatically higher seismic velocities as observed (Fig. 7.11). Notice that the zero-pressure densities of the lherzolite assemblages considered in Figure 7.14 are rather similar below 660 km. This means that the range of starting compositions for the lower mantle covered by models in Figure 7.14 is insensitive to the density constraint.

7.6 THE LOWER MANTLE

According to modified PREM, the lower mantle extends from 660 km down to 2891 km and is characterized by an almost linear increase of density with depth from 4·37 to 5·56 Mg m^{-3}. The only indications of non-linearity occur in the topmost *c*.100 km, where steep seismic velocity gradients (Fig. 7.11) may indicate that phase transitions are spread out, and again in the lowermost 200–250 km where low-velocity gradients occur in the D″ layer (see Chapter 9 for an interpretation).

Not surprisingly, in view of its remoteness, the precise composition of the lower mantle has been the subject of much entertaining and speculative debate. The debate centres principally on whether the lower mantle has the same major-element chemical composition as the upper mantle. The main variable that would affect density is the magnesium/iron ratio, normally expressed as $Mg/(Mg + Fe)$. A second key constraint is that the main lower mantle mineral, Mg–Fe-perovskite, has a large thermal expansion coefficient. So, all other things being equal, the lower the temperature, the higher the density. Taking the combination of minerals given at the bottom of the garnet lherzolite model in Figure 7.14(b) and the upper mantle $Mg/(Mg + Fe)$ ratio of 0·9, a temperature of no more than 1200°C is needed to produce a high enough density at the top of the lower mantle. Since the temperature is almost certainly over 1700°C which, because of thermal expansion, would result in a significantly lower density than observed, an intrinsically heavier atomic mixture is required, suggesting an $Mg/(Mg + Fe)$ ratio of 0.86. There are various other reasons why we believe that there is a small chemical difference between the upper and lower mantle (cf. also Chapter 9). These are connected with:

1. the fact that it is mainly the upper mantle that has been processed to produce the crust;
2. the probability that some differences were 'frozen in' during cooling of the magma ocean following major impacts during the late stages of the Earth's accretion.

Our best estimate is that not only does the lower mantle have a lower $Mg/(Mg + Fe)$ ratio, but also a slightly higher SiO_2 concentration and a lower Ca/Al ratio than the upper mantle. This leaves us with a lower mantle composition approximately as follows (expressed as percentages by mass):

Chemistry		Mineralogy	
SiO_2	49·8	Mg–Fe-Perovskite	81·5
Al_2O_3	3·6	Magnesiowüstite	9·1
FeO	9·7	Ca-Perovskite	5·8
MgO	34·1	Corundum	3·6
CaO	2.8		

The higher ratio of Mg–Fe-perovskite to magnesiowüstite than in the Figure 7.14 models reflects the greater silica concentration proposed for the lower mantle; however, the relative abundances of these minerals might be subject to considerable variation, because for every reduction of SiO_2 by 0·6%, 1% less perovskite and 1% more magnesiowüstite occurs in the model. There is also the possibility that some separation of Mg–Fe-perovskite into the simple oxide forms magnesiowüstite and stishovite may occur. The slightly higher zero-pressure density of this oxide combination (Table 7.4) may mean that they are favoured rather than perovskite in parts of the lower mantle. Another possible lower-mantle mineral has been synthesized in experimental work at pressures above 40 GPa (Irifune *et al.*, 1991). It involves a *combination* of corundum with MgO to produce $MgAl_2O_4$, not in the spinel (low-pressure) structure, but in the calcium ferrite ($CaFe_2O_4$) structure with Mg^{2+} in cubic co-ordination sites. If this high-pressure form were to occur instead of corundum in the lower mantle then our mineralogical model recalculates to give $MgAl_2O_4$ 5·0%, magnesiowüstite 7·7% with the other components unchanged.

In summary, the lower mantle is believed to be a mixture of Mg–Fe-perovskite and magnesiowüstite with subordinate amounts of Ca-perovskite and corundum (or $MgAl_2O_4$ in the calcium ferrite structure), probably with a Mg/(Mg + Fe) ratio around 0·86, somewhat less than in the upper mantle. This chemical difference makes the lower mantle slightly more dense than if it had the same composition as the upper mantle. At the base, the D″ layer is a chemically complex layer which is best considered in the context of the overall chemical and physical evolution of the mantle: this is the subject of Chapter 9.

SUMMARY

1. The major sources of evidence used to determine the composition and state of the mantle and oceanic crust are fourfold:

   (a) chondritic meteorite silicates that provide estimates of the Bulk (silicate) Earth composition;
   (b) the petrology and geochemistry of oceanic basalts, ophiolites and mantle nodules, all derived from the upper mantle;
   (c) seismic velocity and derived density profiles; and
   (d) the results of high-pressure, high-temperature experiments on possible mantle materials.

2. Beneath the top *c.* 6 km of oceanic crust, the upper mantle down to 400 km depth probably has a major-element chemical and mineralogical composition close to that of garnet lherzolite, with *c.* 60% olivine, 30% pyroxene and 10% garnet (Tables 7.2 and 7.3). At shallower levels the aluminous phase, garnet, is replaced by spinel (<75 km depth) or plagioclase (<30 km depth). Significant portions of the upper mantle, particularly near the surface, have been depleted of their (Ca, Al, etc.) fusible components by the removal of a basaltic partial-melt fraction. They now contain a range of depleted (peridotite) compositions (e.g. diorite, harzburgite; Fig. 7.3).

3. Lherzolite may undergo partial melting over a range of depths down to *c.* 150 km, producing with increasing depth progressively more alkali- and magnesium-rich magmas (Fig. 7.5). However, as the normal mantle geotherm does not intersect the dry solidus for mantle lherzolite, the generation of partial melt is due to some combination of:

   (a) dehydration and decarbonation reactions in minerals releasing volatiles that depress melting temperatures;
   (b) enhanced temperatures due to rising hot mantle where temperatures may be elevated by up to 200°C; and
   (c) extensional processes that cause decompression and thus bring deeper material on the normal mantle adiabat up into the *P–T* conditions of partial melting (see Fig. 7.6).

The melts produced have rather low viscosities; small melt fractions (0·1–2%) are easily separated, and rise to the surface rapidly, in 10^5–10^6 years, from the main melt source region at 10–50 km depth.

4. Decompressed material beneath oceanic (extensional) spreading centres forms a broad zone of extensive partial melting (Fig. 7.9). Even here only a total of 10–15% is extracted as melt, and that is in successive batches of small melt fractions. The melt rises to form basaltic ocean crust, while the residues, depleted harzburgite and dunite, remain frozen in the mantle. The basaltic melt fraction reaches a level of neutral buoyancy in a small magma chamber below the ridge axis. Here, as the magma cools, and by crystal fractionation, a layer of gabbro forms on the floor of the chamber. Some of the remaining magma is beneath the zone of periodic magma ejected upwards through dyke-like fissures to form eruptive lava sequences, and the periodic repetition of these processes leads to a layered gabbro sequence (Figs 7.7 and 7.10). With the caveats that many ophiolites are tectonically disturbed and were produced in back-arc basins, their structure reflects this primary melt separation–crust formation process.

5. The Transition Zone is characterized by several important mineral phase changes. An important theme is the compression of large oxygen atoms and the consequent cation co-ordination changes, particularly the step from tetrahedral to octahedral co-ordination for Si^{4+}, which tends to produce relatively incompressible, dense minerals with high seismic velocities (see Figs 7.12, 7.14 and Table 7.4). The 400-km discontinuity is almost certainly caused by the olivine–spinel and orthopyroxene–majorite transformations; various combinations of these minerals persist to 660 km depth where they break down to produce perovskite-structured minerals and dense oxides. A smaller discontinuity at 520 km depth is not yet fully explained; it may be a function of the β-spinel to γ-spinel phase change, or it could mark a compositional change.

6. The lower mantle is a region of smooth seismic velocity and density increase with depth, except for (a) the uppermost c. 100 km, where phase transitions are being completed, and (b) the basal c. 200-km-thick D'' layer (see Chapter 9). The bulk of the lower mantle is believed to have a lower $Mg/(Mg + Fe)$ ratio (c. 0.86) than the upper present-day mantle (0·90–0·91). The main minerals present are Mg–Fe-structured perovskite and magnesiowüstite, together with lesser amounts of Ca-structured perovskite and corundum (or $MgAl_2O_4$ in the calcium ferrite structure).

FURTHER READING

General books:
> Brown *et al.* (1992): general review chapters on experimental petrology and mineralogy of the mantle.
> Floyd (1991): on oceanic basalts, source regions and ocean crust evolution.
> Fowler (1990): various physical aspects of mantle structure and processes.

Advanced journals:
> Eldridge *et al.* (1991): mantle diamonds and recycled sediments.
> Gass (1982); Gass and Smewing (1981): ophiolites and the oceanic lithosphere.
> Kuskov *et al.* (1991): petrological model for the Transition Zone.

Mantle and oceanic crust   McKenzie (1985): melt viscosities, melt fractions and extraction rates.

McKenzie and Bickle (1988): mantle melting processes and geochemical implications.

Advanced books:

Anderson (1989a): on most aspects covered in this chapter.

Coleman (1977): review of ophiolites.

The dynamic mantle

8.1 INTRODUCTION

So far, the mantle has been described largely as if it were static; we have been answering questions such as: What is it made of? What changes across the seismic discontinuities? but not: What processes are operating?

We live on a planet that is dynamic. We know this because, without mountain building and volcanism to rebuild it, the processes of erosion – the scraping of glaciers, the scouring of rivers, etc. – would have reduced the Earth's surface to an almost featureless peneplain within a small fraction of the Earth's age. Erosion is powered by solar energy, which raises moisture that then descends by gravity, but what powers mountain building and volcanism? In the past few decades it has been recognized that they are largely a consequence of the relative motions of tectonic plates. But this in turn poses the problem, what moves the plates?

Almost certainly, the plates are moved by convection (or rather, as we shall see, are a part of convection), a sluggish stirring of the mantle, powered by heat transported out of the Earth. Thus our enquiry must be into a number of related topics: how the solid Earth can flow, what and where are the sources of heat, what form the convection takes, and so on.

This chapter establishes the main ideas, while the following chapter follows them in more detail, and discusses possible geochemical consequences.

8.2 RHEOLOGY OF THE MANTLE

8.2.1 How solids deform

We know that S-, or transverse, waves propagate everywhere through the mantle (except for small pockets of magma at shallow depths) and therefore the mantle must be an elastic solid with rigidity (Section 2.1), so how can it flow? In everyday life there is usually a clear distinction between liquids that flow and solids that are rigid – for instance, we drink liquid coffee out of a solid cup – but this division blurs for materials stressed for long periods at high temperatures.

In a liquid, the molecules are not rigidly bound to one another, so it deforms under the slightest shear stress; we can think of layers sliding past one another like a pack of cards (Fig. 8.1). If the rate of deformation is simply proportional to the shear stress the liquid has **Newtonian viscosity**:

$$\text{shear stress} = \eta \text{ rate of strain} \tag{8.1}$$

where η is the coefficient of viscosity. Such ideal liquids differ one from another only in the value of their coefficient of viscosity, with a thick liquid like syrup having a higher value than a runny one like water.

In a solid, the atoms are held together rigidly, usually in the regular arrangement of a crystal. If a crystal is subjected to a shear force it will deform, but as soon as the force is removed it immediately returns to its original shape.

145

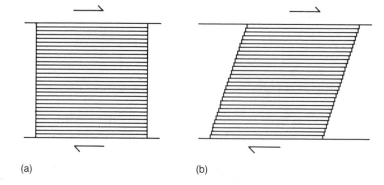

Figure 8.1 Shear in a liquid can be thought of as many layers sliding one past the next, like a stack of cards.

(a)   (b)

Provided it returns completely it is perfectly **elastic** (a meaning that differs from the common idea of an elastic substance being highly extensible). The strain is related to the stress, through the rigidity modulus μ (Section 2.1):

$$\text{shear stress} = \mu \text{ strain} \tag{8.2}$$

Thus, to say a material is rigid is to say it has a non-zero rigidity modulus and resists a change of shape; generally, we also mean its value is large, so that the shear deformation produced by a given stress is small or even negligible. The difference between ideal solids and liquids is that in a solid a constant shear stress produces a constant strain, but in a liquid it produces a constant rate of strain.

Between these two ideal behaviours there are intermediate ones. Real crystals contain many defects, of various types, and these weaken the crystal locally, so that in their vicinity bonds can break more easily than in the perfect parts. This may allow the crystal to deform permanently without overall breakage, or **brittle fracture**. Figure 8.2(a) shows one type of defect, a lattice vacancy. Because of random thermal energy, all atoms in a crystal are vibrating about their mean positions, and if the temperature is high enough, one next to the vacancy may temporarily break its bonds and move into the vacancy; that is, it exchanges places with the vacancy. Repetition of this process allows the vacancy to wander randomly, or diffuse, through the crystal. But if the crystal is under strain, with consequent distortion of the lattice, atoms will tend to jump into a position so as to relieve the strain, and this will bias the diffusion, giving a net flow of atoms in a particular direction.

Figure 8.2(b)(i) shows a much larger type of defect, an edge dislocation; it is a plane of atoms that ceases part way across the crystal. The crystal is deformed from its regular spacing only near the edge of the sheet, hence the name of the defect. Thermal fluctuations can cause the dislocation to jump to left or right (Fig. 8.2(b)(ii)). If the crystal is deformed, one direction will be favoured over the other, and the dislocation, on average, will move to relieve the strain, as was the case with vacancy diffusion. When it has travelled all the way to the side of the crystal the upper part of the crystal has moved one spacing to the right (Fig. 8.2(b)(iii)). When many dislocations, which occur throughout the crystal, move in response to a steady shear force, these offsets add up to a slow but steady deformation like that of a liquid.

This slow, permanent deformation in a solid is **creep**. Its rate depends upon several factors, but the most important ones, other than the nature of the material, are temperature, stress and time. Temperature is important because it is thermal fluctuations that permit atoms to jump, and the higher the tempera-

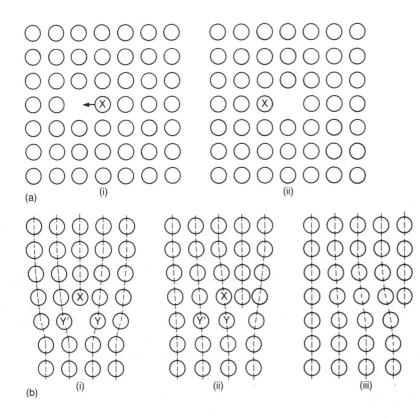

Figure 8.2 Two kinds of lattice defect. In (a) the defect is a vacancy. It can move through the crystal when a neighbouring atom, e.g. X in (i), chances to have enough thermal energy to exchange with it (ii). (b(i)) Shows an edge dislocation ending at X. Given sufficient thermal energy, X may realign with Y or Y'. If the crystal is under shear strain it is more likely to realign with Y (ii). When an edge dislocation reaches the side of the crystal (iii), this is similar to one of the steps of Fig. 8.1(b).

ture the more likely it is that a thermal fluctuation of sufficient energy will occur in a given time. As the melting point is the temperature at which so many bonds are being broken simultaneously that the crystal loses its rigidity, temperature may be usefully expressed as the **homologous temperature**, the actual temperature divided by the melting temperature (both measured in K) (if a material melts partially, the solidus temperature is used in place of the melting temperature). Very roughly, creep becomes significant if the homologous temperature exceeds two-thirds. As all the interior of the Earth exceeds this value, apart from the top few tens of kilometres below the surface, we expect creep to be important. Once the temperature is sufficient for a significant chance of thermal fluctuation to cause a jump, a small increase in temperature greatly increases this chance, so the viscosity is strongly temperature-dependent.

Stress is necessary because it provides, through the strain it causes, the bias that makes the jumps more likely to go one way rather than the other. If the temperature is high enough that jumps are occurring anyway, even the smallest of stresses provides a bias and leads to a net deformation; thus the crystal behaves like a liquid and creep gives it an effective viscosity, though far greater than that of a normal liquid. Increasing the stress increases the rate of deformation, though not always in simple proportion. Diffusion creep is simply proportional to the stress, so that it results in Newtonian viscosity, whereas dislocation creep increases more rapidly (termed power law creep).

Time is important because the thermal fluctuations take time to recur. One difference of creep from elastic deformation, therefore, is that it takes time, whereas elastic deformation is immediate. One common occurrence of creep is the crystalline ice of a moving glacier, which despite being cold has a high homologous temperature, since water melts at 0°C.

147

This simple account of creep should also make clear how the mantle can be both solid and liquid. It is a question of time-scale. At any instant, only a few atomic bonds are broken and the remaining bonds allow the crystal to behave elastically to the passage of S-waves. But if the stress is applied steadily for thousands of years there is an irreversible deformation, or flow.

It should be appreciated that there are many other types of lattice defect and that they respond differently to changes of temperature, stress and other variables. Rheological behaviour is therefore very complex, and in general will display some combination of elastic, plastic and viscous behaviour. However, as a first approximation, the uppermost mantle plus crust that is cool can be treated as an elastic layer, above the remainder of the mantle, which is simply viscous, because it is hot and susceptible to creep.

Finally, a few more rheological terms are explained, for completeness. If an elastic material is subjected to increasing stress, there comes a point at which the elastic limit is reached. Thereafter, the material either breaks **brittlely**, with formation of fractures, or continues to deform **ductilely**. The deformation that occurs depends partly upon the nature of the material, but also upon the rate of stress increase, and the temperature; high rates of stress encourage brittle failure, but high temperature favours ductile flow. The term 'plastic' is often used instead of 'ductile', but is more correctly reserved for inelastic deformation beyond the elastic limit, the amount of which depends only upon the stress, and which occurs immediately.

8.2.2   The rheology of the mantle

Estimates of the rheological properties of the mantle can be deduced from geophysical measurements, supplemented by a combination of theory and laboratory measurements, particularly on olivine, for this is the dominant and weakest constituent in the mantle. However, neither approach is yet able to give definitive values.

The first evidence that the Earth's interior is not rigid came from the discovery of isostasy. As recounted in Section 1.3, it began to be appreciated that mountains have less mass than is apparent at first sight. When surveying was being carried out in northern India, it was found that the Himalayas deflected the plumb-lines and spirit-levels, but less than was calculated for their size. This could be explained if the rocks beneath the mountains were less dense than those at the same level to either side. Since this was found for mountains in general, and whether they were young, or old and eroded, it could not be due to chance. In 1855, both Sir George Airy and Archdeacon Pratt suggested a suitable mechanism to achieve this: mountains essentially float in a denser substratum. Their explanations differed as to whether the highest mountains were formed of the lightest or of the thickest rock (Fig. 8.3). A combination is possible, and we now know that mountains are higher than their surroundings primarily because they are thick and so have deep 'roots' (Airy mechanism), whereas continents as a whole are higher than sea floors primarily because they are composed of lighter material (Pratt mechanism); this is a fundamental difference between oceans and continents, which has consequences that will be discussed later in this chapter.

The concept of isostasy evidently requires a rigid layer over a yielding one: the upper layer must have rigidity to preserve the topography of the Earth, or else

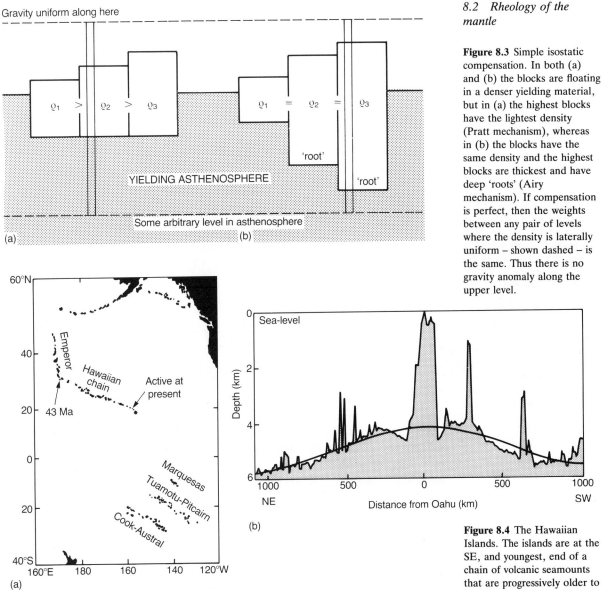

Gravity uniform along here

$\varrho_1 > \varrho_2 > \varrho_3$

$\varrho_1 = \varrho_2 = \varrho_3$

YIELDING ASTHENOSPHERE

'root'

'root'

Some arbitrary level in asthenosphere

(a)　(b)

60°N

Emperor

Hawaiian chain

Active at present

43 Ma

Marquesas

Tuamotu-Pitcairn

Cook-Austral

40°S

160°E　180　160　140　120°W

(a)

Sea-level

Depth (km)

1000　500　0　500　1000

NE　Distance from Oahu (km)　SW

(b)

8.2 Rheology of the mantle

Figure 8.3 Simple isostatic compensation. In both (a) and (b) the blocks are floating in a denser yielding material, but in (a) the highest blocks have the lightest density (Pratt mechanism), whereas in (b) the blocks have the same density and the highest blocks are thickest and have deep 'roots' (Airy mechanism). If compensation is perfect, then the weights between any pair of levels where the density is laterally uniform – shown dashed – is the same. Thus there is no gravity anomaly along the upper level.

Figure 8.4 The Hawaiian Islands. The islands are at the SE, and youngest, end of a chain of volcanic seamounts that are progressively older to the NW (a). The weight of the large volcanic islands has depressed the ocean floor out to about 200 km (based on McKenzie, 1983). (The significance of the broader uprising, or swell, indicated by the smoothed curve, is discussed in Section 9.3.)

mountains would level themselves like syrup poured on a plate, while beneath there must be a layer that yields to permit floating. The rigid and yielding layers are termed, respectively, lithosphere and asthenosphere (other definitions of lithosphere are given in Section 8.4.4).

However, isostatic compensation cannot be as simple as that shown in Figure 8.3, for that would mean many blocks moving up and down independently. Instead, the lithosphere has lateral strength, so that when a localized load is placed on the lithosphere it produces a depression over a larger area. For example, the great volcanic islands of Hawaii are surrounded by a 'moat' that extends out to about 200 km (Fig. 8.4).

Essentially, the lithosphere is elastic because it is cold, while the asthenosphere yields by creep. As we saw in Section 8.2.1, creep is a very slow process, so isostatic adjustment takes a long time, i.e. thousands of years.

149

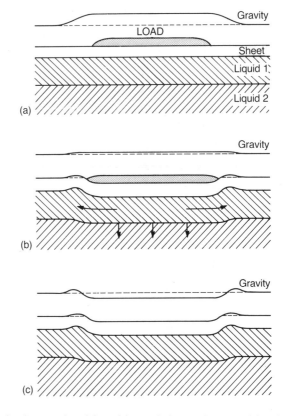

Figure 8.5 Simple model showing the response of an elastic sheet over two viscous liquids, such as plywood over thick oils. (a) Shows the situation immediately after adding the load, which creates a positive gravity anomaly, but this diminishes as the system returns asymptotically towards isostatic equilibrium, with outflow of asthenosphere permitting the sheet to deflect until the load is supported largely by buoyancy, and the anomaly has diminished (b). Abrupt removal of the load inverts the situation, creating a negative anomaly (c). For further explanation see the text.

We can obtain an intuitive idea of how the combined lithosphere and asthenosphere behave by imagining a sheet of plywood covering a layer of very thick oil (Fig. 8.5). For small loads, corresponding with buildings and small mountains, the sheet takes all the weight. Larger loads will cause a depression in the sheet, some of the weight being taken by the sheet and some by the buoyancy of the displaced oil; this corresponds with the example of Hawaii on the ocean floor. For the largest loads, corresponding with ice-sheets extending over thousands of kilometres, the sheet will be flat (except at the edges of the load) and so its strength will have negligible effect, and it will be depressed until practically all of the weight is supported by buoyancy. At the present time, the rock surface of Greenland, below several kilometres of ice, has been depressed into a saucer shape, with its centre below sea-level. Figure 8.11 shows that oceanic spreading ridges are close to compensation on a scale of hundreds of kilometres but not on a small scale. The significance, in Figure 8.5, of the bulges to either the load or the third layer will be explained below.

To deduce the rheological properties of the mantle, we take advantage of the fact that there was a major period of glaciation not long ago. The ice-sheets formed upon northern Europe and North America in the last glaciation were so large that they were supported almost entirely by buoyancy. They formed over a period of about 10^5 years, but melted in about a tenth of that time, beginning roughly 18 000 years ago. The history of glaciation at different locations can be deduced from the extent of glacial deposits, the dating of beaches formed when the sea-level was different, and other evidence. In addition, uplift is still continuing, as measured by surveying (e.g. Fig. 8.6(a)), at about a centimetre a

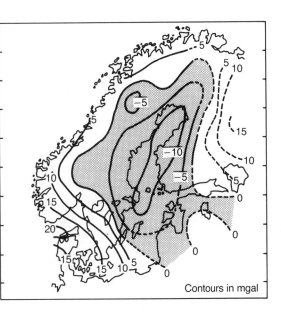

(a)

(b)

Figure 8.6 Uplift of Fennoscandia. (a) Shows the current rate of uplift, contoured, while (b) shows the negative gravity anomaly. Both indicate that equilibrium has not yet been reached. (After Sharma, 1984.)

year at the former centre of the ice-sheets. That neither region has yet reached isostatic equilibrium is also shown by the presence of negative gravity anomalies (e.g. Fig. 8.6(b)), indicating a mass deficit, as expected if the depression caused by the ice-sheet is still returning to equilibrium by the slow inflow of asthenospheric material from below.

To deduce the viscosity values, a model is needed, and an elastic sheet over a liquid with Newtonian viscosity is often used, as in Figure 8.5. The parameters of the model are adjusted to match both the history of sea-level change and its present rate of change. At the centre of the huge ice-sheets the strength of the lithosphere has negligible influence, and the viscosity of the upper mantle is found to have a viscosity of 10^{21} Pa s. The viscosity of the lower mantle is also important, since if it were very large the asthenospheric flow would be mostly confined to the upper mantle, which would form a relatively shallow channel as its depth is only a small fraction of the width of the ice-sheets. In fact, it is found that increasing the viscosity to more than two or three times that of the upper mantle causes the model to fail.

At the edge of the ice-sheet, the lithosphere has a peripheral fore-bulge, because the displaced asthenospheric material has to go somewhere, and it forces up the lithosphere where it is not loaded. Here, because of its curvature, the strength of the lithosphere is important. The response here to the melting of the ice-sheet is more complex, being a combination of the flattening of the fore-bulge and the rise of sea-level due to the melt-water, which occur at different rates. For North America, the actual change is best matched by a lithosphere approaching 200 km thick (Peltier, 1989); this is thicker than in many other places because the part of North America that was glaciated is mostly old continental shield (Fig. 10.2). Again it is found that the viscosity of the lower mantle cannot be more than about twice that of the upper mantle.

151

This result is supported by other studies (see Peltier, 1989), so that the asthenosphere is probably effectively the whole mantle below the lithosphere, and has a viscosity of about 10^{21} Pa s. It used to be accepted that there is a layer of about 100 km thickness immediately below the lithosphere that has a viscosity at least an order of magnitude less than that of the deeper mantle, thought to be due to the temperature approaching the solidus (Section 7.3.2), and this was regarded as the asthenosphere; but it is now considered that the evidence does not support its existence. The simplicity of the result, however, probably partly reflects the limitations of the methods of measurement; viscosity, for instance, may be non-Newtonian and vary laterally (for further discussion see Ranalli, 1987); and neither does it take into account the variation of composition of the continental crust (Section 10.2.4).

The significance of the result is that we expect the whole mantle, apart from the lithosphere, to be able to flow and so to convect if there is sufficient heat to drive it.

8.3 THE THERMAL BALANCE SHEET: HEAT SOURCES AND SINKS

Since convection is driven by the transport of heat, knowing how much heat is being lost by the Earth's interior and where it originates will help us understand the form that convection takes. As the sink, or total heat loss, is the better understood, this will be considered first.

Heat can be transported in three ways: through a material by conduction, bodily within a moving material by convection, or by radiation through space (and through a material too if it is transparent to the radiation). In the top few tens of kilometres of the Earth, conduction is the most important mechanism, for there the rocks are too cool to flow; so most of the heat arriving at the Earth's surface, where we measure it, has been transported by conduction.

In principle, we could measure the heat reaching the surface by measuring the rate at which an object on the surface heats up, but this is far too slow to be practicable (it would take about ten years to boil a kettle, even if all heat loss from the kettle could be prevented). The heat reaching the surface from below is too small to have a noticeable effect upon the temperature there, for it is many times smaller than the heat supplied by the Sun, but it accounts for the rise in temperature with depth, found, for example, on descending into mines. A typical continental value is 20–30°C km^{-1}. If this vertical temperature gradient is measured by placing sensitive thermometers some distance apart, either in a borehole on land or on a probe pushed into sediments on the ocean floor, and the thermal conductivity of a sample of the rock is also measured, the two quantities can be combined to give the heat flow:

$$\text{heat flow} = \text{temperature gradient} \times \text{thermal conductivity},$$
$$\text{in units of joules } m^{-2} s^{-1} \tag{8.3}$$

Details can be found in, for example, Press and Siever (1986).

The results (see Fig. 8.10) show that the heat flow through the ocean floor decreases away from spreading ridges, for reasons explained in Section 8.4.2. Continental heat flow also is age dependent, but depends upon the time since the last orogeny, and the decrease with age is much slower than for the oceans.

Table 8.1 Thermal data.
(a) Heat outflow (as % of total outflow).

Oceanic heat flow (by conduction)	60%
Oceanic hydrothermal convection	10%
Continental heat flow	30%
Total	100% $= 4.2 \times 10^{13}$ W

Source: Slater *et al.* (1980).

(b) Radioactive concentrations and thermal productivities.

Material	U (ppm)	Th (ppm)	K (%)	K/U	heat productivity (μW m^{-3})
Bulk Earth*	0·02	0·08	0·02	10 000	0·014
Gabbro	0·05	0·15	0·08	16 000	0·03
Granodiorite	1·6	6·2	2·1	13 125	1·0
Granulite	0·2	0·7	0·25	12 500	0·13
Carbonaceous chondrite	0·02	0·07	0·04	20 000	0·01
Ordinary chondrite	0·015	0·046	0·09	60 000	0·015
Iron meteorite	negligible	negligible	negligible	—	negligible

* Undepleted periodotite.
Table based on data in Stacey and Loper (1988), Taylor and McLennan (1985) and O'Nions (1987).

(c) Radioactive heat generation (as % of present total heat outflow).

Continental crust (roughly equally granodiorite and granulite), contains about 0·7% of Earth's volume	~10%
Oceanic crust (broadly gabbro), contains about 0·2% of Earth's volume	~0·15%
Mantle (undepleted periodotite), contains about 84% of Earth's volume	~30%
Core (iron), contains about 16% of Earth's volume	negligible*
Total	~40%

* Assuming that the dynamo is powered by growth of the inner core, not radioactivity; see Chapter 6.

These relations of heat flow to age allow values to be estimated for areas where there is little data. Determining heat flow as above only measures heat transported by conduction, but a significant amount is lost near oceanic spreading ridges to circulating water that penetrates to a depth of several kilometres through fractures, reaching close to where the magma is solidifying. There it is heated and rises back to the sea floor, giving rise to vents of highly mineralized brines known as black-smokers, because of the minerals precipitated on contact with the cold sea-water (Section 7.4.1). Because continental heat flow is generally less than that of the oceans, and also because the oceans are larger, continents contribute only about 30% of the global heat outflow, while the oceans contribute 60% plus a further 10% by hydrothermal ridges. The values are given in Table 8.1(a). The total is about 4.2×10^{13} W.

The heat reaching the surface must derive either from heat generated in the interior, or by cooling of the Earth, or a combination of the two. Possible significant generative sources are radioactivity and the settling of denser materials. It is believed that at present the only significant settling is the partitioning of iron into the solid inner core, described in Section 6.5 as the likely energy source of the geodynamo. As we are concerned here with the mantle, this process will be treated simply as a heat input into the base of the mantle (Table 8.1(c)).

The only radioactive elements important as heat sources are uranium (isotopes ^{238}U and ^{235}U), thorium (^{232}Th) and potassium (^{40}K); other radio-isotopes are either uncommon or decay very slowly or both, so they generate heat only very slowly. It was previously believed that the bulk composition of the Earth was chondritic (in which case the current heat production would roughly equal the heat loss, and so it was thought that the Earth was not cooling). We now believe however that the Earth lost some of its more volatile elements during accretion, as discussed in Section 5.2, and these include potassium, evident as the low K/U ratio in the Earth compared to chondrites (Table 8.1(b)), so the amount remaining has to be estimated from the contents of representative rocks (Table 8.1(b)).

Table 8.1(c) shows estimates of the radioactive heat generation for various parts of the Earth, deduced by calculating their volumes, and equating core composition to that of iron meteorite, mantle to undepleted peridotite, oceanic crust to gabbro and continental crust to granodiorite plus granulite. Table 8.1(c) shows that the total heat generated is probably less than half of the total outflow (the values given are rough, but most published estimates range from 30 to 50%), so the Earth must be cooling (required for the growth of the inner core, Section 6.5). Cooling is quite able to provide enough heat: the Earth's thermal capacity is so huge, in comparison to its ability to lose heat by conduction through its surface layers, that it could provide the whole of the present heat flow by cooling at a rate of only 0.2°C in a million years (Turcotte and Schubert, 1982). Of course, if the Earth is cooling it must have been hotter in the past, which carries implications that will be discussed in Chapter 11.

8.4 CONVECTION IN THE MANTLE

8.4.1 General considerations

Convection is the circulation of fluids resulting from density differences, with lighter material rising and denser sinking. The density difference can be compositional, arising from separation of lighter from denser elements, as was described in connection with the core dynamo (Section 6.5) but any such separation in the mantle is believed to be at the most quite minor. The most common cause of density differences is temperature differences, in which case any convection is **thermal convection**.

Convection is often illustrated by the homely example of soup heated in a pan, and though convection in the Earth is very different in form, as we shall see, this example can make clear what drives convection. Suppose that, before the soup is heated, it is confined in vertical tubes, as shown in AB and CD of Figure 8.7.

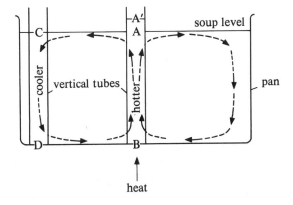

Next, let the temperature distribution that exists throughout the heated soup-pan be established in some unspecified way, hot at the base and up the centre and cooler at the top and sides. Column AB will expand and rise to A'. The pressure at A is now greater than at C, because of the weight of column AA' above it, so if a hole were drilled at A, liquid would flow out sideways. This would tend to lower the height of the column A'B and so reduce the pressure at B, until now equal to the pressure at D. This pressure difference would tend to move liquid in at B if there were a hole at B. And if the liquid flowing into B were hot, then the flow in at B and out at A would continue. The converse argument applies for the cool liquid CD. So removing the tubes would allow the liquid to circulate as the arrows show. This is a simple **heat engine**, for some of the heat energy being transferred from the hot base to the cool top and sides has been converted into mechanical motion. In old-fashioned central heating systems with wide-bore piping this mechanism was sufficient to circulate the hot water without the aid of a pump.

An important quantity that determines the vigour of convection is the dimensionless **Rayleigh number**:

$$\text{Ra} = \frac{\begin{pmatrix}\text{volume coefficient} \\ \text{of thermal expansion}\end{pmatrix} \times \begin{pmatrix}\text{vertical temperature} \\ \text{difference}\end{pmatrix} \times g \times \text{density} \times \begin{pmatrix}\text{depth of} \\ \text{liquid}\end{pmatrix}^3}{\text{thermal diffusivity} \times \text{viscosity}}$$

or, in symbols

$$\text{Ra} = \frac{\alpha \, \Delta T \, g \, \rho \, d^3}{K\eta} \tag{8.4}$$

Increasing the values of the quantities in the numerator encourages convection: α, because it is the thermal expansion that allows an increase in temperature to lower the density and so creates buoyancy; g, because without the pull of gravity the difference of density would not generate a force (convection would not occur in the weightlessness of space); ρ, because doubling the density of the liquid would double the force; d, because increasing the height of column AB in Figure 8.7 increases AA' and hence the excess pressure. The effect of increasing ΔT is obvious, though it is not the actual temperature difference between top and base

155

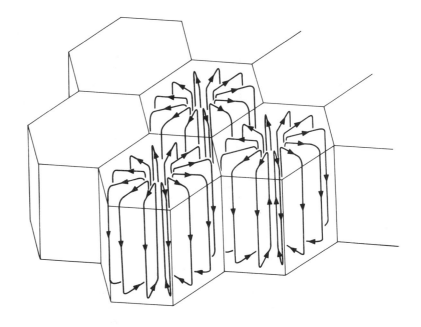

Figure 8.8 Convection at moderately high Rayleigh numbers takes the form of a honeycomb, with upwellings concentrated to the axes, and downwellings taking place round the margins of the cells.

for, as we saw in Section 3.5, in a liquid compressed by the weight above, there is no buoyancy unless the temperature gradient exceeds the adiabat. Thus ΔT is the temperature difference in excess of the adiabatic difference. Increasing the value of the quantities in the denominator discourages convection: the thermal diffusivity, K (given by the ratio of conductivity to thermal capacity), is a measure of the heat in the hotter parts that is lost to the cooler parts by conduction, so tending to reduce the heat available for convection; while increasing the viscosity obviously tends to slow the convection.

The effect of increasing the Rayleigh number is illustrated by considering a liquid layer between horizontal metal sheets, the upper held at a lower temperature than the lower. As the temperature difference is increased from zero, a temperature gradient is set up in the liquid. At first the buoyancy is too slight to produce convection and heat will be transported upwards by conduction. Convection begins when Ra reaches about 5000 and takes the form of horizontal rolls, like a box of Swiss rolls coiled in alternate senses. As Ra is further increased, convection changes to a vertical honeycomb (Fig. 8.8) with hot liquid ascending up the axis of each cell. Convection is now transporting much more heat than conduction. A further increase of Ra to, say, 10^5, concentrates the flows to the axes and surfaces of the cells. When Ra exceeds about 10^6, the pattern becomes less regular, with cells moving about and changing size, and upwelling being further concentrated into narrow columns and downwellings into sheets. Although the geometry of the system has some effect on the vigour of convection, it is unimportant compared to changes of Ra by powers of ten; for instance, in all systems, convection becomes significant when Ra equals a few thousand and irregular over 10^6. It is clearly important to estimate Ra for the mantle.

Approximate values of the quantities in Eq. (8.4) for a mantle dominated by olivine and pyroxene (Section 7.1) are: $\alpha = 3 \times 10^{-5}$ K^{-1}, $g = 10$ m s^{-1} (g is remarkably constant throughout the mantle, see Fig. 3.9), $\rho = 4.66$ Mg m^{-3}, K $= 10^{-6}$ m^2 s^{-1} (Turcotte, 1980). For the whole mantle $d \sim 2.9 \times 10^6$ m (it is not known whether convection crosses the 660-km discontinuity, as discussed in

Figure 8.9 Convection in a tank of glycerine. This example is driven by bubbles of cold carbon dioxide, seen rising at the left, which forms a cold layer above the glycerine and helps to cool it. The dark band at the top is cool glycerine, which about half-way across dives down because of its greater density. Similar behaviour is shown when a heater is used to produce the convection (After an experiment by J. S. Turner, 1973.)

Section 9.2), $\eta \sim 10^{21}$Pa s (Section 8.2.2), while ΔT is several hundred K (Section 8.5). These figures give Ra $> 10^6$, so convection is likely to be vigorous.

The form that the convection takes will depend upon the complexities of the mantle, including:

1. the distribution of heat sources;
2. the distribution of sinks; and
3. how viscosity varies with temperature, stress and other variables. The problem is too difficult for a full solution at present.

One general conclusion is that there must be an **upper thermal boundary layer**, because at the top of the system the flow is horizontal (Figs 8.7 and 8.8) and so it cannot transport heat vertically to the surface; instead, it loses heat by vertical conduction. (Because it is the coolest part of the system, the upper thermal boundary layer will normally be the most viscous; in the case of the mantle (including the crust) it will be rigid.)

The existence of the thermal boundary layer can be demonstrated by a tank of glycerine (Fig. 8.9). A cool layer forms at the surface and, because it is cool, it is viscous and dense, and the layer plunges back into the tank.

It may seem natural to identify the upwelling with the mantle below spreading ridges; the plunging layer with a subduction zone; and the thermal boundary layer with the lithosphere, but we shall see shortly that although there is some truth in this, the actual situation is more complex.

If heat is supplied at the lower surface, there will be a boundary layer there too, for the same reason. The relevance of this concept to the anomalous layer at the base of the mantle, the D'' layer, will be discussed in Section 9.2.3.

8.4.2 Plates and mantle convection

The theory of plate tectonics has been very successful at explaining orogenies and much igneous activity as being due to processes at plate edges. But plate tectonics is primarily a kinematic theory, describing plate movements but not explaining what plates are or why they move. It should be appreciated by now that plate movements are related to mantle convection, though exactly how, has to be made clear.

157

Although a plate can embrace both oceanic and continental parts – the South American plate, for instance, stretches from the west coast of South America to the Mid-Atlantic Ridge (Fig. 8.14) – there is a fundamental difference between the two parts, as mentioned in Section 8.2.2. Continental parts are highly enriched in silica, which makes them intrinsically lighter than both the mantle and oceanic plates, and so they float higher. This buoyancy means that they resist subduction and so continents may be very old; they have therefore experienced a complex history, so that they are not uniform in their properties. Continental plates will be described in Chapter 10. Here, we shall concentrate on oceanic plates.

There are three possible ways that plates might be moved by mantle convection:

1. independently existing plates are carried along on mantle convection currents, like logs in a stream; this might be applicable to continental plates;
2. plates are the upper thermal boundary layer, and so an integral part of the convection, being created where a mantle upwelling reaches the surface, and later returning to it in a downwelling, as illustrated by the glycerine tank experiment of Figure 8.9; this might be applicable to oceanic plates;
3. in the third alternative, the plates are again an intrinsic part of the convection, but they are not simply a thermal boundary layer, because the Earth is not as simple as a tank of liquid.

The reader will have guessed that it is the third alternative that occurs, but attention will be given first to 1 and 2.

Oceanic crust is formed of basaltic material created by partial melting of the mantle, and so it is somewhat enriched in silica and lighter than the mantle beneath (Section 7.3), but for the moment we shall ignore this relatively small compositional difference. Instead, we shall start by assuming that oceanic plate is simply mantle material cold enough to be solid; it starts to form at a ridge axis, where magma reaches to the surface, and thickens as it moves away from the

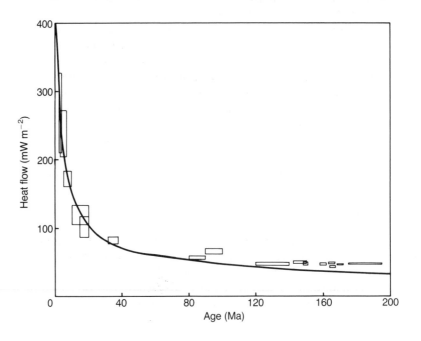

Figure 8.10 Heat flow versus age of ocean floor. The sizes of the boxes denote the errors. A square-root relation is shown by the line. (After Lister *et al.*, 1990.)

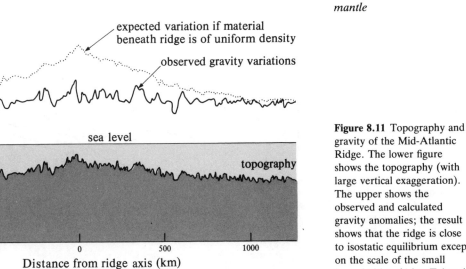

Figure 8.11 Topography and gravity of the Mid-Atlantic Ridge. The lower figure shows the topography (with large vertical exaggeration). The upper shows the observed and calculated gravity anomalies; the result shows that the ridge is close to isostatic equilibrium except on the scale of the small irregularities. (After Talwani *et al.*, 1965.)

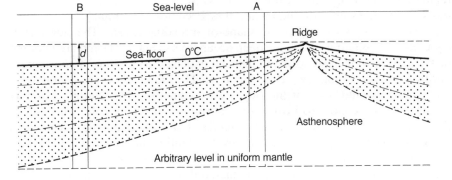

Figure 8.12 Schematic section of an oceanic spreading ridge. The material is assumed to have uniform composition but varies in temperature. If isostasy is maintained, then the weights in any vertical column, such as A and B, are the same. Compare with Figure 8.11.

ridge. A close analogy is the freezing of ice upon a still pond in frosty weather: as it gives up its heat to the cold air, the water freezes. As it thickens, the temperature *difference* between the upper and lower surfaces of the ice remains constant, so that the vertical temperature gradient, and hence the heat flowing upwards, must decrease (Eq. (8.3)). Since the rate at which water can freeze on the base depends upon the rate at which the latent heat can be removed by conduction, the rate of underplating of ice must slow too. A simple calculation shows that the thickness increases in proportion to the square root of the time elapsed.

If we cut through the ice along a line and pull the two parts slowly and steadily apart, then the ice will always be forming anew at the line and thickening as it moves away, its thickness increasing as the square root of its age (see Fig. 8.10). This insight can be applied to oceanic spreading ridges, except that it is not a simple solid over liquid, but rather a change from cold, elastic material to hot, deformable material, without a sharp lower boundary (a derivation can be found, for instance, in Turcotte and Schubert, 1982).

159

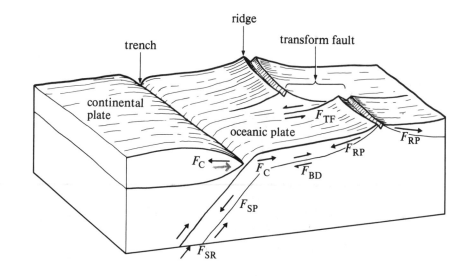

Figure 8.13 Possible forces acting on plates: F_{BD}, basal drag or drive; F_{SP}, slab-pull; F_{RP}, ridge-push; F_{TF}, transform fault friction; F_C, collisional resistance; F_{SR}, slab resistance. (After Forsyth and Uyeda, 1975, by permission of the Royal Astronomical Society.)

This simple model can be tested in various ways. As the plate thickens, the heat flow should diminish, decreasing as the square root of the age (this relationship does not hold near the spreading ridge where heat is also removed by circulating sea-water, as described in Section 8.3). Figure 8.10 shows that this relation is closely obeyed (beyond about 80 Ma the relationship begins to hold less well, perhaps because the mantle flow becomes less simple).

A second test is to predict the shape of spreading ridges. The ridges stand about three kilometres above the deep ocean floors (Fig. 8.11) and are hundreds of kilometres wide, and so they might be expected to produce a considerable positive gravity anomaly. But Figure 8.11 shows that the anomaly is far less than would result if the ridges were composed of material of uniform density. They are close to isostatic equilibrium on a broad scale, which means that the material at depth must be lighter; but this cannot be because it is compositionally different – as is the case for continental mountain chains – for the spreading would carry it away. Instead it is light because it is hot. In fact, the ridges are high simply because their material has expanded due to their high temperature. Equating the weights of columns A and B in Figure 8.12 according to the principle of isostasy (Fig. 8.3) and combining it with the previous result that the thickness of plate increases as the square root of its age, predicts that the depth, d, below the crest of a point on the side of the ridges, is proportional to the square root of its age. This too is found to be closely the case. All ridges have the same shape when their different spreading rates are taken into account.

The plate is colder than the mantle beneath, and therefore it is denser, and this is an unstable situation. But we have now to remember that oceanic crust is not just cold mantle material but, being basaltic, is compositionally lighter; this ensures that at first the average density of the plate, crust plus rigid mantle, is less than that of the yielding mantle but, as the plate thickens by addition of mantle freezing on its base, its average density increases and finally exceeds that of the mantle. Sooner or later, we expect the plate to plunge into the mantle, and give rise to a subduction zone. (A close analogue is seen in the lava lakes of some of the Hawaiian volcanoes. There, a skin of solid lava forms and demonstrates many of the features of plate tectonics (Duffield, 1972), including 'plate' growth at constructive margins – with transform faults – and subduction, though the 'plates' on both sides subduct.) Once an oceanic plate begins to dive it will be pulled down by its own negative buoyancy.

We are now in a position to discuss the various forces that may be acting on plates (Fig. 8.13). Although we cannot measure the forces directly, we can make predictions from the various models and test them against observations.

There are three, main, possible drive forces:

1. If a plate rides passively on the mantle, like a log on a stream, there will be a **basal drive force**, F_{BD}, due to viscous coupling trying to keep the plate moving at the same speed as the convecting mantle beneath it.
2. As described above, old oceanic plate is gravitationally unstable and when it founders its negative buoyancy pulls it into the mantle. This is the **slab-pull force**, F_{SP}.
3. New oceanic plate is formed at the crest of a ridge and as it ages it moves away and subsides, releasing gravitational energy which tends to propel it down the slope. An intuitive grasp of this force can be gained by thinking of lava pouring out of a volcano and running down its sides. Next, imagine that the lava solidifies immediately on leaving the vent and slides down the sides with little friction between it and the slope of the volcano; we now have a solid sheet, fed by the volcano, but still being moved by gravity. In the case of oceanic plate, the low friction is provided by the creep of the hot mantle at depth. This force is called the gravity-sliding force or, since it acts to move the plate away from the ridge, the **ridge-push force**, F_{RP}. The energy to drive it was supplied when the magma was raised to the top of the ridge, which in turn was due to expansion because of its high temperature, as discussed in connection with Figure 8.7.

Any drive forces must be opposed by retardative forces, or else the plates would accelerate indefinitely. The likely retardative forces are also shown in Figure 8.13 and are of two sorts: frictional forces where plates slide past each other, as at ridge offsets (transform faults) (F_{TF}) and also where a subducting plate slides below a surface plate (F_c – for colliding force), and their existence is evident from the occurrence of earthquakes. The second type is due to the viscous resistance where plates move relative to the mantle, such as the subducting slab moving downwards (F_{SR} – slab resistance) and also on the base of the surface parts of plates. There will be a viscous force whenever plate and mantle are moving at different speeds; we have already considered the possibility of a basal drive force, if the mantle is moving faster than the plate, but now we are suggesting alternatively a retardative force if the plate is going faster than the mantle. Fortunately, F_{BD} can stand for either basal drive or basal drag force.

To predict how these various forces move the plates, first they are classified into area and edge forces. The area forces are just the F_{BD} forces, drive or drag, which act over the whole underside of plates, while all the other forces are edge forces, for they are developed only at or near the edges of plates (strictly, both the ridge-push and slab-pull forces are developed over some hundreds of kilometres, but this is small compared with the widths of plates, which are thousands of kilometres across).

The speed at which a plate moves will depend upon the size of the drive forces. Drive and retardative forces balance automatically, for the retardative forces increase with speed (this is usual for retardative forces, and is why cars

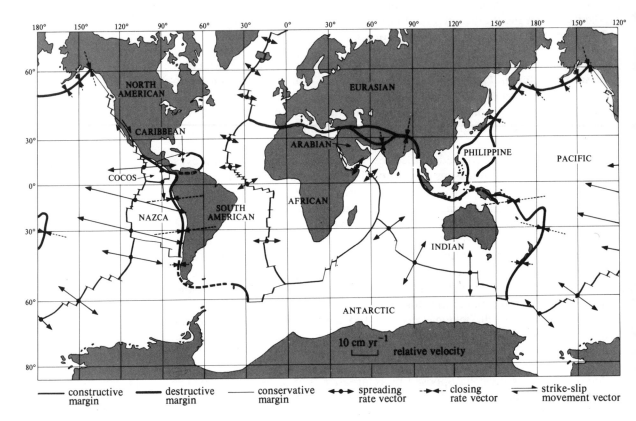

| ——— constructive margin | ——— destructive margin | ——— conservative margin | ◄—►—► spreading rate vector | ─►◄─ closing rate vector | ═══► strike-slip movement vector |

Figure 8.14 Plate map showing the 12 major plates. The arrows show sizes of the relative motions, determined from sea-floor magnetic anomalies and earthquake records. (After Minster *et al.*, 1974, by permission of the Royal Astronomical Society.)

and boats return to a constant speed if the power of the engine has been changed). Then, if the dominant force is, say, basal drive, large plates will go fastest, because they have the highest ratio of area to edge. If the dominant force is slab-pull, say, then plates with a large percentage of subducting edge will go fastest; similarly for the other edge forces. In fact, an adjustment is needed: if a plate were completely surrounded by, say, ridges – as is almost the case for the Antarctic plate, Figure 8.14 – the forces on opposite edges would cancel and there would be no net force. So the effective length is used, the percentage of the edge which has uncancelled force.

To apply this idea, the speeds of the plates are needed. These are not their relative speeds, but their speeds relative to the mantle, because these are what determine the magnitude of F_{BD}, F_{SR}, etc. This means that a marker in the mantle is needed, an unlikely concept considering that we believe the mantle is convecting. However, an approximate marker does seem to exist: this is the hot-spot frame of reference. Hot spots will be discussed in Section 9.2.2, so they will be described only briefly here. Hot spots are sources of volcanism that are not due to the usual plate-edge processes. The best example is the Hawaiian–Emperor seamount chain (Fig. 8.4(a)). The Hawaiian Islands are volcanic, with active volcanoes confined to the SE end, while to the NW are a line of seamounts, whose age increases in proportion to the distance along the chain. They seem to have formed as the Pacific plate moved over an intense source of

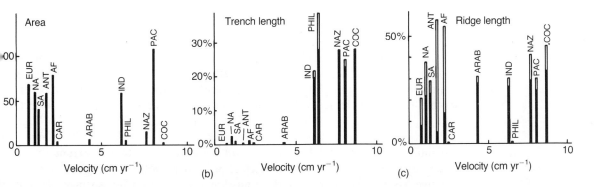

magma or 'hot spot'. A number of other hot spots are recognized (Fig. 9.1), and though they are not stationary, their relative motion is only about a tenth that of the plates. The rate at which igneous activity moves along the hot-spot trace gives the plate's 'absolute', or mantle, velocity. Once the velocity of one plate has been determined, then the velocity of adjacent ones can be found, knowing their relative velocities (Fig. 8.14).

Figure 8.15 shows these absolute velocities plotted against plate areas, and effective percentage of two different types of edge. The basal force cannot be dominant because, for instance, the tiny Cocos plate is going faster than the giant Pacific plate. The strongest correlation is with subducting edge: all the plates with a large effective percentage of subducting edge are moving quickly, and all those with a small percentage are moving slowly. There is a weak correlation with effective percentage of spreading ridge, and no significant correlation with any of the other forces (not shown). Thus the passive, log-carried-along-stream model can be ruled out.

Figure 8.15 Plate velocities versus (a) area, (b) percentage of subducting edge and (c) percentage of spreading ridge edge. In (b) and (c), the height of the bars is the actual percentage, the solid part is the effective percentage as explained in the text. (From Forsyth and Uyeda, 1975, by permission of the Royal Astronomical Society.)

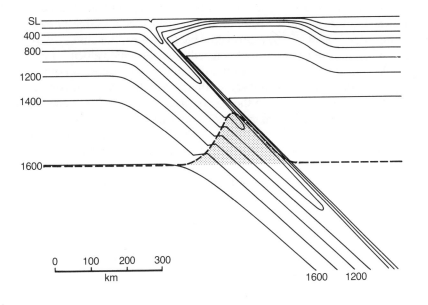

Figure 8.16 Geotherms in a subducting slab (°C). The slab tends to take its geotherms with it as it subducts, but heat conducting in from the mantle causes it to warm up, so they taper. The lower temperature in the slab means that compared with mantle at the same depth, the exothermic 400-km change occurs at a shallower depth, and the part shown shaded is extra-dense compared with normal at the same depth. (After Turcotte and Schubert, 1982.)

163

Why should the slab-pull be the dominant drive force? The main reason is that as the plate dives into the mantle it warms up only slowly, so it continues to be cooler and hence denser than the surrounding mantle, and so the negative buoyancy persists for hundreds of kilometres (Fig. 8.16). And the pull can be increased by phase changes. In normal mantle there is a phase change at a depth of, for example, 400 km (Sections 2.3 and 7.5), but because it is exothermic – heat is released on conversion into the denser form – the phase change occurs at a lower pressure if the temperature is reduced. Thus the change occurs at a shallower depth in the plate (Fig. 8.16), increasing the density contrast at this depth. (The phase change at 660 km is endothermic and has the opposite effect, with implications that will be discussed in Section 9.2.1.) Evidence of a large force acting at subduction zones is provided by gravity (see Note 8). The largest (free-air) anomalies anywhere are found near trenches; to the extent that a trench represents an absence of material it is perhaps not surprising to find large negative anomalies there, but the question then arises why isostasy has not removed it; why doesn't the trench disappear as do the depressions formed by ice-sheets when the ice has melted? The answer is that a force – the slab-pull – continues to act to prevent it.

Slab-pull cannot be the only drive force, for there are plates that have a negligible percentage of subducting edge, yet have relative motion; the separation of the African and South American plates is an instance. But ridge-push from the Mid-Atlantic Ridge is acting in the correct sense to account for their motion. Thus it seems that the plates with a large effective percentage of subducting edge move rapidly, and ridge-push forces act on the remaining plates.

This analysis does not actually identify the dominant forces, but only where the dominant forces act – along a subducting edge and, to a lesser extent, ridges – but we believe that the major force along subducting edges is the negative buoyancy of the subducting plates. Other investigations have extended the analyses. One has looked into whether the area of continental plate is significant, because of the possibility that the greater thickness of continental plate may affect the coupling with the mantle, but it seems not to matter. Another has extended the investigation into the past (Schult and Gordon, 1984), and shown that the similar velocity of all five plates with a large percentage of subducting edge (Fig. 8.15(b)) is probably due to chance; this similarity was once thought to be because the slabs had reached their terminal velocity, the velocity they would sink at if not attached to the remainder of the plate; in other words, that F_{SP} is equalled by F_{SR}. But it seems that retardative forces act on the whole plate. It has also been found that the size of the slab-pull force depends upon the age and rate of subduction, as expected, for these two factors determine the thickness and length of slab and hence its total excess weight. However, these are refinements and do not alter the main conclusion that the major drive force is slab-pull.

A different approach is to examine the direction of stress in the lithosphere, deduced from the fault plane solutions of earthquakes (Note 7). It turns out that the stress is aligned over large areas of plates, and is generally consistent with slab-pull and ridge-push being the dominant forces (Zoback *et al.*, 1989).

We are now ready to discuss why a plate is not simply a thermal boundary layer moving between an upwelling and a downwelling. There are several pieces of evidence, which include the following. Figure 8.14 shows that the Antarctic plate is almost entirely surrounded by spreading ridges, so it must be growing everywhere, with the ridges moving away from the South Pole. If they were over an upwelling, then the upwelling would have to be moving too. And as spreading is close to symmetrical about the ridges, this would require that the African plate, for instance, is moving away from the Antarctic plate at twice the rate of the ridge, which is implausible. Secondly, though spreading ridges are paired on either side of the ridge axis, subduction is unpaired (differing from the Hawaiian lava lake mentioned above). One reason for this is that continental plates have intrinsic buoyancy, so an oceanic plate may subduct beneath a continental one, but not vice versa; and when two continents meet they collide rather than either subducting (these two situations lead to Andean and Himalayan mountain-building processes, respectively, as described in Chapter 10). Thirdly, the uppermost part of plates is too cool to deform by creep, and this gives plates a strength and hence an existence beyond a simple boundary layer. Plates are therefore hard to pull in two, so once a spreading ridge is in existence it remains the weakest point and will tend to continue to be the division between two separating plates. Magmatism occurs there, not because it is above an upwelling, but passively by decompression melting (Section 7.4.2).

In summary, plates are an intrinsic part of mantle convection but, because of their special properties, they strongly determine the form of the convection. It is only a partial exaggeration to say that, rather than convection driving plates, plates drive convection; and the downwellings (subduction) are far more concentrated than the broad, sluggish upwellings, in contrast to the convection in a soup-pan.

8.4.3 Seismic tomography

A development of seismology makes it possible to observe the form of convection in the upper mantle. The variation of the seismic velocities within the Earth depends upon the variations in composition, structure and temperature. In the crust the variation is mainly due to differences in composition and

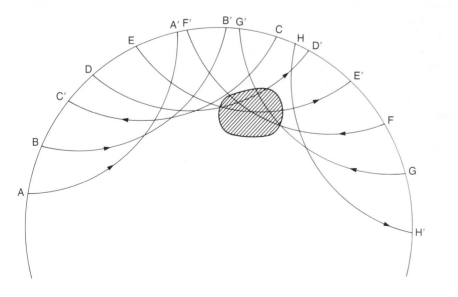

Figure 8.17 The principle of seismic tomography. The shaded region has a slower velocity. If seismic travel times are measured for the rays AA', BB' . . ., and for other rays penetrating to different depths, all those that travel through the region will take longer, from which it is possible to deduce the approximate location of the anomalous region and that its velocity is slow.

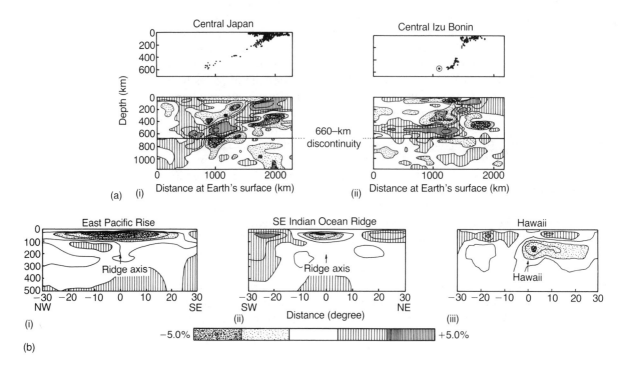

Figure 8.18 (a) Shows vertical tomographic sections using P-waves, of West Pacific subduction zones; (i) shows a section across central Japan and the fast region continues across the 660-km discontinuity (shown by the horizontal line), whereas in (ii) the section across central Izu Bonin is deflected (van der Hilst *et al.*, 1991). (b) Shows sections using S-waves, of the fast Pacific and the slow Indian spreading ridges, (i) and (ii), and, for comparison, the Hawaiian hot-spot chain (iii) (Zhang and Tanimoto, 1992).

structure, reflecting the different rock types, but in the mantle, which is more uniform in composition and structure, the main variation is probably due to temperature differences which are related to convection. Mantle regions that are relatively hot have a lower velocity compared to cooler regions at the same depth. This is because although the higher temperature tends to reduce density by expansion and hence increase the velocity (see Eq. (2.2)), this is more than offset by the reduction of the values of the elastic moduli. As hotter and cooler regions have gradational rather than sharp boundaries, a technique different from those described in Chapter 2 is needed to detect them, that of **tomography**. Figure 8.17 gives a simple example.

The situation in the mantle is obviously much more complex, having both fast and slow regions and being in three dimensions. A further difficulty is that there are not enough earthquakes and seismometer stations to measure as many paths as would be wished, even though P-, S- and surface waves can all be used. As a result, the regions that can be detected are large, being thousands of kilometres across and hundreds deep.

Figure 8.18 shows some recent results. The shading denotes how much the velocity in a region differs from that given by a spherically symmetrical model, such as iasp91 (Section 2.2), which therefore must be accurate so as not to produce spurious anomalous regions. Figure 8.18(a) shows sections across two subduction zones; the subduction zone defined by the earthquakes clearly shows up as a fast (cold) region. In a(i) the zone appears to cross the 660-km discontinuity, whereas in a(ii) it is deflected by it. A recent earthquake, denoted by the dot, was in the deflected part of a(ii). Figure 8.18, b(i) and b(ii), shows sections through the Pacific and Indian Ocean spreading ridges. A low-velocity (hot) region can be seen in the top 100 km, slowest at the ridge axis; it is wider for the Pacific ridge, which spreads faster than the Indian one. This contrasts with a section through Hawaii, a, b(iii), a hot spot, where the low-velocity region extends to at least 200 km (the resolution decreases with depth so the lack of an anomaly may not prove that no anomalous region exists). These results support a passive origin for the volcanism at ridges, and a deeper origin for hot-spot volcanism (Section 9.2.2).

In general, tomographic results show a strong correlation with the present plates at shallow depths, such as fast (cold) under old shields and in subduction zones, and slow (hot) under spreading ridges and other tectonically or volcanically active areas (not shown). This correlation diminishes with depth and also the variation lessens. However, because convection is such a slow process – for example, it has taken the Atlantic roughly 100 Ma to open – the temperature in the mantle is expected to depend on the past as well as the present plate configuration. In fact, there may be a much better correlation with the pattern of plate motions integrated over the past 180 Ma (the limit of accurately known plate motions), (Richards and Engebretson, 1992).

Unfortunately, the tomographic imaging of the lower mantle is more poorly known; velocity variations seem to be smaller than in the upper mantle, but, as yet, results are unable to reveal reliably the hot and cool regions.

8.4.4 The lithosphere: different definitions

The lithosphere was first defined in connection with isostatic compensation (Section 8.2.2), but since then the concepts of thermal boundary layer and plate have been introduced and their use needs clarifying.

In isostasy, the lithosphere is the rigid, and therefore elastic, uppermost layer. As it deflects under load (see Figs 8.4 and 8.5), it can be most nearly equated with the elastic thickness; that is, what thickness (and rigidity modulus) would the sheet have to have to match the deflection? For the example of the Hawaiian Islands (Fig. 8.4) it is 20–30 km, while for the old, continental glaciated areas of North America it was approaching 200 km. The matching assumes a uniform sheet, whereas the properties are changing with depth due to increasing temperature; the elastic thickness is therefore the equivalent uniform sheet.

The thermal boundary layer is that part that moves horizontally and conducts heat vertically. It does not require great rigidity for this to occur, so it is much thicker than the elastic thickness, though, for a uniform composition, one would expect elastic and thermal thicknesses both to be limited approximately by isotherms and so in proportion.

A plate is that part that moves laterally as a single entity, often for thousands of kilometres. An oceanic plate is closely similar to the thermal boundary layer, and in older parts reaches a maximum thickness of about 100 km. The deep structure of continents is less well understood, but older parts are probably 200 km or more thick (the term 'tectosphere' has sometimes been used for all of that part of the crust-plus-mantle that moves with a plate).

Finally, the low-velocity zone (LVZ) (Section 2.3) is sometimes equated with the asthenosphere, and the part above, the **seismic LID**, with the lithosphere. However, the definitions of the LVZ and of the asthenosphere are distinct, being based respectively on seismic velocity and rigidity; and although a decrease of seismic velocities – particularly V_s – and rigidity both result from an

167

increase in temperature with depth, they do so in different ways. Evidence that, in practice, they should be regarded as different comes from the observation that the LVZ is probably absent beneath old continental cratons – such as North America – which nevertheless are able to rebound isostatically.

Despite all of these separate meanings the terms are often used loosely, with, 'lithosphere', in particular, being treated as interchangeable with 'plate'.

8.5   TEMPERATURE WITHIN THE MANTLE

If we knew the temperature at all points within the Earth, then we would be able to deduce the form of convection. Unfortunately, with the exception of the top few kilometres, we have no direct way of measuring the temperature, and so we have to rely on a variety of clues.

Broadly, there will be a steep temperature gradient in the upper thermal boundary layer where heat is transported by conduction, a much smaller gradient within the convecting part of the mantle, and a further steep gradient in the lower boundary layer, which needs to exist to remove heat flowing out of the core. This broad picture has to be modified to take account of the distribution of heat sources, phase changes and the up- and downwellings of convection.

The region below the oceans is the best understood. Estimated heat production within an oceanic plate is so low (Table 8.1(b)) that the temperature within the plate is accounted for almost entirely by the model of plate growth described in Section 8.4.2. In the convecting part beneath, the temperature gradient must exceed the adiabat. As explained in Section 3.5, the adiabat is defined by the condition that, on moving material vertically upwards, the adiabatic cooling due

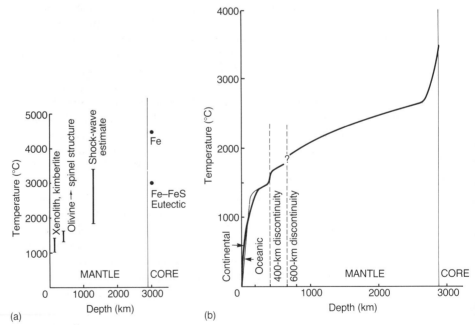

Figure 8.19 Temperatures in the mantle. (a) Shows constraints on temperature at various depths in the mantle (top of core temperatures are from Fig. 6.8, and the other estimates are from Mulargia and Quareni, 1988). (b) Shows schematic temperature profiles. The oceanic geotherm is for plate old enough to have reached close to its maximum thickness; the continental geotherm is that given in Figure 10.10. The profile has large errors, as indicated by the error bars in (a). See text for further details.

168

to expansion exactly equals the drop along the temperature gradient, so that buoyancy remains neutral. The adiabatic temperature gradient is given by the expression $(\alpha g T)/C_p$ (see, e.g. Turcotte and Schubert, 1982), where C_p is the specific heat at constant pressure and the other symbols are as for Eq. (8.4). These quantities can be estimated: near the top of the mantle the adiabat is about $0.5°C$ km^{-1}; it decreases with depth because α, the coefficient of expansion, decreases with pressure, reducing to only $0.25–0.3°C$ km^{-1} in the lower mantle. Convection is such an efficient way of transporting heat in the mantle that the temperature gradient probably only exceeds the adiabat by a small amount. However, these figures for the adiabat do not allow for the effect of phase changes, for these involve latent heat. The exothermic 400-km discontinuty is estimated to add a temperature step of about 160°C, but as the nature of the 660-km discontinuity is unclear (Section 9.2.1) its effect is not known; and if it is a barrier to convection there will be a large temperature difference across it, as discussed in Section 9.2.1.

The thermal structure of a continental plate is different from that of an oceanic plate, partly because of its greater thickness but mainly because of its far greater content of radioactive elements (Table 8.1(b)). In fact, the near-surface crystalline rocks are so rich in these elements that if the whole continental crust – let alone the whole plate – had these concentrations it would produce more than the observed global heat flow. This tells us that the concentration must decrease with depth. It has been found that the surface heat flow is simply proportional to the heat productivity of the surface rocks. Since this relationship seems to hold despite erosion of the surface, it requires the relationship:

$$q_s \quad = \quad q_m \quad + \quad c\,H_s \exp\,(-d/h_r) \tag{8.5}$$

|surface|sub-lithospheric|lithospheric heat|
|heat flow|heat flow|production|

where d is the depth and c and h_r are empirical constants dependent on the thermal history of the area; the longer the time since the area was tectonically active, the cooler it is. Heat production therefore decreases with depth, roughly halving for every 5 km increase in depth, and so becoming very small at the base. This fall off of thermal productivity means that the temperature gradient also decreases with depth, until the base is reached. Examples of oceanic and continental temperature profiles in the uppermost mantle are given in Figure 8.19(b).

Estimates of the temperature at particular depths can be deduced in a variety of ways. Especially useful are xenoliths (Section 7.2.2); if they have been transported sufficiently quickly from depth they will not have had time to re-equilibrate, and so they preserve the mineral assemblages corresponding with their source region, from which the temperature and pressure (and hence depth) can be deduced. Best known are those in kimberlites, which can be transported from a depth of 150 km to the surface in 2–3 days, and sometimes include the high-pressure mineral diamond, but xenoliths can also be carried up from lesser depths in pyroclastic eruptions that form tuffs. Further, the 400-km discontinuity is the depth at which olivine converts to spinel structure; from laboratory measurements this occurs at a temperature of about 1500°C at that depth (Section 7.5.1). An upper limit on the temperature of the mantle is provided by the temperature at the top of the core. In Chapter 6 it was explained that the molten outer core is mostly iron but contains a light diluent, most probably oxygen or sulphur; oxygen raises the melting temperature, whereas sulphur lowers it, introducing large uncertainties in estimates of the core temperature; the temperature of the top of the core was discussed in Section 6.6 and summarized in Figure 6.8. These various pieces of information have been assembled in Figure 8.19(a), and a possible mantle temperature profile shown in (b).

Lateral temperature differences are even less well known than radial ones. Some estimates can be made from the variation of seismic velocity found using seismic tomography (Section 8.4.3). Conversion from velocity to temperature is uncertain, but slabs have been deduced to be 300°C cooler than adjacent mantle, from the deflection of the phase change and hence the seismic discontinuities, as described in Section 8.4.2 (Vidale and Benz, 1992).

8.6   THE EARTH AS A HEAT ENGINE

At the beginning of this chapter it was explained that the reason why the Earth's surface has not been levelled by erosion is because of tectonic and associated igneous processes, and that the mechanical forces that build mountains and cause earthquakes result from convection – the mantle functioning as a simple heat engine. To round off this chapter, we look at how much heat is converted into mechanical energy, drawing on Verhoogen (1980).

Earthquake strain energy.   Earthquakes occur when rocks have been strained by shear to failure, the strain energy being released by elastic rebound (see Note 7). About 90% of earthquakes occur along plate boundaries due to relative plate motions, which we have seen are powered by convection. Although many thousands of earthquakes occur each year, most of the energy is dissipated by the relatively few large ones. The annual average is about equal to one of the largest size (magnitude 8·6), giving a power of about 3×10^{11} W, roughly 1% of the global heat transported to the surface.

Uplift of mountains.   Mountain building involves the lifting of rocks, by which process mechanical energy is stored as gravitational potential energy (later, when erosion breaks down the rocks, this energy is released to help to transport the debris to lower altitudes). Assuming that uplift equals erosion, estimated to be 7·5 km^3 per year, with an average rock density of 2·5 Mg m^{-3}, and that the material moves from the average land height of 1 km to sea-level, the energy is stored at a rate of 7×10^9 W. This is a very small proportion of the global heat transported.

Plate tectonics.   From the point of view of this chapter, the major consequence of convection is the creation and movement of plates. It may be thought, therefore, that we need to include the energy required to move them. But this is not so, for the plates are an integral part of the convection (Section 8.4.2) and are therefore simply part of the 'working substance' of the heat engine, the material that expands and contracts to develop mechanical energy, like the steam of a steam engine. Work is done only when the convective flow encounters opposition, as at plate boundaries, giving rise to earthquakes and mountain building, which have been considered.

In summary, the heat that arrives at the Earth's surface derives from *Summary* radioactive decay, growth of the inner core which releases gravitational energy, and cooling. It is transported through the mantle by thermal convection, which is a form of heat engine. A by-product of this process is the formation (and later destruction) of the oceanic plates and, where plates collide, surface forces are developed that cause earthquakes and mountain building. The proportion of the heat energy that goes into these two mechanical processes is no more than 1%, but small as it is, it gives us our dynamic planet (though at a cost in lives lost in earthquakes and volcanic eruptions).

SUMMARY

1. The Earth's surface is active, with significant relief despite erosion, because of tectonic and igneous processes. These mostly occur at plate margins.
2. The total amount of heat arriving at the Earth's surface is mainly conducted through the plates, supplemented by hydrothermal convection at spreading ridges, plus minor processes. The total amount is about $4 \cdot 2 \times 10^{13}$ W.
3. Most of this heat is generated by radioactive decay in the mantle and crust, plus heat from settling in the core which leads to inner core growth, but 25–50% of the total comes from cooling of the Earth.
4. Except for the cool, outer few tens of kilometres, the mantle is everywhere hot enough to deform by creep, a very slow form of adjustment that involves only a small proportion of the crystalline bonds being broken at any instant; this permits the mantle to behave rigidly on a short time-scale, permitting the transmission of seismic S-waves, but viscously on a long time-scale, permitting convective flow.
5. The rheology of the mantle is not well understood, but it may be approximated by a rigid, elastic layer up to nearly 200 km thick – depending on the age and nature of the plate – over a mantle with Newtonian viscosity of 10^{21} Pa s, which increases by only about a factor of two in the lower mantle.
6. Heat is transported within the mantle by convection, except for the upper and lower boundary layers (and possibly across the 660-km discontinuity), where transport is by conduction.
7. The dominant force on plates is the negative buoyancy of subducting slabs, 'slab-pull', which drives the plates, with ridge-push and basal drag having lesser roles.
8. Plates are an integral part of convection but the physical integrity of plates and the intrinsic buoyancy of continental plates means that they play a greater role in the form of convection than is the case for a simple thermal boundary layer in a homogeneous liquid.

   Subduction determines the downwellings, and spreading ridges are not the result of upwellings, but the cause of them.
9. Temperature in the mantle is poorly known, but there is a steep gradient in the plates, and the lower boundary layer, with the remainder of the mantle (with the possible exception of near 660 km depth) being only a little above the adiabat, apart from phase changes.
10. Earthquakes and mountain building are the result of mechanical forces developed by the mantle acting as a heat engine, but involve only about 1% of the heat transported.

FURTHER READING

General journal:

McKenzie (1983): mantle convection.

General books:

Bott (1982): an account of the Earth's interior, largely from the physical standpoint.

Turcotte and Schubert (1982): contains chapters on rheology, elasticity and heat, giving the underlying physics.

Verhoogen (1980): an account of heat sources and sinks and the Earth as a heat engine.

Advanced journals:

Olson *et al.* (1990): review of large-scale mantle convection.

Romanowicz (1991): seismic tomography.

Advanced book:

Anderson (1989a): includes many of the topics of this chapter.

Evolution of the mantle

9.1 INTRODUCTION

It is not enough to describe what the Earth is – what it is made of, and what state it is in – for the Earth is dynamic, and so *processes* are also important. But neither is it sufficient to stop at its processes, for these imply change. The Earth's supply of energy is decreasing as radioactive isotopes decay away, so it is cooling, and one consequence is the growth of the inner core. There are also chemical changes, for the formation and destruction of oceanic plates does not simply cycle mantle material, but also separates elements, first at spreading ridges (Section 7.4), and then at subduction zones (Section 10.3.2), so that the material returned to the mantle differs from that at its source region. Igneous activity also can occur away from plate margins, giving rise to oceanic islands and plateaux, and to continental flood basalts; these have already been touched on in connection with the hot-spot reference frame (Section 8.4.2). All this activity changes the composition of the mantle so that it is chemically heterogeneous.

While the topics of this chapter describe second-order effects compared with those covered in the previous two chapters, they provide an insight into some of the more subtle mantle processes which are important in its long-term evolution. But, by the same token, they are harder to unravel, and there remains much uncertainty, despite vigorous investigation. First, we deal with mainly physical evidence as to whether the 660-km discontinuity is a barrier to mantle convection, and in Section 9.3 we examine the consequences and evidence for the development of isolated geochemical reservoirs in the upper and lower mantle.

9.2 DOES CONVECTION CROSS THE 660-KM DISCONTINUITY?

9.2.1 The nature of the 660-km discontinuity

Chapter 8 deliberately left vague the effect of the 660-km discontinuity on convection. This is because, despite being a puzzle for at least two decades, it is not known whether convection crosses the discontinuity with little difficulty, and this is linked to the nature of the discontinuity. The viscosity of the lower mantle is low enough and the quantity of heat to be removed from it (including that supplied by the core) is large enough for convection to be needed for its removal. Therefore convection is expected to occur without interruption throughout the mantle, unless the 660-km discontinuity is a barrier.

There is little doubt that there is a phase change from $(Mg,Fe)_2SiO_4$ in the spinel structure to $(Mg,Fe)SiO_3$ in the perovskite structure + $(Mg,Fe)O$ as magnesiowüstite across the discontinuity, but is this the only change? There has been discussion whether the seismological observations can be fully matched by a simple transition, and in Section 7.6 arguments were put forward for a lower $Mg/(Mg + Fe)$ in the lower mantle, implying an increase in the mean atomic weight. If there is a compositional change is it large enough alone to be an obstacle to convection? The minimum compositional density difference needed to prevent convection has been claimed to be from as little as 0·1 to as much as 3%.

An alternative approach is to look for evidence of whether convection crosses the discontinuity. The maximum depth of earthquakes beneath subduction zones is about 700 km, but their absence below this depth does not prove that the plates penetrate no deeper, for it may be that a slab has merely, for example, become too hot to sustain the necessary shear stress; as the nature of such deep earthquakes is poorly understood it is difficult to decide what conditions are needed for them to occur.

The 660-km discontinuity is expected to provide some resistance to convection, because the phase change would be expected to occur deeper in the plate than the adjacent mantle, decreasing the buoyancy, in contrast to the 400-km discontinuity (see Section 8.4.2). Fault-plane solutions (Note 7) show that where earthquakes occur in the lowermost upper mantle, the slab is in compression – unlike nearer the surface – suggesting that the slab is encountering some form of resistance. There is also evidence that the discontinuity is up to 30 km deeper where slabs reach it. Attempts to detect slabs below the 660-km discontinuity have been ambiguous, but recent seismic tomographic evidence suggests that while *some* slabs are deflected by the discontinuity, others are able to penetrate it (Fig. 8.18).

If the 660-km discontinuity is a total barrier to convection, then there must exist separate convective systems above and below it, and heat then has to transfer from the lower to the upper mantle through a conductive thermal boundary layer. This would require a large temperature gradient, resulting in a temperature differences across the boundary of 500–1400°C. As a result, the homologous temperature at its base would be much higher, and this would be expected to decrease the seismic velocities and increase the attenuation, as found for the low-velocity zone in the upper mantle (Section 2.3); such effects have not been observed.

9.2.2 Hot spots and plumes

A **hot spot** is any volcanic activity that occurs away from plate margins, or exceptional activity at margins. The best-known example is the Hawaiian–Emperor seamount chain (Figs 8.4 (a) and 9.1). Radiometric dating shows that activity has progressed along the chain and currently is confined to the most south-easterly island (plus a submarine volcano developing to its south-east). This may be explained if the Pacific plate moved over a particularly hot spot in the mantle, hence the hot-spot concept. Similarly, hot spots have been identified with the source of continental flood basalts, such as the Lower Tertiary Deccan lavas (Réunion plume), while Yellowstone is a currently active one (see Fig. 10.17).

A feature of hot spots is that their volcanism produces alkali basalts, different from the tholeiites of spreading ridges. Figure 7.5 shows that this requires a source region that is deeper in the upper mantle. Thus when a hot-spot trace crosses a spreading ridge, the characteristic alkalic signature appears in the ridge tholeiites, for hundreds of kilometres on either side of the hot-spot crossing. Moreover, the compositional difference continues for some time after the hot spot has moved away from the ridge. This accounts, for instance, for the composition of Icelandic lavas, whereas their great volume arises because a

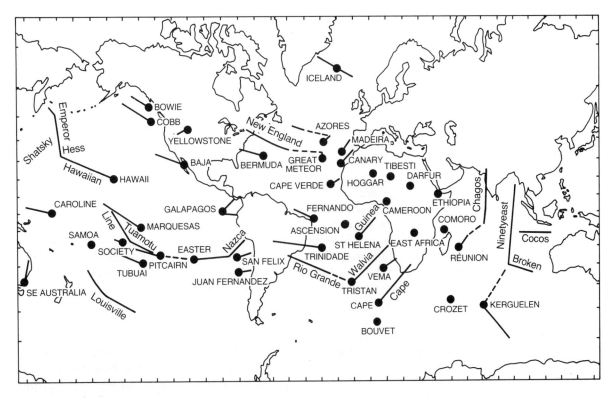

Figure 9.1 Hot spots. The blobs show the location of present activity, while for the most widely accepted instances of hot spots, the lines show past positions. (After Crough, 1983.)

much larger volume of mantle than normal has been partially melted at a higher potential temperature (Fig. 7.6(b)). The greater height of hot-spot volcanoes relative to ridges is also support for a deep origin, for a taller magma column is needed to generate the extra lift.

There is no agreement on the total number of hot spots, but a few dozen are widely accepted (Fig. 9.1). Neither is it fully understood what causes hot spots. The most widely supported theory is that they are the surface expression of **plumes**, columns of hot mantle material, rising because of their higher temperature and lower density. (A familiar example of a plume is smoke rising from a cigarette in still air.) According to White and McKenzie (1989a and b), the column is about 100 km across, but when it first reaches the lithosphere it spreads out laterally to form a mushroom-like head which may be up to 2000 km across. The mushroom causes a wide uplift, apparent in the cross-section of Hawaii (Fig. 8.4(b)), but the hot-spot volcanism is limited to a smaller region nearer the centre. Observations of ancient flood basalts caused by plume activity (e.g. the Deccan and Columbia River basalts – Fig. 10.17) show that this central region can still be some hundreds of kilometres across. It is the large amount of heat available in a plume, resulting from its large volume and excess temperature of about 200°C, that may allow several million cubic kilometres of magma to erupt in only a few million years. This is partly a consequence of P–T conditions in the upper mantle approaching the dry solidus for lherzolite (Figs 7.5 and 7.6(b)), but also due to the larger depth extent of the mantle that is enabled to produce melts. The large extent of the plume head also explains how the geochemical signature of hot spots can appear in ridge basalts several hundred kilometres from the hot spot (see Section 9.3). After the prolific 'head' volcanism, activity is reduced, giving rise to a hot-spot trace as the plate moves over the plume stem. However, it is possible that pulses of magma may

175

subsequently ascend the plume as single waves, to give rise to renewed bursts of high activity.

The question then arises, at what depth do the plumes originate? The most popular source is the base of the mantle, identified with the D'' layer.

9.2.3 The D'' Layer

It was pointed out in Section 2.3, that in the lowermost 200 km of the mantle the seismic velocities increase less rapidly with depth than in the rest of the lower mantle. This might be because of a higher temperature gradient there, or because of compositional differences. For example, Anderson (1989, 1992) has argued that, since the core–mantle boundary represents the largest density contrast in the Earth, it might act as a 'natural collection point for any light material leaving the core or dense material settling out of the mantle'. Thus if D'' is an 'unclean' chemical boundary between the core and mantle, it may be strongly enriched in iron oxides and silicates relative to the rest of the mantle: $FeSiO_3$, perovskite, and FeO, wüstite, could be important. D'' may also contain some early refractory phases condensed in the Earth (Section 5.2) which would include Ca–Al–Ti minerals such as perovskite itself, $CaTiO_3$, and melilite, $Ca_2Al_2SiO_7$. It may even be the repository of any deeply subducted ocean crust that penetrates the 660-km discontinuity as a narrowly focused high-density *sinking* plume (Section 9.3). In short, D'' is likely to be a chemically complex boundary layer, but it is almost certainly also a lower thermal boundary layer at the base of the mantle, because of the substantial quantities of heat coming from the core (Section 8.4).

There is good evidence from seismic rays refracted around the core that the D'' layer is strongly heterogeneous laterally on a scale of 150–500 km, while smaller irregularities have been shown up by PKKP rays (Fig. 2.6) which reflect from the underside of the core–mantle boundary. There are also 'bumps' on the core–mantle boundary up to 6 km high and 20 km across which, because of the low viscosity of the core, can only be supported dynamically, that is, by core upwellings and downwellings; this means that heat is delivered unevenly to the core–mantle boundary, just as it is to the Earth's surface. An important suggestion is that as this heat is delivered from the core, the boundary layer heats up and becomes progressively more buoyant. Unlike the upper boundary layer, it has a lower viscosity than the rest of the mantle, so instabilities form. But these make little progress up into the more viscous mantle above until they have grown by amalgamation to have a buoyancy sufficient to rise, a process that may take about 100 Ma. They then rise as plumes, at about 10 cm per year, comparable with the velocities of descending slabs, but much faster than the general mantle convection.

A case has been made for a superplume in the mid-Cretaceous, when there was a 50% or more increase in the production of oceanic crust, at ridges and plateaux, primarily in the Pacific. Many consequences have been attributed to it. For instance, it coincided with the Cretaceous magnetic quiet zone, when the Earth's magnetic polarity remained normal for 40 Ma (Fig. 6.3); this could be because the cooling of the core–mantle boundary by the plume removed sufficient energy from the core to prevent reversals. Another possible consequence is the increased delivery of carbon and other nutrients to the surface, causing higher temperatures through the greenhouse effect, and increased production of hydrocarbons (Larson, 1991a and b).

Our understanding of the nature of the 660-km discontinuity, the D″ layer and of plumes is interconnected, for the many pieces of evidence relate to more than one of the three. The 660-km discontinuity is probably an obstacle to convection, but whether it is a total barrier is less clear. The D″ layer is undoubtedly anomalous, possibly compositionally different, and almost certainly a thermal boundary layer from which convective flows ascend. The high Rayleigh number of the mantle means that these flows are likely to be concentrated in thin sheets or columns, but this does not ensure that they reach the upper mantle without interruption.

In their ascent, plumes are likely to entrain mantle material and so become larger, and it has been argued that the largest hot spots would need to have ascended through the whole mantle to have reached their size. The amount of heat transported by plumes, estimated as 6–10% of the global total, matches the heat delivered by the core to the mantle whereas, if only heat crosses the 660-km discontinuity, it would give rise to many more plumes, which would carry the majority of the heat. The distribution of hot spots on the global scale seems to be associated with broad geoid anomaly highs (Note 8) which also suggests a deep, lower-mantle origin. But arguing against a lower-mantle (and hence, presumably, a D″ origin) is the fact that hot-spot spacings, often less than 1000 km, are too close for this to reflect the spacing at their sources on the core–mantle boundary (where it would be less than 500 km, because of the small radius of the core). Moreover, seismically anomalous mantle has not been detected more than a few hundred kilometres below hot spots, and it does not always have a roughly circular head. It also has to be borne in mind that not all volcanic activity away from plate margins matches that of Hawaii. Some lines of volcanoes have erupted simultaneously (as the Line Islands and Cameroon volcanic line), and not all continental flood basalts may need a source below the lithosphere (see Anderson *et al.*, 1992).

A possible solution to this conflicting evidence is that no single explanation accounts for all hot-spot activity. Sources may be at different depths. In addition, the 660-km discontinuity may be only a partial barrier to convection. Arguably only some, the heaviest, and/or most rapidly descending slabs can penetrate through it, and only the largest rising plumes can rise through this barrier. Support for such **hybrid convection**, involving partial or intermittent mixing across the 660-km discontinuity, comes from computer modelling of mantle convection incorporating an endothermic phase change at 660 km, and from seismic tomographic investigations below the Pacific (Fig. 8.18(a)) and geochemical arguments for large-scale mantle heterogeneity, which are the subject of the next section.

9.3 GEOCHEMICAL HETEROGENEITY IN THE MANTLE

In Chapter 7 we discussed oceanic basalts as though they all arise from chemically similar lherzolite sources. The major-element variations between basalt types are easily reconciled with this view for they are a product of the *P–T* conditions under which the melts were produced (Fig. 7.5). However, the trace- 177 element story is rather more complicated, and much attention has been focused on **incompatible elements**, so called because they are strongly partitioned into the first melts to form, and also on initial **radiogenic isotope** ratios, which give information on the time-integrated geochemical history of the melt zone (details below). The advantage of studying oceanic basalts is that they provide a direct window on the geochemistry of upper-mantle source regions and are uncontaminated by melting of the continental lithosphere. The snag is that only very young rocks, reflecting the recent state of the mantle, are available in oceanic regions. As we shall see, some of the most intriguing geochemical evidence for heterogeneities comes from oceanic islands remote from ridges, which have magma sources deep within the upper mantle: these are the **ocean-island basalts** or **OIBs**. In some senses, however, they have distracted attention from the remarkable geochemical consistency of the more voluminous igneous products at spreading ridges which account for 90% of mantle volcanism. (Mid-) ocean ridge basalts have acquired the acronym **MORB** since it happens that some spreading ridges are in the *middle* of oceans. Remember that all these magmas are produced in the upper mantle, so it is the geochemistry of upper-mantle sources that we are considering.

9.3.1 Incompatible trace elements

A convenient method of comparing the incompatible trace-element abundances of different basalt suites is to plot them in order of increasing compatibility (left to right in Fig. 9.2) and to normalize them to chondrites (i.e. close to Bulk

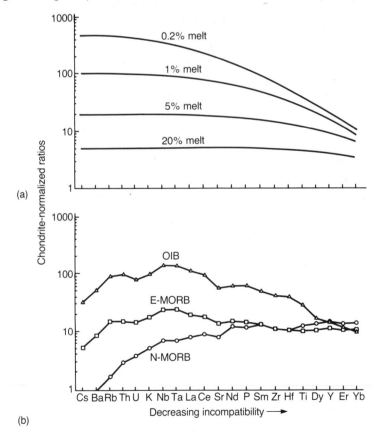

Figure 9.2 Incompatible trace-element ratios relative to chondrites in mantle-derived magmas (a) showing the trends expected for 0·2 to 20% melting of primitive mantle, and (b) averages for two types of ocean ridge basalts (MORB) and for ocean-island basalts (OIB) recently derived from the upper mantle. (Data compiled by Floyd, 1991.)

178

Earth). Thus a source region that is neither enriched nor depleted relative to chondrites will have ratios of *one* for all these elements and we assume that this characterized the primitive mantle. Extraction of melt will preferentially remove incompatibles from the source, and the elements on the left will be proportionately the most extracted. Thus the first melts from primitive mantle would show steadily increasing values from right to left (see Fig. 9.2(a)), and the degree of melting would determine the extent of the enrichment relative to chondrites.

Figure 9.2(b) shows that typical modern basalts are, as expected, enriched in almost all incompatible elements, yet none of the curves matches the theoretical curves of (a). Although all three curves are similar for the least incompatible elements, with an enrichment factor of about 10, they differ from each other and the theoretical curves above for the most incompatible elements. Normal, or N-type MORB, for example, actually has less Cs and Ba than chondrites. The explanation is that, before the MORBs were extracted, their source regions were strongly depleted by *previous* extraction of small melt fractions (cf. Section 7.3.3) during the long history of the upper mantle. As Figure 9.2(a) shows, small melt fractions are extremely efficient in scavenging the available incompatible elements from their source regions. So ocean-crust formation has removed these elements from the mantle, leaving the depleted N-MORB source regions that we must infer from Figure 9.2(b).

What about the other oceanic basalts in Figure 9.2(b)? There are smaller depletions of highly incompatible elements in E-type, or enriched (relative to N-type) MORB, which represents the products of thicker melt zones affected by plumes, such as in Iceland. Similarly, the deeper OIB alkali basalt source regions achieve even higher values for most elements. Notice here that there are two possible interpretations: *either*:

1. the E-MORB and OIB magmas are tapping mantle source regions that are less depleted in the highly incompatible elements; *or*
2. they are being produced from geochemically similar sources to the N-MORBs but with smaller degrees of melting.

Since E-MORBs and OIBs tap deeper mantle than N-MORBs, (1) implies that the upper mantle is chemically zoned in incompatible elements. For (2) this need not be the case, because the presence of the plume allows a large volume of the mantle to produce small melt fractions. Isotopic data provide a further dimension in this debate.

9.3.2 Isotopic heterogeneities

A wealth of research effort has been invested over the past 20 years in unravelling the distinctive features of basalt source regions using radiogenic isotope geochemistry. The advantages over simple trace-element analysis are that:

1. the geochemical history of the source; and
2. contributions from different sources may be investigated.

The historical aspect arises as follows: for a particular radioactive isotope parent–daughter pair, the more of the daughter that is detected at the time the magma came from the source region, the more of the parent must have been present in the source(s) tapped *on average*, prior to melt formation. As we have seen (Fig. 9.2(a)), melting processes can cause dramatic changes in element *ratios*; for example, the Cs/Yb ratio is higher by a factor of 10 in a 1% mantle

Figure 9.3 Strontium isotope (a) and neodymium isotope (b) evolution diagrams for a chondritic Bulk Earth (CHUR = chondritic uniform reservoir) showing the effect of a simple melt extraction event on the reservoir at 3 Ga before present. Of course, upper-mantle depletion to form oceanic, then continental crust (cf. Section 10.3), will have been a continuous process throughout the Earth's history. Notice that the depleted source region for oceanic basalts evolves with time towards lower ^{87}Sr/^{86}Sr and higher ^{143}Nd/^{144}Nd ratios, as reflected in present-day MORB sources.

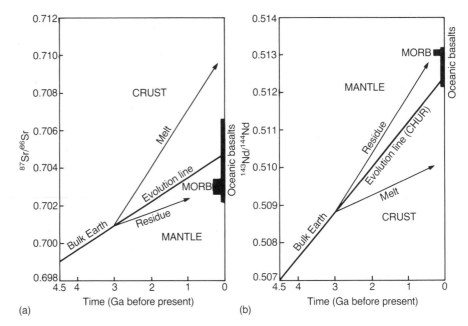

(a)   (b)

melt as compared with its source. (This arises because of a 100 times enrichment of Cs and a 10 times enrichment of Yb compared with the source.) Another advantage of isotopic methods is that, as the ratio between two *isotopes* of the same element is *not* changed by melting, the melt reproduces the isotopic ratios of the source region. Various techniques are in use (see Note 4) and we shall concentrate on **Rb–Sr** and **Sm–Nd isotopic methods**.

From isochron diagrams for oceanic basalts we can determine the so-called **initial ratios** (^{87}Sr/^{86}Sr)$_{init}$ and (^{143}Nd/^{144}Nd)$_{init}$ (see Eqs (N4.6) and (N4.7), Note 4) which represent the isotopic ratios of the mantle source region at the time of extraction. Dealing first with strontium isotopes, we wish to compare the strontium isotope ratios of recent magma source regions with each other, and particularly with the Bulk Earth value, which in a homogeneous chondritic mantle would, today, have reached 0·7047 (Fig. 9.3(a)). The range of values actually found in modern oceanic basalts is 0·7022 to 0·7066 (Fig. 9.3(a)), but with the majority less than 0·7047. This reflects:

1. the fact that, over time, mantle source regions have been more depleted in Rb than Sr (due to the greater incompatibility of Rb, Fig. 9.2) resulting in its greater concentration into the crust; and
2. not surprisingly, that the mantle is isotopically heterogeneous.

In contrast, Nd is *more* incompatible than Sm. Thus a 'depleted' source region will have lost less of the parent isotope, ^{147}Sm, than its daughter, ^{143}Nd, and so becomes more radiogenic with time than an enriched source region (cf. Fig. 9.3(b)): the opposite behaviour to the Rb/Sr system. The model Bulk Earth value of (^{143}Nd/^{144}Nd)$_{init}$, 0·51264 for the present day, is therefore *lower* than those for the majority of modern oceanic basalts (0·5133 down to 0·5123), reflecting the preferential incorporation of Nd into the crust. Again, the range of Nd-isotope ratios testifies to considerable mantle heterogeneity.

The Sr and Nd isotope variations in mantle source regions, deduced from modern oceanic basalts, are summarized in Figure 9.4. The array of N-MORB values, defined by the Mid-Atlantic Ridge, the East Pacific Rise and the Indian

Ocean, indicates sources that, for some time, must have had low Rb/Sr and high Sm/Nd ratios relative to Bulk Earth. Just to reinforce this conclusion, if the highly incompatible elements had been depleted *recently*, involving the preferential removal of Rb and Nd, then the Sr and Nd isotope ratios would not have had time to change and would have evolved with Bulk Earth. Notice that some oceanic islands have Sr and Nd isotope ratios that overlap with MORB, whereas others are closer to Bulk Earth. Few show higher Rb/Sr (hence higher $^{87}Sr/^{86}Sr$) and, to a lesser extent, lower Sm/Nd (hence lower $^{143}Nd/^{144}Nd$) than Bulk Earth. Dealing first with the OIBs that lie close to N-MORB (e.g. the E-MORB of Iceland and the OIBs of Ascension, the Canaries and Easter Island), despite their relatively high abundances of incompatible elements (from Fig. 9.2(b)), the isotopic data indicate a time-averaged source depletion relative to Bulk Earth. For these island suites, the ambiguity noted at the end of Section 9.3.1 remains; *either*:

1. the suites are tapping mantle sources that were enriched in incompatibles prior to or during partial melting; *or*
2. smaller degrees of melting in sources similar to those for N-MORB are involved.

Note that for model (1), radiogenic isotope data have advanced us significantly – if deep upper-mantle sources are richer in incompatible elements than shallower sources, enrichment was a *recent* event. This has led to the proposal that enrichment by mantle plumes, possibly rising from the lower mantle and carrying higher levels of incompatible elements, may be a trigger for OIB production. This process appeals to a source outside the upper-mantle reservoirs that we have been considering; the consequences are discussed further in Section 9.4.

Similar alternatives apply to the higher Sr, lower Nd isotope ratio part of the array in Figure 9.4. Some of the isotopic variability may be introduced by the subduction of oceanic crust, which is preferentially enriched in radiogenic

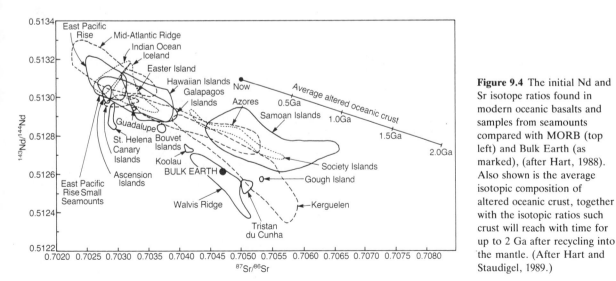

Figure 9.4 The initial Nd and Sr isotope ratios found in modern oceanic basalts and samples from seamounts compared with MORB (top left) and Bulk Earth (as marked), (after Hart, 1988). Also shown is the average isotopic composition of altered oceanic crust, together with the isotopic ratios such crust will reach with time for up to 2 Ga after recycling into the mantle. (After Hart and Staudigel, 1989.)

181

strontium because of its interaction with sea-water. But this cannot explain much of the variation in Figure 9.4, and another alternative is that continental lithospheric *mantle* (cf. Chapter 8), which has been isotopically isolated for much of the Earth's history, may become negatively buoyant beneath continental collision zones, and may then be incorporated into the mantle flow. Xenoliths of such mantle are known to carry both enriched and depleted isotopic characteristics relative to Bulk Earth. There is also the possibility that plumes of previously untapped material, carrying Bulk Earth isotopic ratios, rise and mix with the depleted upper mantle sources, in which case the main trend in Figure 9.4 is a mixing line between depleted upper mantle and Bulk Earth isotopic ratios. However, on model (2) (small melt fractions), much of the isotopic variability could be generated if melts do not reach the surface but are stored as Rb-and Nd-enriched zones, which then evolve isotopically, and become available for later melting.

At the time of writing, the debate about these alternatives is continuing. While all these processes may well contribute to a varying degree, the idea of plumes of undepleted material (i.e. close to Bulk Earth isotopic ratios) rising from the lower mantle has become popular for the following reasons. Firstly, the geographical distribution of enriched OIBs, roughly between latitudes 20° and 50° S in the SE Atlantic and SW Indian oceans, suggests a *large* province with enriched mantle geochemistry. This is the so-called **DUPAL anomaly** (after Dupré and Allègre, 1983), which has since been traced in MORB samples from the SW Indian Ridge. It is not unique, and a similar but less distinct anomaly has been identified in the South Central Pacific (the Magellan Seamount area). Secondly, according to some researchers, these large-scale geochemical anomalies seem to coincide with regions of anomalously low seismic velocity in the lower mantle, with hot-spot clusters and relative highs on the geoid (see Section 9.2), all of which indicate ascending flows (Olson *et al.*, 1990). This brings us to a similar conclusion to that reached in Section 9.2, that some limited input of lower-mantle material to the upper-mantle source regions for basalt magma genesis may be occurring. This provides a useful but not unique interpretation of the geochemical data.

9.4 SYNTHESIS

We are now in a position to draw together the threads of this discussion. Clearly, while the upper-mantle source regions of the majority of modern oceanic basalts are essentially garnet lherzolite in their petrology and major-element geochemistry, they are strongly depleted in incompatible elements relative to Bulk Earth (Fig. 9.2(b)). This is consistent with their extraction in small melt fractions from a convective, well-mixed regime at depths along the solidus, as indicated in Figure 7.5. These melts rise to form the ocean crust, thereby removing incompatible elements from the mantle. The reason why they are not returned to the mantle at subduction zones is because the incompatible elements then enter the partial melts that contribute to the continental volume at destructive plate margins (Section 10.3). Thus, by a two-stage partial melting process, the continental crust has become the repository of these elements, including Rb and Nd, and this is why MORB sources, in particular, are displaced from Bulk Earth (Fig. 9.4). In fact, to produce such non-radiogenic strontium and radiogenic neodymium isotope ratios, much of this depletion must have occurred early in the Earth's history (Chapter 11).

Geophysical data given in Section 9.2 indicate that the upper-mantle convective regime extends across the Transition Zone to 660 km depth, but less clearly below this. If the upper and lower mantles are at least partially isolated chemically, they should have become geochemically different in trace-element concentrations (we already know that this is probably the case for major elements – Section 7.6). Mass-balance calculations for trace elements confirm the expected differences. For example, if the whole mantle were involved in convection, then the highly incompatible elements now in the continental crust would represent some 30% of the Bulk Earth inventory, leaving 70% distributed through the mantle. In contrast, if only the upper mantle (33% of the total) is required to furnish the continental incompatible elements, 90% will occur within the continents and 10% will be distributed through the upper mantle. The strong depletion of MORB source regions (Fig. 9.2) supports the second model. So the picture that we have is of a lower mantle containing close to Bulk Earth concentrations of highly incompatible elements, and an upper mantle with just 10% of those concentrations. Detailed modelling suggests that there may be a small transfer from the lower to the upper mantle. The input rate for incompatible elements from the lower mantle requires exchange of only 2–5% of the upper mantle mass per Ga (see O'Nions, 1987, for details).

Isotopic data (Figs 9.3 and 9.4) illustrate the homogeneous Rb- and Nd-depleted nature of MORB sources, which contrast with the probable heterogeneous sources of OIBs. It was argued in Section 9.3.2 that while the ageing of trapped small melt fractions provides one explanation of the OIB data (our model (2)), some OIB sources may have been enriched in incompatible elements immediately prior to or during partial melting (our model (1)). Again, one obvious source of isotopically undepleted material close to the Bulk Earth composition is the lower mantle. Indeed, Nataf (1989) has argued that narrow, intense density anomalies can partly pass across the upper/lower mantle boundary promoting hybrid convection (cf. Section 9.2.4). The isotopic and trace-element signature of such a mantle plume may be modified by mixing with further partial melts in the upper mantle to provide a spectrum of OIB and E-MORB magmas. There is a coincidence of the emergence of such 'enriched' magmas with some of the geophysical evidence for upwelling below 660 km, and with primordial ^{3}He anomalies most probably from the lower mantle (O'Nions, 1987), that further strengthens the case for such plumes.

But this is unlikely to be the whole story, for crustal recycling in the form of subducted oceanic basalts, ocean-floor sediments and foundering fragments of dense continental lithosphere are all likely to occur. We know that the intense eclogite metamorphism that these rocks suffer increases their density to the point that they probably accumulate above or even penetrate across the 660-km discontinuity. However, if these geochemically distinctive materials became part of the main upper-mantle convective regime, their isotopic heterogeneity will be short lived compared with the time-scales over which they were created.

In crude terms, Figure 9.5 summarizes the characteristics of upper-mantle melt sources as we now understand them. Of course, the picture we have drawn is necessarily simplified and is under continuous revision; no doubt it will be refined further during the lifetime of this book. Hopefully, those readers interested only in the broad picture will find the above discussion adequate, while those wishing to probe further will find it a useful starting point.

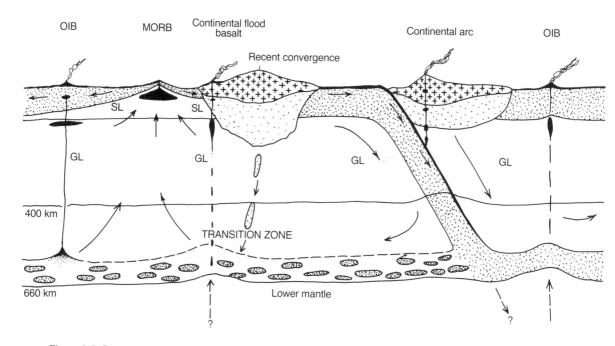

OIB   MORB   Continental flood basalt   Continental arc   OIB

Figure 9.5 Cartoon illustration of the upper *c.* 800 km of the Earth, indicating possible sources for OIBs, MORBs, continental flood basalts and arc andesites (the latter two to be discussed in Chapters 10 and 11). Arrows indicate the directions of relative lithospheric plate motion and (simplistically) the convective regime in the upper mantle. Possible regions of material exchange between the upper and lower mantle are indicated by broken arrows. Zones containing a small amount of melt are shown solid; sub-continental mantle lithosphere is lightly stippled; sub-oceanic mantle lithosphere is heavily stippled. GL = garnet lherzolite; SL = spinel lherzolite. (After Davies *et al.*, 1989 and Floyd, 1991.)

SUMMARY

1. The 660 km discontinuity in the Earth's mantle is most likely to act as a partial barrier to convection. Seismic data suggest that while some subducting plates are deflected, others penetrate this barrier, a view supported by computer modelling of hybrid convection. Similarly, some rising hot plumes may originate in the lower mantle, penetrate the 660 km discontinuity, and supply both heat and material for oceanic hot spot and continental flood basalt volcanic activity (Fig. 9.1).

2. The D″ layer is a thermal boundary layer where heat is transferred from the core to the mantle. It may be chemically different from the rest of the mantle, perhaps an 'unclean' chemical boundary layer, and almost certainly it gives rise to thermal instabilities in the mantle which may be the source of plumes reaching the surface.

3. Analysis of the trace-element and isotopic characteristics of mantle-derived magmas suggests that:

   (1) the upper mantle is extremely depleted in the most incompatible elements relative to Bulk Earth (Fig. 9.2), and that this is a consequence of continental crust formation; and

   (2) the upper and lower mantle are largely chemically isolated, possibly with the exchange of a few per cent of the upper mantle mass per Ga.

   Thus the lower mantle has a chemical composition close to Bulk (silicate) Earth, whereas the upper mantle may retain only 10% of the highly incompatible element concentrations of Bulk Earth.

4. The trace-element and isotopic contrasts between MORBs and OIBs (Figs 9.2–9.4) reflect either:

(a) source differences, with the OIB sources being enriched in incompatible elements prior to or during partial melting; and/or

(b) different (small) degrees of melting in geochemically similar sources.

Model (a) is consistent with magma mixing following enrichment of OIB sources by upwelling plumes from the lower mantle, though recycled lithospheric material may also play a significant role. Figure 9.5 presents a simplified summary of the upper mantle, based on the available geophysical and geochemical data.

FURTHER READING

General journals:
Vink *et al.* (1985): hot spots.
White and McKenzie (1989a): plumes.
Advanced journals:
Hart (1988): geochemical heterogeneities in the mantle.
Hergt and Hawkesworth (1992): remobilization of continental mantle lithosphere.
Hill *et al.* (1992): relation of plumes to mantle tectonics.
Larson (1991a) and (1991b); Larson and Olson (1991): superplumes.
Olson *et al.* (1990): review of mantle convection and implications for evolution.
O'Nions (1987): convection and geochemical mass balance in source regions of source regions.
Stixrude and Bukowinski (1990): the lower mantle and D" mineral phases.
White, R.A and McKenzie, D.P. (1989b): plumes.

10 The continental crust

10.1 THE NATURE OF THE EARTH'S CONTINENTAL CRUST

Occupying 41% of the Earth's surface but only 0.7% of the Earth's volume, the continental crust is the most accessible and best studied part of our planet. But because it is by far the most complex part in both its physical and its chemical nature, much remains enigmatic. It contains a wide range of rock types from the relative purity of many sedimentary rocks to the complex chemical mixtures of igneous granites and metamorphic granulites, two of the most common rock types. In thickness it is also variable, ranging from as little as 20 km below regions of **plate extension**, such as the Basin and Range Province of the western USA, up to 70 km beneath the Andes and 90 km beneath the Himalayas, which are both regions of **plate convergence**. Overall, however, the continental crust has an average thickness of 36 km compared with just 6 km for the oceanic crust (Section 7.4). Much of the complexity of the continental crust stems from its long history, around 4000 Ma, during which time its rocks may have experienced many tectonic and/or volcanic episodes. The history of the continental crust is examined in Chapter 11; here we concentrate on a description of the crust and crustal processes as they are today, aiming to answer these questions:

1. What are the major surface and in-depth features?
2. What rock types are involved, how did they originate and become differentiated both vertically and horizontally?
3. To what extent can these features and rock types be accounted for by the tectonic and magmatic processes we observe today, which occur principally in zones of plate convergence and extension?

As with earlier chapters on the mantle and oceanic crust, our picture of the composition and evolution of the continental crust depends on a combination of:

1. seismic data to reveal the layered structure;
2. experimental petrological results that help us to understand the pressure–temperature–time evolution of metamorphic and igneous rocks; and
3. observations on the composition and physical state of crustal rocks at the surface, especially in relation to tectonic setting.

Near-surface rocks are relatively cool, and pressures are much lower than in the mantle, and therefore the rheological behaviour of crustal rocks subjected to stress is very different to that of rocks at depth. Instead of slow creep, near-surface rocks may fracture, resulting in faulting and earthquakes. Both horizontal and vertical movements may occur, the latter allowing deep crustal rocks to be exhumed by erosion, or shallow rocks to be buried. So we start our examination of the crust with a description of the materials present and their strength. Later sections of this chapter take a more 'dynamic' view of the relationship between tectonism and magmatism in the different modern environments. Then in Chapter 11 we will extend this picture to appreciate the way in which the continental crust has grown and changed with time as the rate and style of geological processes have evolved.

10.2 CRUSTAL STRUCTURE, COMPOSITION AND RHEOLOGY

10.2.1 Major surface features

The continents and ocean floor contrast strongly in both topography and age. Figure 10.1 shows that the topographical range extends from the highest Himalaya (maximum 8848 m) to the Marianas ocean trench (10 912 m), but within this range the crustal levels are distributed bimodally, with the continents (average height 0·12 km) being distinct from the ocean basins (average depth 3·8 km). This bimodal topography occurs because the continental lithosphere

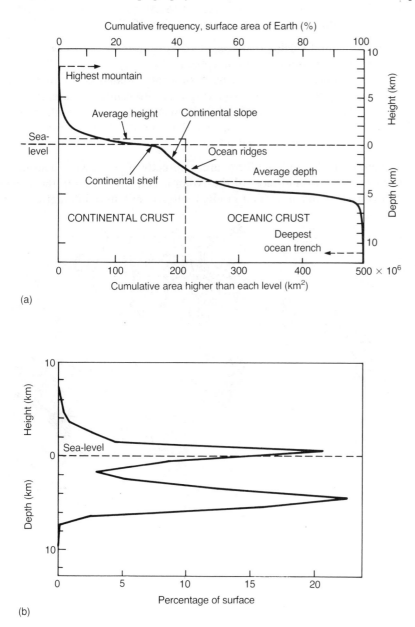

(a)

(b)

Figure 10.1 The distribution of levels in the Earth's solid surface (a) on a cumulative basis and (b) showing the percentage of the crust lying in each 1 km elevation interval (known as a hypsometric curve). The total area of the Earth's surface is about 510×10^6 km^2, the average height of continental crust above sea-level is 120 m and the average depth of ocean basins below sea-level is 3·8 km. (Data from Cogley, 1984.)

187

Continental crust contains a much greater thickness of low-density silica-rich rocks than does the oceanic lithosphere with its thin basaltic crust and so, because of isostatic compensation in the asthenosphere (Section 8.2), the continents stand proud. Oceanic areas principally take the form of large flat-bottomed basins that are traversed by 2 km high ridge systems (Figs 8.12 and 10.2). Around their margins, the ocean basins have **trenches** adjacent to destructive plate margins – as around most of the Pacific – or broad **continental shelves**, up to 200 km wide, where subduction is not taking place – as around most of the Atlantic. For simplicity, we shall refer to these as the **active** and **passive** edges of the continents.

The continental crust is more complex because most of it is older than oceanic crust, and it has frequently been reprocessed (Sections 10.3 and 10.4). At the surface, there are three main components (see Fig. 10.2):

1. exposed continental **shields** that consist of Precambrian (>570 Ma) crystalline igneous and high-grade metamorphic rocks;
2. continental **platforms**, with a flat-lying or gently folded cover of younger metasedimentary rocks, usually overlying more Precambrian basement; and
3. young, mainly Tertiary (<70 Ma) **fold mountains** that may contain older, deformed, metasedimentary rocks, and that almost always contain young igneous rocks, both volcanic and intrusive.

The continental shield and platform areas directly abut passive continental edges, such as those around most of the Atlantic (Fig. 10.2). Here the *surface* transition from ocean to continent is gradual, often over tens of kilometres of the

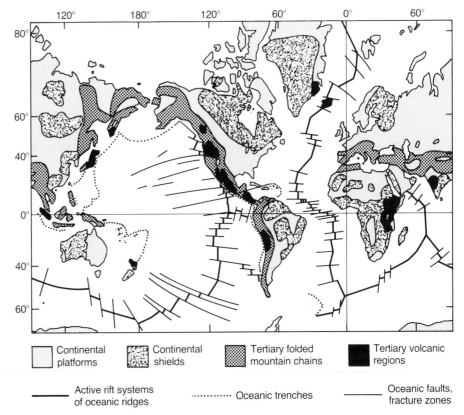

Figure 10.2 Major features of the Earth's continental crust in terms of their geological age and surface structure. See also Figure 8.14 for plate definitions. (After Wyllie, 1971.)

Continental platforms Continental shields Tertiary folded mountain chains Tertiary volcanic regions

Active rift systems of oceanic ridges Oceanic trenches Oceanic faults, fracture zones

188

continental slope. Passive margins also have characteristic continental shelf zones which were produced by marine erosion during the recent (Pleistocene) glacial period of lower sea-levels. These are the margins of former intracontinental extensional basins that developed to form ocean basins (see Section 10.4). In contrast, active margins such as the Pacific 'ring of fire' are marked by volcanic mountains. These include continental volcanic arcs, such as the Andes, and ocean island arcs, such as the Aleutians and Philippines of the northern and western Pacific (Fig. 10.2). Another prominent young mountain belt forms a line extending east from the Mediterranean through the Alps and the Himalaya. Here, igneous activity is less important, but the degree of folding and deformation is extreme – huge slices of continental crust have been thrust over one another, leading to considerable crustal shortening. So there are marked geological contrasts between the products of ocean–continent or ocean–ocean convergence (volcanic arcs), and those of continent–continent convergence (folded and deformed mountains) – see Section 10.3 for further details.

10.2.2 Vertical structure and principal rock types

The pattern of continental crustal thickness (Fig. 10.3) tends to reflect the surface crustal components (Fig. 10.2), for the depth of the Moho mirrors the surface topography in an exaggerated way. Thus, the thickest crust occurs under the continental mountain ranges as would be expected from the Airy model of isostasy (Section 8.2). Above the Moho, the continental crust has a layered structure similar to the oceanic crust (Fig. 7.7), but the thickness and junctions between the layers are often far less well defined. However, Figure 10.4 shows that a four-layer crust is common. Even the most sharply defined boundary, the Moho, where velocities increase from $c.7·5$ to 8 km s^{-1} (cf. Chapter 2), is not always a simple two-dimensional plane but may be a zone varying from less than 1 km to several kilometres in thickness. The seismic discontinuity in the mid-crust, at depths ranging from 10 to 30 km, is usually known as the **Conrad discontinuity** after its discoverer. It can be traced, again often as a diffuse zone under many continental regions, separating the two upper crustal layers from the two layers of the lower crust.

The high-velocity basal layer varies most in thickness between different tectonic environments (Figs 10.2 and 10.4). In most continental shield and platform areas it is 5–10 km thick. But it is not necessarily thicker in areas of thick crust for, whereas it may reach 20 km in island or continental arcs (IA in Fig. 10.4), it is often only a few kilometres thick in young **orogenic belts**, regions formed by recent continental convergence. Both overall and lower crustal thicknesses are diminished beneath continental rifts, and the lower layer is

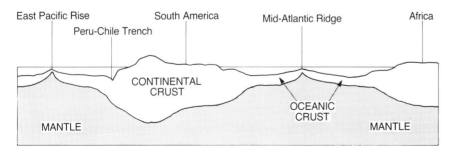

Figure 10.3 Relative thickness of oceanic and continental crust in a cross-section from the East Pacific Rise to Africa. Note the crustal thinning under ocean ridges and the thickening under continents. (Schematic diagram with approximate vertical exaggeration of 1 cm = 25 km.)

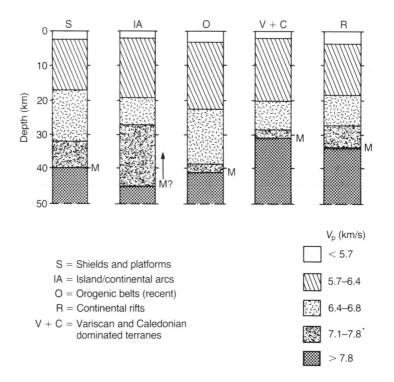

Depth (km)

S = Shields and platforms
IA = Island/continental arcs
O = Orogenic belts (recent)
R = Continental rifts
V + C = Variscan and Caledonian
dominated terranes

V_p (km/s)

☐ < 5.7
▨ 5.7–6.4
▨ 6.4–6.8
▨ 7.1–7.8˙
▨ > 7.8

Figure 10.4 Vertical structure of the continental crust, as revealed by seismic refraction and wide-angle reflection surveys, according to tectonic province. See text for description and comments on the probable rock types in each layer. M = Moho. (After Mooney and Meissner, 1991.)

particularly thin beneath the Variscan and Caledonian crust of western Europe (Chapter 11) where the last tectonic event was also extensional. The depth of the Conrad discontinuity (the 6·4 km s^{-1} V_p boundary in Fig. 10.4) and particularly the depth of the Moho have been confirmed by deep seismic reflection studies that have become increasingly fashionable since the mid-1980s. In summary, the important distinguishing features that relate to tectonic environment are the overall crustal thickness and the thickness of the high-velocity lowermost crustal layer.

Turning to the likely compositions of crustal layers, the top layer is usually the most variable, with the uppermost few kilometres of material ranging from relatively unmetamorphosed volcanic or sedimentary rocks (V_p generally less than 4·5 km s^{-1}) to medium-grade metasediments, such as quartzites and greenschists (V_p generally less than 4·5–5·5 km s^{-1}). Beneath this shallow layer, the main part of the upper crust (second layer in Fig. 10.4) is often regarded as 'granitic' in composition. However, acid plutonic rocks of the granite family vary considerably from intermediate silica **diorites** (about 55–60% SiO$_2$) to the rather less common **granites** *sensu stricto* (>70% SiO$_2$). It is more likely that the *average* composition of the upper crust is **granodiorite** or **tonalite** (i.e. quartz diorite) with an average P-wave velocity of 6·25 km s^{-1} (see Sections 10.2.3 and 10.3).

The composition of the lower continental crust is less well known because of the relatively few places where such rock types are exposed. Historically, it was regarded as basaltic, but this idea became less popular with the realization that a basaltic lower crust would transform into dense **eclogite** (Section 7.2.1) because of the pressure, and such rocks would have much too high a seismic velocity, around 8·4 km s^{-1}. However, rocks with less mafic compositions but in the eclogite **metamorphic facies** (i.e. formed under the same conditions of

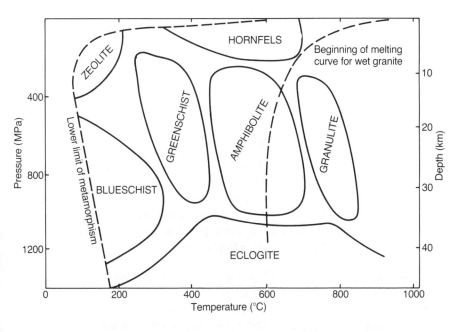

Figure 10.5 Experimentally
determined pressure–
temperature scheme for
metamorphic facies in which
characteristic minerals occur
across a range of possible
rock compositions. For
example, felsic and mafic
granulites both contain high-
temperature (Ca) feldspars
and high-temperature (Mg)
pyroxenes, but in different
proportions, along with other
minerals. Also shown is the
beginning of the melting
curve for wet granite. (See
text for further details; also
Thompson, 1992 for a more
advanced analysis of
metamorphism.)

pressure and temperature – Fig. 10.5), could be present, but still would have
rather a high seismic velocity. A second possible rock type for the lower crust is
granulite, which forms over a range of high-pressure and high-temperature
conditions (Fig. 10.5). Note that the term 'granulite' refers to a metamorphic
P–T facies rather than to a specific rock composition. Thus granulites may vary
from felsic to mafic, with different proportions of their constituent minerals,
mainly magnesium-rich pyroxenes and calcium-rich feldspars. Experimentally
determined P-wave velocities on granulite samples at 1000 MPa pressure have
produced values between 6·4 and 7·5 km s^{-1}, increasing towards the most mafic
samples. Yet a third possibility is that much of the lower crust lies in the field of
amphibolite-facies metamorphism. This would imply rather lower temperatures
(Fig. 10.5) and the presence of relatively hydrous rocks with amphibole and
subordinate feldspar. Amphibolites have appropriate seismic velocities (up to
7·1 km s^{-1}) for much of the lower crust, but only if their compositions lie within
the intermediate–mafic range (45–60% SiO_2).

Support for the occurrence of mainly granulite with some amphibolite-facies
rocks in the lower crust comes from their occurrence in:

1. the most deeply eroded crustal regions of Precambrian age;
2. xenoliths in volcanic vents brought up by magmas passing through the lower
   crust; and
3. experimental data indicating that, during the formation and emplacement of
   granite magmas into the upper crust, the lower crust will be depleted of water
   and a silicate melt fraction, thus leaving only the refractory pyroxenes and
   feldspars of granulite (Section 10.3).

An analysis of the depths at which crustal granulite xenoliths formed, by
geobarometry (involving the use of pressure-sensitive mineral chemistry),
suggests that felsic–intermediate granulites give way to mafic granulites in the
deep crust (Bohlen and Mezger, 1989). Indeed, the addition of mafic magmas to
the lower crust, which then crystallize in granulite-forming *P–T* conditions,

191

increasingly is recognized as an important control on magma genesis in the crust (see later sections).

In summary, the continental crust has a four-layer seismic structure: 1. a thin sedimentary and metamorphic layer overlying; 2. a major upper crustal zone where the most likely average composition and rock type is granodiorite. The lower crust (3) and (4) is mainly granulite, but of felsic to intermediate composition giving way to more mafic rocks at depth.

10.2.3 Chemical composition of the crust

Early attempts to determine the crustal chemical composition used weighted averages of all the varied rocks that have been collected at the Earth's surface. But this is invalid on two counts. Firstly, the deeper layers were both poorly known and under-represented and, secondly, little attention was paid to the oceanic crust. Accurate estimates became possible only when the granulitic nature of the lower continental crust was recognized and when the ocean crust was explored by drilling.

Table 10.1 gives a breakdown of crustal composition by layer and by location. The surface processes of crystalline rock weathering, erosion, transportation of dissolved and solid material, followed by deposition in selected environments lead to extremely diverse sedimentary rock compositions. A useful approach is to make use of natural sampling processes by intercepting material transported by rivers. The average particulate compositions being carried by the world's major rivers are quite similar, and have the chemistry of fine-grained (i.e. well-mixed and homogenized) shales. To this must be added the soluble component which ends up mainly in limestones, leading to the average given in Table 10.1 for the sedimentary–metamorphic layer, which comprises 8% by mass of the continents. The main upper-crustal layer is available for sampling in areas where metasedimentary rocks have been removed by erosion and, today, this is commonly granodiorite. However, estimates of crustal compositions are model dependent and it is impossible to assess true compositions without anticipating later arguments. Thus, rather than opt for a granodiorite layer, in which SiO_2 would be 67% and K_2O nearly 3%, we have moved some distance towards a tonalite composition. This reflects the fact that tonalites are particularly common in early Precambrian upper crustal zones, and that most of the continental crust was formed at this time (Chapter 11). Lower-crustal compositions are even more elusive, but rocks described as mafic granulites and amphibolitic gneisses are taken as typical (Table 10.1). Their incompatible-element concentrations are usually low because of previous partial-melt extraction.

Combining these estimated compositions for the different continental layers by weighting them according to their estimated abundances from seismic data (Fig. 10.4), produces the total continental average in the penultimate column of Table 10.1. Previous estimates had SiO_2 values ranging from 57 to 65%, with K_2O at 1·1 to 4·2%. Our estimate agrees quite well with that of Taylor and McClennan (1985 – last column, Table 10.1) which is based on 75% Archaean average crust plus 25% 'model' andesite, the latter to reflect post-early Precambrian additions to the continents. This model has a rationale based also on trace-element data (as discussed below). Notice that whereas the overall composition of the ocean crust (Section 7.4 and Table 10.1) is mafic (basaltic), *the overall composition of the continental crust is intermediate (i.e. 55–64% SiO₂)*

Table 10.1 The chemical composition of the Earth's crust and its component layers. Data are in per cent by mass of the eight principal major-element oxides rounded to 100% in each column (excluding volatiles). This approximation is greatest in the top layer of continental and oceanic crust, both of which have CO_2 fixed in limestones. (Adapted from Ronov and Yarnoshevsky, 1969; Taylor and McLennan, 1985.)

% mass	Oceanic crust				Continental crust				
	Layer 1	Layer 2	Layer 3	Total oceanic	Sedimentary -metamorphic	Tonalite granodiorite	Granulites	Total continental	Taylor and McLennan continent
	3	16	81	100	8	44	48	100	
SiO_2	50·2	50·2	49·9	50·0	56·0	64·2	54·4	58·9	57·3
TiO_2	0·7	1·2	1·5	1·4	0·8	0·6	1·0	0·8	0·9
Al_2O_3	14·0	16·0	17·3	17·0	16·8	16·3	16·1	16·2	15·9
FeO	6·9	9·2	10·2	9·9	6·1	5·8	10·6	8·1	9·1
MgO	3·7	6·9	8·4	8·0	3·5	2·6	6·3	4·5	5·3
CaO	20·6	13·2	9·9	10·8	12·7	4·6	8·5	7·1	7·4
Na_2O	1·4	2·2	2·7	2·6	1·8	3·4	2·8	3·0	3·1
K_2O	2·5	1·1	0·1	0·3	2·3	2·5	0·3	1·4	1·1

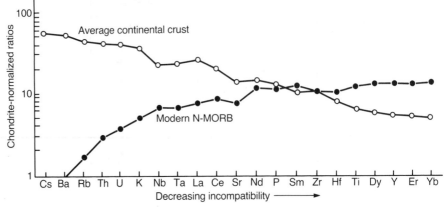

on almost all estimates. Compared with Bulk Earth (Table 7.1) this represents a very considerable differentiation by partial-melting processes to produce the Earth's silica-rich continental crust. Moreoever, comparisons between the undepleted upper-mantle analysis of Table 7.1 and the oceanic and continental crust analyses of Table 10.1 show that the continental crust is by far the more strongly enriched in lithophile elements.

A remarkable property of geochemical systems is that wherever major-element enrichments have been produced, incompatible trace elements are usually even more enriched. So it is with continental chemistry, which shows about 40–50 times enrichments of the six most incompatible trace elements with respect to chondrites and, therefore, Bulk Earth (Fig. 10.6). This is amplified in Table 10.2, where the concentrations of some important trace elements in

Table 10.2 The concentrations of some critical trace elements in the Earth's crust and mantle. All data in parts per million. (Sources: Taylor and McLennan, 1985; O'Nions, 1987; Sun and McDonough, 1989)

	1 Undepleted mantle	*2* Ocean crust (N-MORB)	*3* Continental crust	*4* Continental enrichment
Rb	0·7	1	32	46
Th	0·08	0·15	3·5	44
U	0·02	0·05	0·91	46
Sr	21	150	260	12
Nd	0·97	10	16	17
Sm	0·32	3·1	3·5	11

undepleted mantle, continental and oceanic crust are shown together with the continental enrichment factor. Two important conclusions follow from these data. Firstly, since the continental crust is about 0.5% by mass of the Bulk (silicate) Earth, enrichment factors of 40–50 imply that 20–25% of the Earth's total content of these elements now resides in this thin continental layer. Secondly, unlike modern N-MORB (Section 9.3), continental trace elements follow the predicted pattern of stronger enrichments in the direction of increasing incompatibility.

Reference to Figure 9.2(a) reveals that, starting with a mantle containing Bulk Earth trace-element concentrations, an absolute maximum melt fraction of only 2% is allowable to generate the continents. However, we know from Section 9.3.1 that the *modern* upper-mantle N-MORB source region must be strongly depleted in highly incompatible elements relative to Bulk Earth. We now see that this is a consequence of continent formation, most of it early in the Earth's history (Chapter 11), which scavenged these elements to give the modern mantle concentration patterns.

In summary, there are two observations on continental geochemistry for which we must account. *Firstly*, we shall need to explain how continental trace elements have come to resemble those of small fraction mantle melts which must be essentially mafic (i.e. basaltic). Given an initial upper mantle with Bulk Earth geochemistry, these magmas would have had a time-average geochemical pattern resembling the continental rather than the modern N-MORB data in Figure 10.6. *Secondly*, we need to explain how the overall major-element chemistry of the continents has come to be intermediate rather than mafic. Indeed, the most likely source of intermediate magmas is by 10% or more melting of mafic material (i.e. large melt fractions). We shall start to resolve this apparent paradox by examining modern processes of magma generation in zones of crustal tectonic activity (See Sections 10.3 and 10.4).

10.2.4 Continental lithosphere rheology

As explained in Section 8.2, the rheological behaviour of a material depends upon several factors, notably temperature, magnitude of the stress, and duration of the stress. Within the continental lithosphere, the stresses are usually not large, but they do become magnified at plate boundaries. The increase of temperature with depth causes the rheological behaviour to vary. Near the surface, where the temperature is low, the material is brittle so it extends or compresses by displacements along faults. Below this it is elastic and so the stress causes it to develop a strain which it will support indefinitely, unless the stress is increased. Deeper again, the higher temperature ensures that creep permits the material to be ductile. Now suppose that the whole lithosphere sheet is extended or compressed by a steady horizontal force: all three zones will deform by the same amount but, because the brittle and ductile parts yield, no stress can be developed in them, and so the force is opposed only by the stress developed in the elastic zone (Fig. 10.7, solid line). Notice how the stress has been *concentrated* or amplified, into this elastic zone, just as it would be into, say, a sheet of rubber sandwiched between clay. If the stress becomes too large the elastic zone will not be able to sustain it and **whole lithosphere failure** will ensue (Fig. 10.7, dashed line).

In the actual lithosphere, the development of these rheological zones is more complicated because there are layers of different materials. Space does not allow us to introduce more advanced models, and so the interested reader is referred to Park (1988). However, an important concept we shall require here is that of **critical stress**, which is the value of overall lithospheric stress required to produce whole lithosphere failure in 1 Ma. Critical stress varies with temperature: the higher the temperature, the smaller is the stress needed for failure to occur. This is illustrated in Table 10.3, where we have divided the lithosphere

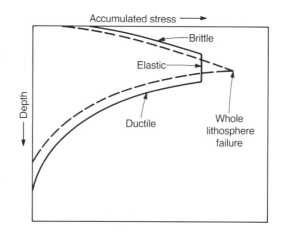

Figure 10.7 Schematic representation of stress accumulation across an elastic layer (solid line), when zones above and below are yielding by brittle and ductile deformation. As time or stress magnitude increases, the regions of brittle and ductile deformation coalesce (dashed line) and the previously elastic layer starts to fail. (After Park, 1988.)

into four main age provinces with typical values of heat flow. Continental heat flow declines with increasing age, and so lithosphere strength increases. The youngest tectonic provinces are therefore most likely to fail. Notice also that rocks deform more easily under extension than compression, and this is reflected in the two sets of critical stress values.

How do these critical stress values compare with the actual stresses developed in the lithosphere? Most of the forces are directed horizontally and are the same ones that drive the plates (Fig. 8.13). These forces produce a strain in the form of plate movement; any resistance leads to a *within-plate stress* being developed, but otherwise the forces are dissipated at plate boundaries. The maximum likely value of net within-plate stress is thought to be *c.*25 MPa whereas horizontal stress magnitudes at a convergent boundary may rise to *c.*80 MPa. So where two continental plates, moving in opposite directions, converge and are not subducted, permanent deformation will occur in lithosphere of average to above-average heat flow (Table 10.3). This results in the uplift of fold mountains

Table 10.3 Critical stress values and heat flows required for permanent deformation (whole lithosphere failure) of the continental lithosphere in 1 Ma.

Heat flow (mW m^{-2})	Critical extensional stress (MPa)	Critical compressional stress (MPa)	Usual types of lithosphere
100	8	25	tectonically active
75	20	40	<250 Ma average age
60	65	100	average continent (e.g. Palaeozoic crust)
40	>200	>200	Precambrian shields

(Section 10.3.4). In contrast, within-plate stresses rarely cause permanent compressional deformation but are capable of extending young lithosphere. Such stresses may tear apart continents to cause first the formation of sedimentary basins and then, later, new oceans (see Section 10.4).

In summary, the important points are:

1. that stress magnitudes approaching 80 MPa may operate across lithospheric convergence zones where they cause uplift of fold mountains in zones of continental collision;
2. that within-plate compressional stresses are rarely adequate to cause permanent deformation; but
3. that within-plate extensional stresses will cause permanent deformation in continental lithosphere of above-average heat flow if they are maintained for more than 1 Ma (Table 10.3).

10.3 PROCESSES IN ZONES OF LITHOSPHERIC CONVERGENCE

10.3.1 The contrasting styles of continental accretion

Earlier, we noted that continental thicknesses in volcanic and folded mountains (Fig. 10.2) range up to more than twice the average value of 36 km. Here the distinction between 'Andean' ocean–continent boundaries (Fig. 10.8) and 'Alpine–Himalayan' continent–continent boundaries (Fig. 10.9) is discussed in terms of their plate-tectonic evolution. The simplest means of generating continental material occurs at the island arcs of ocean–ocean convergence zones, as around the modern western Pacific, and these arcs eventually are swept up by larger continental masses (Section 10.3.3). The *magmatic* processes (Section 10.3.2) in continental Andean-style arcs are analogous, and there is a progressive evolution from their occurrence around the margins of a closing ocean (Fig. 10.9, Stage 1), through to Himalayan-style mountains once the ocean is closed (Fig. 10.9, Stage 4). Crustal shortening is very evident in such regions, and igneous activity, though present, is less significant than at the earlier 'Andean' stage.

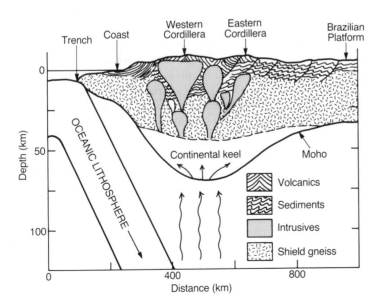

Figure 10.8 Schematic east–west cross-section through the central Andean ocean–continent convergence zone, illustrating possible features of crustal structure, including the thick continental 'keel'. Arrows indicate direction of magma and volatile streaming up from the downgoing plate and mantle 'wedge' between this plate and the continent. The keel itself consists of mafic, mantle-derived intrusive rocks which thicken the crust from beneath, together with some pre-existing deep crustal material. (After Brown and Hennessy, 1978.)

197

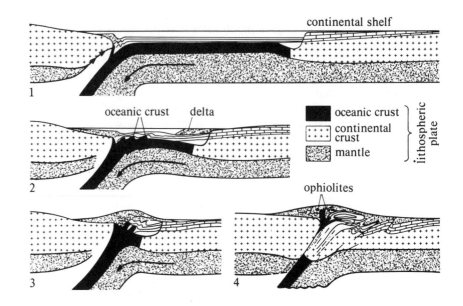

Figure 10.9 Sequence of events leading to a continent–continent collision with subduction on one margin only. Note that sediments accumulated on the ocean floor are caught up in the collision, together with ophiolitic fragments of oceanic lithosphere. Outside the field of view, to the left, is a continental magmatic are that would be active during stages 1–3. (After Toksöz, 1976.)

The magmatic arcs of active continental margins are characterized by andesitic volcanoes and vast linear 'granitic' batholiths, thousands of kilometres long and perhaps 100 km wide, parallel to the margin, together with variable amounts of clastic sediments. These are not 'fold' mountains in the Alpine–Himalayan sense; instead, the crust thickens mainly by igneous intrusion, both mafic intrusions from the mantle into the continental keel and by intermediate–felsic intrusions at higher levels (see Section 10.3.2). Some extension of the crust occurs across the zone of active magma emplacement, thinning the pre-existing gneissic basement and allowing space for the new intrusions. Overall, it seems that the crust thickens at ocean–continent, and similarly at the island arcs of ocean–ocean convergent boundaries, more by *vertical additions* than by shortening due to *lateral compression*. Thus *new material is being added to the continental crust at so-called destructive plate boundaries*, largely by magmatic processes. The large volumes of continental erosion products that accumulate mainly in the shallow ocean basins may be accreted back to the continents at their leading edges as a fore-arc **accretionary prism**, though some may be subducted and thereby form a return flow of continental material into the mantle. Magmatic activity through subduction can last only so long as the oceans themselves and, once a **continental suture zone** of the Alpine–Himalayan style has formed, activity must cease, probably to be renewed at a different location around some younger ocean. Magmatic arcs are developed around shrinking oceans prior to suturing, and therefore it follows that they tend to flank one or both sides of the intracontinental fold mountains. Many examples of ancient sutures are now recognized (see Chapter 11), for example, the Ural mountains, a great north–south mountain chain across central Russia, and the Caledonian–Appalachian mountains of Scandinavia, Britain, eastern Canada and USA (a single mountain chain pre-dating the modern Atlantic).

10.3.2 Magma genesis and temperatures in the continental lithosphere

Here, our first aim is to develop a model for magma production at destructive plate margins. As for the mantle (Chapter 7) we use the results of melting experiments in relation to pressure–temperature variations in the lithosphere.

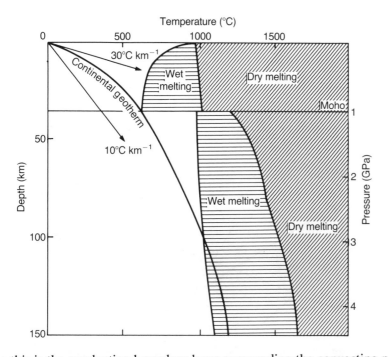

Figure 10.10 Illustration of the variation of temperature with depth, the geotherm, in typical modern continental lithosphere with average heat flow. Higher temperature gradients near the surface reflect the selective concentration of radio-isotopes of K, U and Th in the upper crust. Lower gradients below 125 km depth reflect the onset of mantle convection. Also shown are the beginning of melting curves for wet and dry felsic continental crust, above the Moho, and for wet and dry mantle lherzolite beneath (see Figs 7.5 and 7.6). Increased density of shading indicates the temperature intervals across which wet and dry melting takes place to produce granitic (*sensu lato*) magmas in the crust and basaltic magmas in the mantle. (After Brown, 1985 and Wyllie, 1992.)

Since this is the conducting boundary layer surrounding the convecting mantle, in a province with average heat flow (thickness 125 km, basal temperature 1360°C), the average thermal gradient is $c.11°C\ km^{-1}$. Now the gradients measured in mines and shallow continental boreholes are much higher than this, typically 20–30°C km^{-1}. The reason for this discrepancy is that continental temperature gradients are strongly augmented by the high levels of **heat-producing radio-isotopes**, isotopes of the incompatible elements K, U and Th that occur in the shallow crust. So the **continental geotherm** is strongly curved in the crust, giving higher temperatures near the surface than would occur with a linear 11°C km^{-1} gradient (Fig. 10.10).

Also shown in Figure 10.10 are the wet and dry melting curves for the production of felsic (granodiorite–tonalite) magmas in the crust and mafic magmas in the sub-continental mantle lithosphere. Notice that the continental geotherm barely grazes the field of crustal melting, and since all the geological evidence points to the fact that lower crustal rocks are dominantly dry, refractory granulites (Section 10.2.2), it is extremely unlikely that crustal melting will occur spontaneously. However, a rising body of hot magma from mantle depths may perturb the geotherm significantly, inducing partial melting in the crust. So, *one simple and widely accepted model for crustal melting is that mafic magmas must already be rising from subcrustal depths*. However, in the particular case of continent–continent convergence zones, crustal temperatures may be increased to melting conditions by friction and abnormal radiogenic heating, where sedimentary layers rich in K, U and Th are thrust over one another in the brittle deformation zone (see Section 10.3.4).

The need to account for the continuous supply of mantle-derived magmas to oceanic and continental magmatics arcs throws the spotlight on to the deeper part of Figure 10.10. Here we see that unless large amounts of volatiles are present, temperatures are not high enough for the mantle to melt. Of course, Figure 10.8 gives the game away and demonstrates that the key to understanding arc magmatism lies in pursuing the fate of subducted oceanic lithosphere as it is

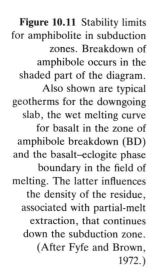

Figure 10.11 Stability limits for amphibolite in subduction zones. Breakdown of amphibole occurs in the shaded part of the diagram. Also shown are typical geotherms for the downgoing slab, the wet melting curve for basalt in the zone of amphibole breakdown (BD) and the basalt–eclogite phase boundary in the field of melting. The latter influences the density of the residue, associated with partial-melt extraction, that continues down the subduction zone. (After Fyfe and Brown, 1972.)

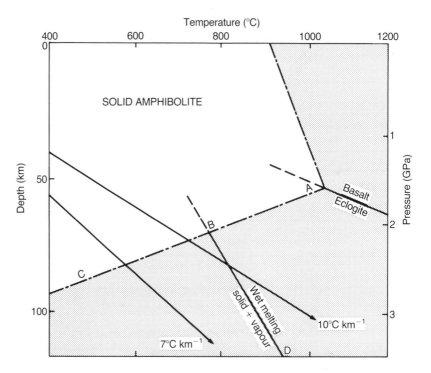

drawn down beneath the overriding plate edge. Typically, this comprises an upper 5–10 km of mainly basalt–gabbro crust, overlying depleted peridotite. The latter is unlikely to melt because it is already the residuum after partial melting in an oceanic extension zone. But what about the ocean crust? It is characteristically hydrothermally altered by interaction with sea-water, and so it contains many hydrous minerals such as zeolites and amphiboles. As temperatures rise during subduction, metamorphism will ensure that most of the remaining pyroxene–feldspar mineralogy is converted to amphibole. So the problem now reduces to the question of amphibolite stability: it breaks down to release water in the shaded area of Figure 10.11. The prominent change of slope on the amphibole breakdown line occurs where the basalt crosses into the eclogite metamorphic facies (Fig. 10.5), with a marked increase in material density. Since we are plunging a cold oceanic plate into the mantle, only the high-P, low-T part of the diagram is relevant. The downgoing plate is heated by conduction from the surrounding hotter mantle and by friction at its boundaries, giving a thermal gradient down the subduction zone of 7–10°C km^{-1}. Now the line BD in Figure 10.11 indicates where, with increasing temperature (left to right), wet melting of amphibolite will occur, returning mafic melts. At high thermal gradients, amphibolite dehydrates across AB, and the products are a basaltic andesite melt plus a depleted eclogitic residue. At lower thermal gradients, where amphibolite dehydrates across BC, the products are water vapour and mafic eclogite. The water vapour penetrates the overlying mantle wedge between the subducted plate and overlying crust (Fig. 10.8), carrying with it mobile chemical species, mainly the most incompatible elements scavenged from the dehydrating ocean crust. The mantle wedge is relatively hot and dry, and therefore volatile addition depresses melting temperatures and promotes wet melting at 100–150 km depth (Fig. 10.10).

The presence of a subducting oceanic plate can therefore initiate melting in two ways, either by generating a melt from its own substance – which seems to occur locally when young, hot oceanic crust is subducted – or by releasing a hydrous component which initiates melting in the mantle wedge, the more normal situation. In either case, the bulk of the subducting plate continues its descent as dense refractory eclogite, but the melt phase is buoyant, and rises to invade the crust. Some of these magmas reach the surface in the form of the most 'primitive' are basalts. However, the majority are trapped at the crust–mantle density boundary where they effectively *underplate* the crust with incompatible-element-rich mafic material. In the process they thicken the crust, form the keel in Figure 10.8, and release volatiles and latent heat due to crystallization. The continued addition of hot material to the base of the continental crust is exactly what is required to promote remelting at higher levels in the thickening pile, so generating intermediate and felsic melts. By and large, these melts carry the pre-formed continental incompatible-element signature towards the surface where some are erupted as andesites, silicic ash flows and ignimbrites. But the majority of these magmas are trapped where their buoyancy is reduced at a few kilometres depth to form granodiorite and tonalite intrusions – which are subsequently revealed by downwards erosion of the volcanic superstructure. Their much higher viscosity than for mafic magmas means that these relatively silicic magmas must achieve at least a 10% melt fraction, probably as much as 40% for the most silicic granites, before they can migrate from their deep crustal source regions to the shallow zone of emplacement. They leave behind depleted mafic granulite residues in the deep crust.

In summary, using modern subduction environments as an example, we have seen how the separation of tonalitic upper crust from mafic granulites of the lower crust may have occurred throughout the Earth's history. Moreover, we can begin to understand how the continental trace-element signature is that of small melt fractions from the mantle when the upper crust itself was produced from large melt fractions. But why does the crust have an overall intermediate composition when the material added is mafic? The probable answer is that some of the depleted residues in the lowermost crust become denser than the mantle beneath, partly because of eclogite-facies metamorphism, as well as removal of a light melt. So this material founders into the mantle and is lost, thus preserving the relatively silicic nature of the continents.

The products of arc magmatism vary considerably in space and time; for example, in the Peruvian Andes (e.g. Pitcher *et al.*, 1985) magmatic cycles repeated over almost 100 Ma have followed a progression with time from intermediate to more felsic magmas, both within individual cycles and between successive cycles. Critical elemental ratios, such as K/Na and Rb/Sr, increase with time and also with distance from the trench *across* the magmatic arc. Isotopic data provide useful indicators of magma genetic processes, and Figure 10.12 illustrates that the earliest magmatic suites from North Chile have isotopic characteristics, depleted relative to Bulk Earth, resembling those of modern sub-oceanic MORB sources (Fig. 9.4). This signature is consistent with the magma genesis picture developed above, that the Nd and Sr either were recently derived from the mantle wedge, which would have to be part of the well-mixed upper mantle, or were in the fluid phase derived by dehydration of MORB-like oceanic crust. These early arc rocks are the most mafic, and cannot have been exposed to crustal Rb/Sr or Sm/Nd ratios for very long, or more mature $^{143}Nd/^{144}Nd$ and $^{87}Sr/^{86}Sr$ ratios would have developed.

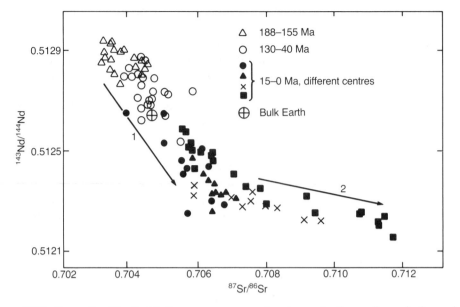

Figure 10.12 Neodymium and strontium isotope initial ratios for intrusive and volcanic rocks taken from a transect across the North Chilean Andes. For discussion of the two trends and interpretation of the data see text. (After Rogers and Hawkesworth, 1989.)

With time, the North Chilean magmas evolve towards the radiogenic Sr and less-radiogenic Nd ratios that are more typical of mature crustal rocks, but in two stages. The earlier trend 1 probably represents the mobilization of Precambrian mantle lithosphere as magmatism progressed east and involved older sub-continental mantle zones. Such lithosphere is capable of preserving much higher Sr and lower Nd ratios from a time when the upper mantle was less strongly depleted and, moreover, can develop such ratios over time if Rb/Sr is high and Sm/Nd is low. The second trend in Fig. 10.12, towards higher Sr ratios, is a crustal melting and contamination (magma mixing) trend, whereby old, isotopically mature rocks *within* the continental crust become involved in producing evolved, silicic magmas. Such mature isotopic trends only tend to occur in arc magmas clearly associated with thick continental crust. The 'ponding' of mafic magmas in the deep crustal keel of such areas (Fig. 10.8), with subsequent partial melting of the crust and mixing between magmas, appears to be an important means of generating evolved intermediate and felsic magmas. This is the **MASH hypothesis** of Hildreth and Moorbath (1988), where the acronym stands for melting–assimilation–storage and homogenization.

Finally, what about the *rate of new magmatic additions* to island and continental arcs? Crustal thickening typically occurs to the extent of between 10 and 30 km during a 100 Ma active cycle, and we shall use an average of 20 km. A characteristic arc width is 100 km, and the total length of active destructive margins is close to 40 000 km. Thus we have a total volume of $20 \times 100 \times 40\,000 = 80$ million km^3 which has added over 100 Ma at *c. 0·8 km^3 per year* (other modern estimates vary between 0·5 and 1·1 km^3 per year). Since the continental volume is about $7·5 \times 10^9$ km^3 it would have taken nearly 10 000 Ma to produce the continents at this growth rate, more than twice the time available. Yet our growth rate is likely to be an over-estimate because we know that continental erosion products are recycled into the mantle via subduction zones, and we postulate that the lowermost eclogitic crust is also removed. Of course, we have focused only on one site of continental growth, subduction zones, and in Section 10.4 will consider magmatic additions in extension zones and above mantle plumes. But even then the firm conclusion

10.3.3 Continental evolution by terrane accretion

So far we have been dealing with processes that occur at plate margins where the motion of the oceanic lithosphere is more or less at right angles to the trench where subduction takes place. But this is by no means universally valid, as a glance at the modern plate vectors adjacent to Colombia (NW corner of South America) and coastal North America reveals (Fig. 8.14). One of the best-studied areas is coastal California where strike-slip motions on the San Andreas Fault are carrying entire crustal blocks northwards relative to continental North America at several centimetres per year (Fig. 10.13). This has led to the fascinating forecast that Los Angeles will have become a new suburb of San Francisco a mere 30 Ma into the future. So an important aspect of continental assembly involves the rearrangement and, ultimately, the welding of continental blocks, including island arcs to a pre-existing continental margin. Such crustal blocks are known as **terranes**; they are usually bounded by faults and have a geological history distinct from adjoining terranes. Terranes come in all shapes and sizes. For example, the whole Indian subcontinent could be regarded as an individual great terrane that, over 80 Ma, has moved about 6000 km northwards from its origin in southern latitudes where it was once joined to Australia and Antarctica. On a much smaller scale, volcanic seamounts and island arcs become accreted to continental margins during ocean closure, and some of these small terrane fragments may have moved very great distances. Taken to its absurd limit, at a rate of just 10 cm per year a wandering terrane could migrate 5000 km across the Pacific in just 50 Ma.

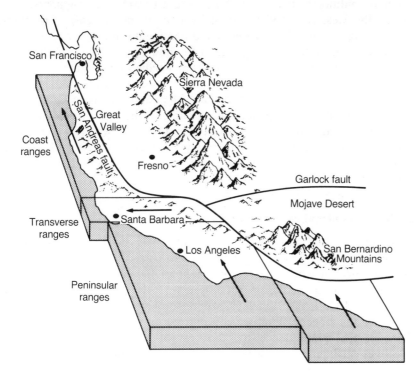

Figure 10.13 Map of the San Andreas fault system in California, illustrating the general northwards movement of lithospheric blocks along the west coast of North America. This is caused by rotation of plate vectors in the NE Pacific associated with intersection of the Pacific ocean ridge with the leading edge of the westwards-migrating North American continental mass.

203

Given this introduction, there is 'little wonder that the continents are patchwork agglomerations of terranes' (Howell, 1985). For example, it is now known that over 70% of the North American Pacific seaboard from California (Fig. 10.13) north to Alaska is made up of discrete terranes, each of which is 'exotic' in relation to its neighbours. Individual terranes are identified by having:

1. *abrupt changes* in rock sequences across major faults, implying very different geological histories in now-adjacent crustal blocks;
2. discontinuities in the *fossil* record: and
3. markedly different *palaeomagnetic directions*, implying derivation from quite separate latitudes.

Most of the terranes of coastal North America appear to have accreted to the ancient cratonic margin during Mesozoic/early Cainozoic times. Strike-slip motions apparently have been perpetuated on the Pacific margin of North America throughout the last 100–120 Ma, such that the north-west corner has become a 'graveyard' for continental fragments. Many of them even originated on the far side of the Pacific; for example, blocks carrying Asian faunas are found as far as 500 km inland from the Pacific coast!

To account for the long history of terrane accretion in NW America we start in the early Mesozoic with a very large palaeo-Pacific ocean. This would have carried large amounts of old oceanic lithosphere waiting offshore to be subducted; intra-oceanic island arcs would have developed and the ocean ridge would have been more central than today. But as the modern North Atlantic opened, North America started to override the Pacific Ocean. The whole of the eastern Pacific has now been overridden and subducted; hence an oceanic transform fault has reached the continental margin, forming the San Andreas Fault. During this process, the eastern Pacific was cleared of older arcs, oceanic plateaux and continental fragments. So how did continental fragments, originally from the western Pacific, eventually end up in the east? Fragments separated during the *initial extension* of the continental lithosphere to form a new ocean, must lie on the *opposite side* of the new ocean ridge from the main continental mass (in this case, on the eastern Pacific plate). Later, any such separated fragments would become swept up by the Americas, moving west. Finally, if terrane accretion due to 'closure' of half an ocean is so significant in the geological framework of North America and elsewhere, we would expect to find similar terrane accretion areas in ancient suture zones where two continental plates have converged. This is precisely what is now being discovered, providing a new understanding of geological discontinuities in intracontinental mountain belts.

10.3.4 Intracontinental convergence zones

The formation of fold mountains (Fig. 10.9) may be thought of as the culmination of lithospheric convergence at a particular site, for the destruction of an ocean forces a wholesale readjustment of the plate regime. Here we examine aspects of the collision process across the world's youngest fold mountains (Fig. 10.14), the Himalaya, where 450 km of crustal shortening has occurred since the Eocene. First we concentrate on the suture zone itself and then look at the consequence for the rest of Asia. Most contemporary compressional earthquakes occur at the Himalayan frontal thrust which is just the latest in a series of suture lines between different terrane blocks across the

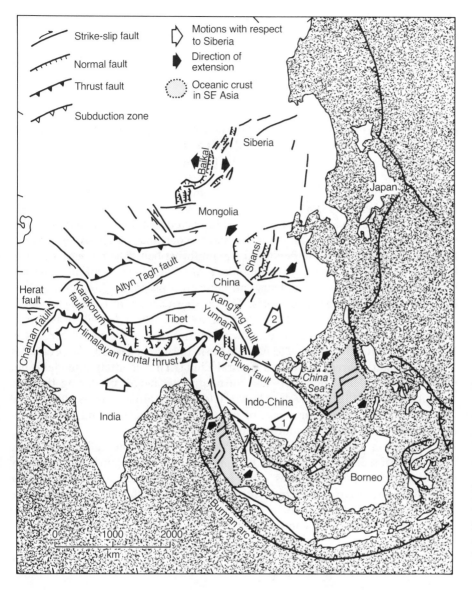

Figure 10.14 Map of tectonic relationships in eastern Asia, illustrating the complex network of faults and block motions initiated by the collision and indentation of India into Asia. The numbers associated with the SE directed arrows indicate phases of easterly block intrusion: 1 at 50–20 Ma before present, and 2 since 20 Ma ago. (After Kearey and Vine, 1990, from an original by Tapponier *et al.*, 1982.)

Tibetan Plateau. This is a 400 km (N–S) by 2000 km (E–W) zone of compressed arc terranes, representing almost 100 Ma of continental accretion to the northern margin of an ancient ocean, known as **Tethys**, which used to separate the Eurasian and Indian cratons. Seismic data show that the expression of surface thrusts can be detected to a depth of *c.* 10 km, consistent with the probable thickness of the brittle deformation zone today (cf. Section 10.2.4). The seismic Moho varies from 50 to 75 km depth with sharp vertical offsets corresponding with the continuation of surface thrusts, extrapolated downwards. This suggests that, during continental collision, blocks extending through the entire thickness of the crust may be carried over each other either in a brittle fashion or as ductile **shear zones**. Moreover, the *thrust planes extend into the upper mantle;* and their persistence to great depth may reflect the initially cold surface of the underthrust blocks. Later, when thermal equilibrium across the crustal section has been re-established, ductile processes in the lower crust may

205

start to erase the deep boundaries between overthrust blocks. Notice that easily melted metasedimentary rocks may have been carried to great depth where high temperatures may induce wet melting (Fig. 10.10). So while early accreted terranes contain classic subduction-related magmatic products, the youngest post-collisional igneous rocks are the High Himalayan leucogranites, which are pure *crustal melts*.

Following the accretion of successive arc terranes, the Indian subcontinent itself started to impinge on Asia *c.* 50 Ma ago. Although the closure rate has gradually diminished, the process is still not complete as is clear from continuing compressional and strike-slip seismic activity, and measurements of recent uplift rates (0·7 cm per year) in the Pakistan sector. The compressional stresses induced by this long-term high-velocity convergence have caused extensive fracturing in the upper crust and deformation across the whole of eastern Asia (Fig. 10.14). Thus, while thrust faulting is restricted to a relatively narrow zone, strike-slip faulting occurs parallel to the Indian coastlines and, more important, towards the east within and beyond the main thrust belt. The network of faults has been modelled successfully (Tapponier *et al.*, 1982) in terms of the impact of a rigid block (India) which indents and deforms a plastic medium (Asia), in a process termed **indentation tectonics.** The northern continent is being squeezed out to the east, almost like toothpaste; Indochina which initially was forced eastwards, changed its direction of movement towards the southeast (arrow 1 in Fig. 10.14), as India continued to migrate north during the last 20 Ma, producing a second easterly directed block in Central China. The indentation model is not without its problems, however, and an alternative model predicting that the various modes of this widely distributed deformation are a response to internal forces within a semi-viscous sheet of variable thickness (England and House-man, 1986) carries more favour in some circles.

Focusing on the uplifted Tibetan Plateau, one of the most fascinating discoveries of recent years has been that, instead of compression, there are many extensional earthquake sources. This has been attributed to a tendency for uplifted mountains to *spread under their own weight*. So the overall near-surface strain across the Tibetan Plateau (Fig. 10.15(a)) is the response to two forces; compression due to N–S continental plate convergence, and radial extension

Figure 10.15 (a) Pattern of horizontal stress components in the shallow continental lithosphere of the India–Asia convergence zone (Indian coast and the Tibetan Plateau shown in outline). Compressional stresses due to convergence are indicated with solid arrows, while stresses due to topographical contrasts are indicated with open arrows. Earthquake focal plane solutions follow the standard format with 'closed eyes' representing compression which occurs around the plateau margins, and the four-sector and 'open-eye' solution representing strike-slip and extensional movements which occur within the plateau (after England, 1992). (b) Schematic model for the decay of mountainous topography and roots with time, due to near-surface normal faulting and flow in ductile layers within the lower crust and upper mantle.

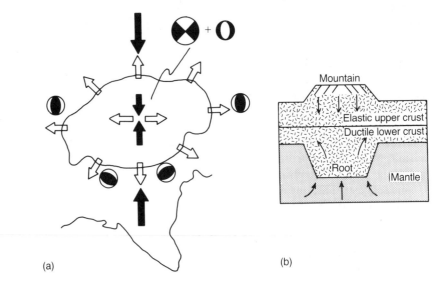

(a)

(b)

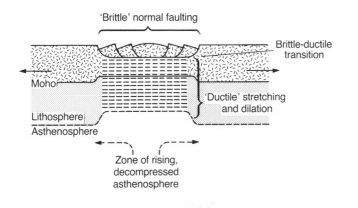

'Brittle' normal faulting

Brittle-ductile transition

Moho

'Ductile' stretching and dilation

Lithosphere
Asthenosphere

Zone of rising, decompressed asthenosphere

Figure 10.16 Lithospheric thinning by active extension. Laterally directed horizontal forces (solid arrows) and/or an upwelling mantle plume (dashed arrows) are responsible for initiating extension. Propagation then depends on the existence of the horizontal forces, while asthenospheric upwelling is promoted by decompression because of extension. Note that the depth of the brittle–ductile transition depends on the temperature profile through the lithosphere, hence on the heat flow. (After Wernicke, 1985.)

resulting from the spreading of the Tibetan Plateau. This results in a combination of strike-slip and normal (i.e. extensional) faulting within the upper brittle zone of the plateau, and compression in the surrounding regions (cf. Fig. 10.16). Surface erosion also contributes to the decay of mountainous topography, reducing its overall height. But what is happening deeper in the crust? The effect of increased weight is to cause lateral ductile flow (Section 10.2.4) away from the region beneath the mountains. In turn, this reduces the crustal root thickness beneath the mountains (Fig. 10.15(b)), replacing it with inflowing mantle. So, after compression ceases in convergence zones, we may anticipate that brittle extension in the shallow crust, and ductile flow in the deeper crust and sub-continental lithosphere will tend to reduce continental thicknesses. Once the crust approaches its 'equilibrium' thickness (35–40 km) we may speculate that the temperature of the deepest crustal regions becomes too low to allow further thinning to occur.

In summary, we have seen that continental collision, the final stage of lithospheric convergence, is often a lengthy process, involving thickening on thrusts and deep ductile shears that penetrate the entire crust and into the mantle lithosphere. Deformation within relatively yielding crust means that the effects of collision are widespread, over thousands of kilometres. Surface collapse and deep ductile flow within the crust then tend to reduce both the surface and basal topography of the continents towards an equilibrium thickness, a process believed to take *c.* 100 Ma after convergence ceases.

10.4 PROCESSES IN ZONES OF LITHOSPHERIC EXTENSION

The importance of continental extension zones as sites of lithosphere accretion and of crustal growth only began to be appreciated during the 1980s. These include uplifted continental areas undergoing either 'passive' extension (i.e. due to gravitational decay – e.g. Fig. 10.15(b)) or 'active' extension above upwelling mantle (Fig. 10.16; cf. also Sections 7.4.2 and 9.2). In either model it is envisaged that stretching of the lithosphere is accommodated by brittle normal faulting in the upper crust with ductile flow in the deeper lithosphere. Decompressed asthenospheric mantle flows in towards the base of the extended lithosphere, with consequences for melt generation that we shall examine shortly.

207

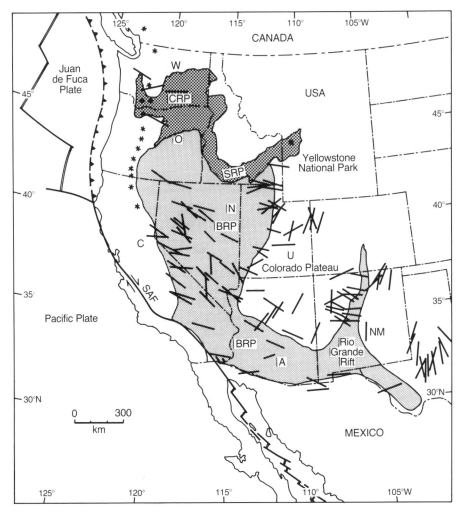

Figure 10.17 Location of the Basin and Range extensional province (outlined in pale-grey tone) in relation to the current plate boundary between the American and Pacific plates (from north to south, the Juan de Fuca plate subduction zone with Cascades volcanoes marked by asterisks, the San Andreas transform fault, SAF, and the Cocos–Pacific ridge transform system of the Gulf of California). The short black lines represent current axes of *least* principal horizontal stress, parallel to maximum extension, as derived from earthquake focal mechanisms, *in situ* stress measurements and a variety of geological data. Throughout the Basin and Range Province, these indicate extension in a WNW–ESE direction. The darker tone labelled CRP and SRP indicates the limits of the Columbia River Plateau basalts and those of the Snake River Plain. (Dot–dash lines are state boundaries: C, California; O, Oregon; W, Washington State; N, Nevada; U, Utah; A, Arizona; NM, New Mexico.) (After Zoback *et al.*, 1981.)

Among the best-documented areas is the **Basin and Range province** of the western USA (Fig. 10.17). This is the dissected margin of the uplifted Colorado Plateau, and gets its name from its parallel ridge-and-valley topography. These are aligned in a WNW–ESE direction, and are the result of extensional faulting caused by tension perpendicular to this direction. Extension is almost certainly due to continental **'back-arc' spreading,** resulting from subduction of the Pacific plate beneath North America. This peaked between 10 and 20 Ma ago when vast quantities (2×10^5 km^3) of 'flood' basalt lavas were erupted in the north of the province: the Columbia River basalts and those of the Snake River Plain. Total extension across the Basin and Range is estimated at 50–100%, meaning that the crust is now 1·5 to 2·0 times its original width. This defines the **stretch factor** (β) as 1·5–2·0. This area is also a zone of high heat flow. Values obtained today reach a maximum of 90 mW m^{-2} in the centre of the Basin and Range, falling to the continental average of 60 mW m^{-2} on the flanks. Almost certainly the heat flow was even higher during the most active period of extension and magmatism.

Now in Section 10.2.4 (Table 10.3), we noted that if heat flow is sustained at 70 mW m^{-2} or higher, the maximum likely within-plate extensional stress of 25 MPa will cause whole lithosphere failure. This prognosis is borne out exactly by observations of contemporary basins where we find that failure is not

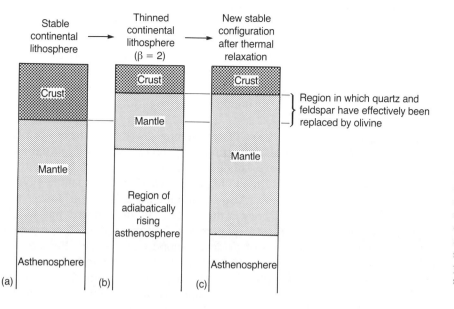

Figure 10.18 Stages in the evolution of a continental lithosphere profile which undergoes symmetrical thinning (stage b) with a stretch factor, β, of two, and then relaxes thermally (stage c) back to its original thickness. Note that the overall effect is to replace some of the original crustal mineralogy with mantle olivine. Topographical anomalies produced by the stretching and relaxation processes are ignored, for simplicity.

occurring in Precambrian shield and some Palaeozoic areas with heat flows below the threshold value. In contrast, there are several younger extension zones (for example the Rhine, Baikal and East African Rifts) that will continue to propagate until the stress has been relieved or accommodated. Figure 10.18 gives a simple illustration, for a stretch factor of two of the lithosphere thinning process. Progressing from (a) to (b), the lithosphere thickness reduces, allowing hot asthenospheric mantle to reach much shallower depths. This should strike a chord, for there are strong similarities with extension across ocean ridge zones (cf. Figs 7.9 and 7.10) where decompression initiates partial melting and the formation of basalt magmas. In the same way, continental extension zones propagating rapidly (i.e. with high **strain rates**) at high heat flows and/or under high stresses will be characterized by basaltic magmatism. If the stresses are maintained, a new ocean basin may form. For example, plate driving forces connected with the subduction of oceanic crust in the floor of the Persian Gulf beneath Iran have pulled Arabia away from Africa during the last 20 Ma: the result is a new ocean, the Red Sea. However, if the strain rate starts to diminish, as has happened in the Basin and Range, the picture in Figure 10.18(c) will develop, where the lithosphere approaches its original thickness but the crustal segment is now thinner. Effectively, some of the relatively weak quartz–feldspar mineralogy has been replaced by stronger olivine, so the overall effect of extensional deformation has been to strengthen the lithosphere, making it less easy to deform: a process known as **strain hardening**. In contrast to zones of rapid extension, continental extension zones propagating slowly (i.e. with low strain rates) will tend to strain harden *during* extension, because there is very little thinning (cf. Fig. 10.18 (b)). Such zones tend not to achieve stretch factors exceeding 1·5.

To summarize this discussion so far. Given that extensional stresses of 25 MPa occur in some continental areas, a heat flow exceeding 70 mW m^{-2} can lead to fast strain rates and these are often associated with stretch factors above 1·5. Against this background it might seem surprising that there are not more areas of young continental crust (Fig. 10.2) currently extending. Quite simply, the

209

reason is that most intraplate regions are in compression and 'extension is limited almost entirely to thermally uplifted zones' (Zoback *et al.*, 1989), perhaps implying that most are active rifts overlying upwelling mantle plumes. Nevertheless, Zoback *et al.* (1989) showed that maximum horizontal stress, compressional or extensional, is commonly subparallel to the direction of absolute plate motion, thus reaching the important conclusion that 'the forces driving the plates also seem to dominate the stress distribution in the plate interior'.

Returning now to the question of continental growth, we have seen from Figure 10.18 that extension at high strain rates generates mantle partial melts. As at ocean ridges and above subduction zones, these will be small melt fractions, basaltic in composition and highly enriched in incompatible elements relative to their mantle source. Also as at modern subduction zones, these magmas may not necessarily reach the surface to be erupted but may be emplaced at depth, either underplating or freezing within the extending continental crust. Several lines of evidence support this model. For example, high-resolution teleseismic and gravity data from the large silicic calderas of the western USA (mostly lying within the extending area of Fig. 10.17) indicate that relatively dense basaltic magmas rise continuously from the mantle, adding heat and mass to large, chemically zoned sub-volcanic magma chambers. Crustal melting and magma mixing processes generate mainly rhyolitic magmas that are occasionally erupted to form silicic ash flow tuffs. The process of crustal growth and intracrustal melting is therefore strikingly similar to that in subduction-related magmatic arcs (Section 10.3.2), and there are likely to be large 'granitic' intrusions developing beneath these calderas.

In conclusion, there is ample evidence confirming that mantle melts are added to the continental crust, where extension initiates passive melting beneath the lithosphere. The existence of an active mantle plume (see Section 9.2.2) augments this process by raising the mantle temperature, but is not an essential prerequisite. Just how extensive this continental growth mechanism is in volumetric terms is difficult to assess because the variables (thinning rate, mantle potential temperatures, etc.) are poorly known on a global scale. Almost certainly it is subordinate to growth by magmatism at subduction margins at the present day, but this was not always the case. Even in the recent geological past there have been periods of major flood basalt eruption within the continents – the Columbia River basalts (Fig. 10.17), for example, which, alone, added $c.\ 0{\cdot}1\ km^3$ per year to the continents for several million years. This is over 10% of the estimated continental growth rate from all subduction zones today (Section 10.3.2) and involves just one small area in North America, albeit one of excess magma production probably related to an energetic mantle plume. At various times in the geological past, particularly in the early Precambrian, extensional magmatism within the continents, crustal underplating and the generation of continental flood basalts probably were the dominant mechanisms leading to crustal growth, a topic to which we shall return in Chapter 11.

SUMMARY

1. The Earth's continental crust comprises:

    (a) ancient Precambrian stable shield areas;

(b) flat-lying platforms with a cover of gently folded metasedimentary rocks overlying more shield (or crystalline basement); and

(c) younger mountain belts (Fig. 10.2). The latter include the continental magmatic arcs that characterize ocean–continent convergence zones (e.g. the Andes), and fold mountain belts which mark the line of continent–continent convergence (e.g. the Himalayas).

2. Rheological studies reveal that the continental lithosphere is sufficiently strong that only where compressional stress magnitudes approach 80 MPa, as in convergence zones, does permanent deformation occur in the form of thrusting and uplift of fold mountains. Compressional stresses within plates are usually insufficient, but extensional stresses within plates (≥ 20 MPa) may cause permanent thinning and rifting of the continental lithosphere if they are maintained for at least 1 Ma in a zone of above-average heat flow (Fig. 10.16). Pressure-release melting will then produce basaltic magmas from the mantle and significant contributions to the volumetric growth of the continents may arise in this way.

3. Continental crust varies from 15 to 20 km thick in extension zones to a maximum of 90 km thick in convergence zones. Once the stresses causing deformation have ceased, the crust returns, in about 100 Ma, to its normal thickness of *c.* 35 km; important processes are: underplating thinned crust, and erosion, gravitational collapse and deep ductile flow affecting thickened crust (e.g. Fig. 10.15).

4. The crust comprises two major seismic and compositional layers beneath a variable thickness of metamorphic and sedimentary rock cover (Fig. 10.4). On average, the upper crust is granodiorite to tonalite (quartz diorite), both in terms of rock type and chemistry. The lower crust comprises mainly high-temperature, refractory granulite-facies metamorphic rocks, with felsic to intermediate compositions, giving way to more mafic compositions at greater depth.

5. Chemically, the overall composition of the crust is intermediate in silica content and must have formed largely by the accumulation of tonalite–diorite magmas produced as large melt fractions from a basaltic parent. In contrast, the enhanced abundances of incompatible trace elements now found in the continental crust resemble those of small melt fractions derived from the mantle (i.e. basalts themselves).

6. Today, new continental material is being added at destructive plate boundaries mainly in the form of mafic mantle-derived magmas. These are produced by partial melting in the mantle wedge overlying subducted ocean crust which releases volatiles during prograde metamorphism, so fluxing the melt zone (Figs 10.8 and 10.11). Rising basaltic melts are trapped at the crust–mantle density boundary where they underplate the crust with material rich in incompatible elements, and also release latent heat of crystallization. This promotes crustal partial melting (Fig. 10.10) and perhaps magma mixing, generating intermediate and felsic magmas which are erupted (e.g. andesites) or are emplaced (as granodiorite, etc. batholiths) in the upper crust. Thus, the silicic upper crust is separated by magmatic processes from dense mafic granulite residues. In order to preserve an intermediate bulk continental composition, the lowermost dense crustal rocks must become tectonically delaminated back into the mantle. Additions to the continents by magmatism at island and continental arcs is the main expression of continental growth today (*c.* 0.8 km^3 per year, further discussion in Chapter 11).

211

7. Material is also added at ocean–continent boundaries where any pre-existing unsubductable continental fragments riding on an oceanic plate, known as terranes, are swept on to the margins of the advancing continent (Fig. 10.13). Abrupt terrane boundaries are common in intracontinental convergence zones where a whole ocean has disappeared. Note that terrane accretion does not increase the *overall* continental volume.

8. Continental convergence is a complex process involving thickening on thrusts and deep ductile shears that penetrate the entire crust and into the mantle lithosphere (Fig. 10.9). Deformation within relatively yielding crust, such as during indentation tectonics, means that the effects of collision are widespread, often over thousands of kilometres (Fig. 10.14).

FURTHER READING

General books:

England (1992): evidence for continental deformation at plate margins.

O'Nions (1992): geochemical and isotopic constraints on crustal melt additions.

Park (1988): reviews of lithospheric rheology and continental tectonics.

Taylor and McLennan (1985): most of the chemical and magmatic aspects of this chapter.

Advanced journals:

Cogley (1984): topographical definition of the continents and the significance of continental shelves.

Coney *et al.* (1980): the concept of terrane accretion.

Griffin and O'Reilly (1987): nature of the lower crust and the Moho.

Hildreth and Moorbath (1988): crustal melt generation, magma mixing and assimilation processes.

Huppert and Sparks (1988): significance of basaltic magma emplacement for crustal partial melting.

Mooney and Meissner (1991): seismic structure of continental crust.

Zoback *et al.* (1989): stress distribution within continental lithosphere.

Advanced books:

Atherton and Naggar (1990): reviews of granite geology and magmatism.

Coward *et al* (1987): review papers on various aspects of continental extension.

Thorpe (1982): reviews of andesite geology and magmatic processes.

Evolution of the Earth's continental crust

11.1   THE FRAMEWORK OF CRUSTAL EVOLUTION

In Section 10.3 we explained how the Earth's crust grows, today, principally by a two-stage fractionation process, first at spreading ridges, then at subduction zones. If the principle of **uniformitarianism** – i.e. that the processes observed today are similar to those of the past – is correct, then why is it necessary to discuss the topic again? The reason is that present-day processes might not have operated in the remote past, primarily because of the much greater heat production during the Earth's early history. Accretionary impacts, short-lived radioactive decays and the gravitational energy released by core formation produced much larger amounts of heat than today (cf. Section 5.3). It is believed that this promoted extremely vigorous mantle convection so that basaltic crust was created and destroyed without continent formation. So it is small wonder that there is no evidence of any crust preserved from the first few hundred million years of the Earth's history.

As shown in Figure 11.1, because of radioactive decay, present-day heat production even from long-lived isotopes is about a factor of five less than when the Earth formed. The rate of internal and surface tectonic processes probably has decreased in proportion to this decreasing heat production and this may explain why the postulated modern rate of continental magmatic additions in Section 10.3.2, i.e. $0 \cdot 8$ km^3 per year, is less than that needed to produce the entire mass of the continental crust in c. 4000 Ma. The primary objective of this

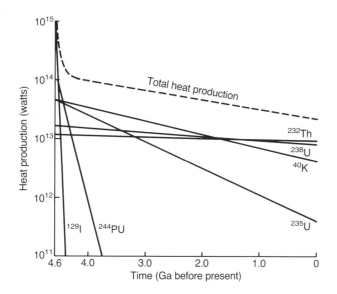

Figure 11.1 Heat production from the important radioactive isotopes incorporated at the Earth's formation. K, U and Th values have been extrapolated back from present-day average abundances in the crust and mantle, while values for short-lived radio-isotopes are estimated from nucleosynthesis theory and decay products in the Allende meteorite. The curvature in the total heat production line during the first 500 Ma is also an estimate reflecting the abundant heat sources available at that time (see text); note that the vertical axis is logarithmic.

chapter, therefore, is to examine the geological record from continental crust of all ages for clues about changing tectonic styles and patterns of crustal growth.

Although most workers agree about the Earth's long-term thermal decay, there is far less agreement about the *mechanisms* and *past rate of continental growth*. So far we have used the term 'crustal growth' to indicate just the addition of new material to the continents, without taking account of material being recycled, removed by erosion or lost by delamination. Over the last 20 years, a plethora of models, describing the change of continental volume with time, has been published, and some of the most strongly supported variants appear in Figure 11.2. Early in this debate it was recognized that the erosion level of the oldest cratonic areas has remained close to the present level and that younger Precambrian terranes still carry their original flat-lying sediments, arguing strongly for constant sea-levels, known as *continental freeboard* (referring to the loading line of ships). Thus one possibility, model (1), requires that since about 3500 Ma ago, volume additions to the continents have been balanced by equivalent losses of material, yielding a **no-growth model**.

In contrast, proponents of **growth models** (2) and (3) argue that a constant freeboard actually requires growth of the continental volume because of the decline of heat flow from the mantle which, it is argued, has had the overall effect of deepening the ocean basins. A deepening that would lower the freeboard level by about 200 m, as would have occurred during the last 2500 Ma, requires an increase in the crustal volume of 10–40% to balance it. The fragmentary nature of the evidence for earlier land–sea levels means that similar calculations cannot be applied simply to the period before 2500 Ma ago, thus opening the way for a group of popular models which propose major episodic continental growth just prior to this time: model (2) in Figure 11.2.

Isotope geochemical data bear on this problem; for example, zircon ($ZrSiO_4$), a minor mineral component of crustal granites, is important in unravelling early crustal history, because it is resistant to weathering, so that it is concentrated as a detrital mineral in sedimentary rocks, especially quartzites. The Mount Narreyer quartzites of Western Australia, themselves about 3500 Ma in age, found fame during the 1980s for producing zircons with U–Pb ages (Note 4) over 4000 Ma, notably one grain dated at 4276 ± 6 Ma. Thus one of the granitic terranes

Figure 11.2 The changing volume of the Earth's continental crust according to three models discussed in the text: 1. no growth since *c.* 3500 Ma ago (cf. Armstrong, 1981), a model in which additions to the continents since that time are balanced by subductions; 2. progressive growth of the continental crust since *c.* 4000 Ma ago but with a period of major continental stabilization at 2500–3000 Ma ago (cf. Taylor and McLennan, 1985; this model allows for some continental recycling into the mantle, but additions have continued to exceed subtractions; 3. a progressive growth model in which the volume of the continents at any one time is the minimum necessary to satisfy isotopic residence age data for sediments (cf. O'Nions, 1992).

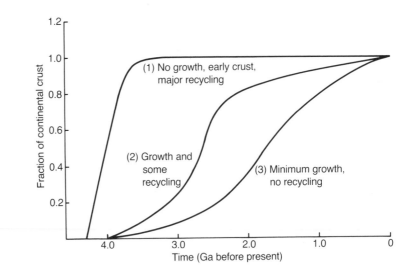

214

contributing to the Narreyer sediments was being eroded 800 Ma after it formed. Given that the oldest real dates on *in situ* terrestial material are 3960 Ma, from the Acasta gneisses of the Slave Province, Canada, it is quite remarkable that the ancestral *c*. 4300 Ma granite complex had survived for so long without being resorbed into the mantle. Taken alone, this might constitute evidence favouring early crustal growth (model (1)) but, significant as this old zircon date is, only 21 of the first 242 zircons analysed produced ages over 4000 Ma, the rest dating much closer to the sediment age, which is generally the case for Archaean rocks. In contrast, sediments with a wide range of Proterozoic ages frequently contain a numerically important zircon population dating from 2500–3000 Ma, evidence that suggests there was a period of rapid continental growth starting about three billion years ago – hence model (2) in Figure 11.2.

Sedimentary rocks of all ages contain minerals with isotopic indicators telling us how long the material from which they formed has been part of the crust (i.e. the time since extraction from the mantle). This is known as the **crustal residence age**, and it turns out that the *mean* residence age for modern sediments is only 1700 Ma. If no material is returned to the mantle, this requires that 1700 Ma is the average age of the sediments, many of which therefore must be relatively young. Thus model (3) in Figure 11.2 was constructed by applying similar logic to sedimentary rocks from the geological record. Clearly, this is a *minimum* growth curve, for any loss of older continental material to the mantle means that more crust must have grown early in the Earth's history. Even model (1) could yield sediments with a 1700 Ma residence age today if recycling of the crust into the mantle and its replacement has proceeded sufficiently rapidly throughout the Earth's history. As we saw in Chapter 10, some additions to the continents are taking place today, whereas there is no good evidence for equivalent resorption, so the balance seems to favour crustal growth models (2) and (3).

A final choice between these models must await Section 11.5, after further discussion of the past rates of crustal recycling and growth (Sections 11.2–11.4). The evidence can be found only in the continental crust because all rocks older than about 200 Ma are continental. They are divided into three major periods.

Archaean	pre-2500 Ma	(Gr. *arkhios*, ancient)
Proterozoic	2500–570 Ma	(Gr. *proteros*, former; *zoos*, living)
Phanerozoic	post-570 Ma	Gr. *phaneros*, visible; *zoos*, living)

Rocks from the Archaean period are relatively sparse, occurring only in widely separated continental nuclei or shields. However, most Archaean suites comprise areas of high-grade metamorphic **gneisses** of tonalitic composition surrounded by compressed low-grade volcano-sedimentary sequences – classically termed **greenstone** belts (Fig. 11.3). For Archaean times, the term **permobile** (rapid motion) is often used to describe the inferred intensity and dynamism of tectonic mechanisms, which may have been plate-like in operation. The later Precambrian, or Proterozoic, period apparently was characterized by markedly different geological and tectonic conditions. The small-scale gneiss–greenstone structures, (hundreds of kilometres across) became trapped within large, stable continental platforms with lateral dimensions up to thousands of kilometres. Extensive sedimentary sequences covered an older crystalline rock basement and linear '**mobile belts**' developed both within the stable blocks and around their margins. These were the sites of intense deformation and igneous activity. Evidence for modern plate-tectonic processes can be traced back into the Proterozoic and may be responsible for some of the craton–mobile belt

215

Figure 11.3 Archaean granite-greenstone terrane in the East Pilbara region of Western Australia. Light areas are granite gneiss domes and the finely striped regions between are greenstones. Archaean rocks are surrounded by Proterozoic sediments and volcanics (dark areas), while a cover of young Tertiary sediments appears in the south-west and north-east corners of this photo. North is at the top and the image is 150 km across. (ERTS image No. 1148–01282.)

configurations. But both the scale and intensity of tectonic processes have changed, and strict uniformitarian concepts cannot be applied to earlier geological times. To reiterate, the main aim of this chapter is to understand the history of the crust and the history of surface conditions on our planet, leading to an evolutionary model for the Earth as a whole and the crust in particular.

11.2 ARCHAEAN AND PRE-ARCHAEAN EARTH HISTORY

11.2.1 Archaean geology

Before speculating about conditions in the pre-Archaean we need the clues about tectonic processes found in the Earth's earliest crust. Within the last decade the wide variety of Archaean sedimentary and igneous rocks, and of different tectonic patterns, has begun to be appreciated (see Nisbet, 1987; McCall, 1991 for reviews). The recognition of early thrust tectonics has suggested that most Archaean cratons consist of complex terrane aggregates in the style of some modern plate margins and suture zones. However, there are important differences; for example, extensive linear magmatic belts, now found above subduction zones, did not develop, and there are few major andesitic suites comparable with those erupted today.

Sodium-rich granite–granodiorite–tonalite (granite *sensu lato*) complexes are the major components of Archaean geology (e.g. Fig. 11.3). They occur in both low-grade and high-grade terrains, the latter typically being 'grey' tonalite gneisses that have been deeply buried in the crust and then exhumed, such as the famous 3750-Ma-old Amîtsoq gneisses of West Greenland. These gneisses envelop and intrude a 3800-Ma-old belt of volcano-sedimentary rocks, the Isua 'supracrustals', so called because they were laid down as a layered sequence in shallow water, presumably on an older granitic crustal foundation. Similarly, the major Archaean greenstone terranes, developed from 3600 Ma onwards, are supracrustal; there is evidence that some were deposited on granitic (*sensu lato*) and others on mafic, probably ocean, crustal basement. The relationship between granites and greenstones is complex (Fig. 11.3 and 11.4); in some cases the granites are the old high-grade gneissic foundations for the volcano-sedimentary sequence, and in other cases they are low-grade intrusives. The classic 'gregarious' batholiths of the Zimbabwe Craton (see Fig. 11.11 for location) are regarded as the products of fold interference due to multi-phase deformation (Figs 11.4(b–d)). Similar conclusions have been reached in Pilbaran (Australia) research where studies of the Shaw batholith margins – westernmost pale mass in Figure 11.3 – demonstrate the importance of low-angle thrust, or sheared interactions between greenstones and their gneissose foundations, again implying significant horizontal and vertical intracrustal motions.

The low-grade greenstone sequences vary considerably in thickness, composition and degree of metamorphism. Much of the material was volcanic in origin, but it includes clastic sediments such as immature molasse and greywackes, the proportions of which increase up-sequence. Mafic, and even ultramafic, lavas

11.2 Archaean and pre-Archaean Earth history

Figure 11.4 (a) Sketch map of the Archaean granite–greenstone pattern in the Zimbabwe Craton (after McGregor, 1951). (b)–(d) Three stages in the possible structures of the pattern in (a) in which (b) a greenstone sequence is first deposited on a gneissose basement and intruded by granite plutons, (c) the greenstones become folded, exhibiting greater deformation than the relatively strong granite masses, and new plutons are emplaced during regional compression, and (d) a second phase of deformation, approximately at right angles to the first, causes fold interference producing complex granitic domes surrounded by more linear deformed metasedimentary zones (After Snowdon, 1984.)

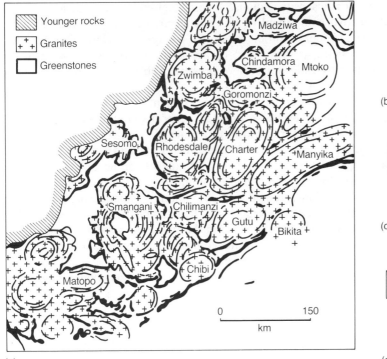

(a)

(b)

(c)

(d)

217

Table 11.1 Lithological proportions of volcanic rocks in some Archaean greenstone belts (reproduced from Taylor and McLennan, 1985, which contains original sources).

Composition	Australia		Canada		Southern Africa	
	Pilbara Block, N Western Australia	Yilgarn Block, S Western Australia	Superior Province, Central Canada	Slave Province, NW Canada	Barberton, South Africa	Zimbabwe Craton
Ultramafic	4·3	20	0·7	—	24·3	9·8
Mafic	59·3	62	54·0	49	72·0	60·2
Andesitic	7·8	5	31·4	35*	—	22·5
Felsic	28·6	13	13·9	16	3·7	7·5

* Includes 80% basaltic andesite.

(typically magnesium-rich **komatiites**) predominate over felsic volcanics, and the eruptive suites are frequently bimodal, mafic–felsic (Table 11.1). Komatiites, named after the Komati River of the Barberton (part of the Kapvaal Craton, Fig. 11.11), are remarkably magnesium-rich (*c.* 20% MgO) low-silica lavas that are almost entirely confined to the Archaean. They require extremely high temperatures to form, around 1600°C in the modern Earth (Fig. 7.5). They are dominated by olivine and clinopyroxene, both often sharing a dramatic skeletal platy texture, known as **spinifex texture**, produced by rapid post-eruptive cooling of hot, low-viscosity liquids. These high-magnesian lavas are an important, distinctive component of the Archaean geological scene, for they speak of a hotter upper mantle, to which we return below. Basalts and komatiites predominate in the lower, earlier parts of greenstone sequences, giving way upwards to andesites (where present, Table 11.1) and more felsic volcanics.

Associated with the upper, more clastic sedimentary greenstone sequence there is often a group of cyclic sedimentary deposits comprising deep-water shales, shallow-water immature sandstones, some of them conglomerates containing detrital gold and uraninite, and *b*anded-*i*ronstone and chert (fine-grained silica) *f*ormations, or **BIFs**. BIFs are usually taken to indicate low levels of atmospheric oxygen, for iron is much more soluble in its reduced, Fe^{2+}, than its oxidised, Fe^{3+}, state. Fine-grained, shallow-water sediments suggest rather tranquil conditions. Indeed, the widespread development of algal reefs in the form of **stromatolites**, such as those found in the sheltered subtidal waters of the modern Bahamas and Western Australia, confirms that greenstone sedimentary basins were developed on stable continental crust.

Finally, we return to the high-grade terranes that often have been metamorphosed in the granulite facies (Fig. 10.5). This produces a recrystallized anhydrous mineralogy in which orthoclase–plagioclase–orthopyroxene ± quartz is the prevalent assemblage. Such rocks, known as **charnockites**, are particularly common in the South Indian craton, where chemical data on minerals sensitive to the pressure–temperature conditions of metamorphism have given values of 830 ± 100 MPa and 760 ± 40° C (Harris *et al.*, 1982). The significance of these results, now duplicated from other Archaean cratonic areas, is that the continental crust was at least 30 km thick, with temperature gradients as high as 25°C km^{-1}, enough to promote crustal melting in hydrous rocks (Fig. 10.10). Other rock types in these high-grade areas are metavolcanic amphibolite sequences, and layered peridotite–gabbro–anorthosite complexes that may represent fractionated crustal magma chambers.

11.2.2 Physical and tectonic conditions

So far we have established that there was continental crust of 'normal' thickness after 4000 Ma ago, some of it emergent from the oceans, but we have no idea of its extent. We also know about the dramatic vertical and horizontal movements of crustal blocks and, compared with the present day, in the Archaean there were:

1. more tonalitic (sodic) batholiths;
2. less andesite and more bimodal volcanic suites;
3. komatiites.

Since surface magmatism and tectonics is ultimately tied to the thermal structure of the Earth's interior it is helpful to consider the likely effect of higher heat production in the early Earth. In physical terms this can have two effects, both of which probably applied to the Archaean mantle:

1. the temperature in the convecting mantle was higher everywhere by the same amount, in other words the adiabat was elevated (e.g. Fig. 11.5); and
2. convection was more vigorous, mainly because with higher temperatures the viscosity was less.

The most significant consequence of (1) is that the adiabat came closer to the surface, thus decreasing the thickness of the lithosphere. The amount of partial melting would also have been greater; indeed it is argued (cf. Section 5.3.3) that the early mantle was extensively molten, perhaps with a surface magma ocean. As the lithosphere developed, magmas would have reached the surface from various different depths. However, magma is more compressible than rock, so the deepest mantle melts that have sufficient buoyancy to escape upwards are komatiites from *c.* 150–200 km depth. An obvious consequence of (2) above, decreased viscosity, is that Rayleigh numbers were higher leading to more vigorous, chaotic convection, which itself tends to increase the rate of cooling. So there may have been large lateral temperature variations, vigorous plumes and probably more, smaller convection cells in the mantle than today.

Now we will consider how much magma would reach the surface if there were such things as ocean ridges in the Archaean. Let us assume a potential temperature of 1430°C, 150°C higher than today (see mid-Archaean adiabat in Fig. 11.5); in these conditions the 'ocean crust' has to be no less than 20 km thick because of the larger melt fractions being produced. This has fascinating implications for the freeboard argument since a thicker ocean crust must be more buoyant, as we see in Iceland today. A world almost covered with thick oceanic crust need only have a small volume of continental crust (cf. Section 11.1) to maintain freeboard. But such buoyant ocean crust introduces another problem for Archaean plate tectonics: it would be less easily subducted. Thus the essential driving force for sea-floor spreading, the sinking of cold, dense lithosphere, would be *reduced* at a time when the Earth was producing *more* heat, which would tend to *increase* spreading rates.

219

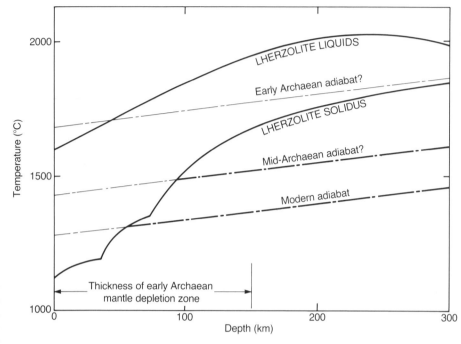

Figure 11.5 Depth–temperature plot for the upper 300 km of the Earth extracted from Figure 7.5 and showing dry solidus and liquidus for mantle lherzolite together with ancient and modern mantle adiabats. Note that because of the increased thickness of the Archaean melt zone, the depleted lithosphere would have also been thicker than today. This may explain why Archaean cratons rapidly developed thick mantle roots (see text and Bickle, 1986).

There are various ways of resolving this paradox; for example, perhaps the Archaean oceanic crust was more komatiitic, thus denser and reinstating the slab-pull force, but this would be inconsistent with the majority of the magmas produced by decompression melting which are basaltic. Physically, the overall effect of a thick, buoyant oceanic crust would be to reduce subduction rates, so mantle temperatures would rise and convection would change to a more plume-like regime and/or smaller plate dimensions (e.g. Fig. 11.6) which produces a greater overall length of subduction margin. In an elegant analysis, I. H. Campbell *et al.* (1989) show that rising plumes in a hotter Archaean mantle will generate komatiite liquids in the hot axial region with basalts in the cooler surrounding zones, including the plume head. They believe such liquids could furnish the lower parts of greenstone sequences in the style of modern flood basalt provinces, which result from plumes (Section 9.2.2). So it seems that the plate-tectonic regime was different from the present, including more vigorous plume activity. Thermal plumes, in some cases perhaps associated with back-arc basins (Fig. 11.7), initiated greenstone belt volcanism. Horizontally directed plate-tectonic forces produced interfolded structures (cf. Fig. 11.4) and terrane accretion to build Archaean cratonic masses from smaller plates and plate fragments. So, to summarize, 'the Archaean was a time of hot spot tectonics and unusual ocean basins' (Schubert, 1991).

How did the the important tonalite gneiss belts originate on this model? Although often regarded as the primitive 'scum' of a differentiated planet,

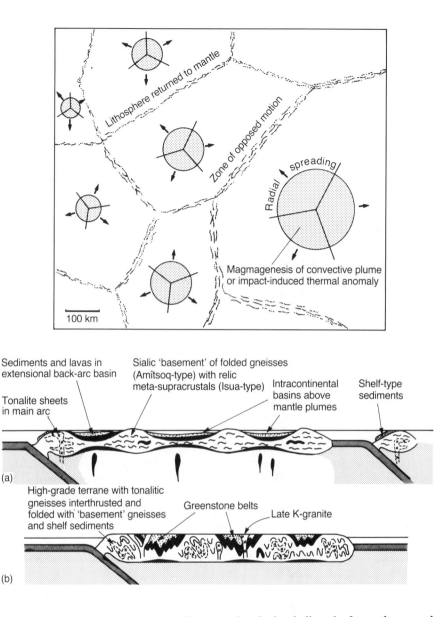

Figure 11.6 Radial spreading from hot spots to produce a polygonal plate pattern, is a style of oceanic tectonics that may have characterized much of the Archaean period. (After Drury, 1978.)

Figure 11.7 Microcontinent model for the genesis of Archaean granites and greenstones. Diagram (a) represents the mobile stage where folded gneisses and island-arc magmas form the domes. Volcanic and sedimentary proto-greenstone materials accumulate in back-arc and other continental basins fed by mantle plumes. In (b) the microcontinents have aggregated to form the structures now observed. (After Windley, 1984.)

Archaean tonalites are no more likely to be derived directly from the mantle than their modern equivalents (Section 10.3). So are they related to subduction and/or crustal underplating? Significant geochemical features emphasizing differences between ancient and modern tonalites, granites, etc. are: K_2O/Na_2O ratios below 0·5, compared with 1 for post-Archaean samples and much more fractionated rare-earth element profiles (denoted by La/Yb ratios), coupled with low Yb in the Archaean samples, (Fig. 11.8). This rare-earth element (REE) pattern is highly distinctive for it implies that the crust-forming magmas were in equilibrium with residual garnet and hornblende. This indicates that the subducted Archaean ocean crust melted, itself, whereas modern crust dehydrates, releases volatiles and the overlying mantle wedge melts. Referring to Figure 10.11, the thermal gradient must have been *above* 10°C km^{-1} to generate *basalt* melts along AB in the diagram, before amphibole started to break down. As today, rising basalt magmas would underplate the thickening crust, promot-

221

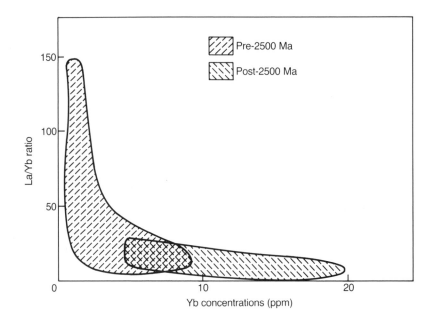

Figure 11.8 Slope of the rare-
earth element profile (light-
heavy as given by the La/Yb
ratio) plotted against Yb
concentration in 644 granite
(*sensu lato*) samples, 319 from
the Archaean and 325 from
Proterozoic and Phanerozoic
complexes. Over 95% of each
sample group plots in the
fields shown – for original
data see Martin (1986). Data
are chondrite-normalized,
which is the conventional way
of expressing rare-earth data.
(Note: La = Lanthanum; Yb
= Ytterbium.)

ing the development of large-volume tonalite magmas, but generally, because of the newness of the developing continents, with shorter crustal residence times than today (see Section 11.5). Today, the rising volatiles have higher K/Na ratios than subducted basalts because K is more easily mobilized by fluids and thus is imported into the continental superstructure. As a test of this model, we find anomalous, Yb-poor, low K/Na contemporary magmas in south Chile, where unusually hot, young (15 Ma old) crust is being subducted. So the geochemistry of Archaean tonalites supports a crustal origin by subduction-related magmatism where oceanic crust was partially melted in a relatively hot environment.

The third important member of Archaean igneous suites (Table 11.1), the felsic volcanics, are chemically indistinguishable from the tonalites. Again, this suggests a simple origin by crustal melting in regions of high heat and material flux from the mantle. The greater prevalence of bimodal suites in Archaean greenstones probably indicates that hotter mafic–ultramafic magmas were capable of penetrating right through the crust, rather than, as today, ponding at depth, fractionating, mixing with crustal melts, and forming eruptive andesites.

Where do all these speculations on Archaean tectonics leave us? The geological record reflects processes in a hotter mantle than today, yielding higher-temperature mantle plumes that penetrated the continents and may have floored the oceans. Oceanic plates were heavy enough to subduct, yielding continent–forming magmas from their own substance. High-MgO oceanic rocks promoted early subduction. Such rocks are much less easily remelted than basalt, so if the pre-4000 Ma crust was all komatiite there would have been little continental growth. Additionally, the highly vigorous mantle convection during the first 500 Ma may have acted against the development of permanent continents. Any continental rafts that formed were inherently unstable, hence the uncertain start of the growth curves in Figure 11.2. After 4000 Ma ago, the thermal regime allowed microcontinental plates to develop, and horizontal lithospheric forces ensured that they accumulated and thickened into cratons.

Finally, from Figure 11.5 we see that continent formation must have depleted a great thickness (≥150 km) of the upper mantle. Abundant isotopic and geochemical evidence indicates that this refractory lithosphere became stabilized and frozen beneath the Archaean cratons, where it has remained isolated from the convective mantle regime. Studies of kimberlite xenoliths from South Africa by Richardson *et al.* (1984), show that xenoliths brought to the surface just 90 Ma ago became part of the sub-continental lithosphere about 3200 Ma ago. They yield mineral equilibration conditions of 1150°C at a source depth of 180 km, well below even the modern adiabat (Fig. 11.5)! The Archaean Earth really was a place of contrasts that we are still only just beginning to understand.

11.2.3 Archaean atmospheres and primitive life

The Earth's present atmosphere is due primarily to volcanic outgassing of the mantle. Any gaseous envelope which condensed around the Earth when it first formed has been either extremely diluted, or totally replaced, because the noble gases are very strongly depleted as compared with cosmic abundances (Fisher, 1985). The most likely explanation is that the giant impact which formed the Moon totally removed any previously formed atmosphere and oceans. In recent years, a consensus has developed that the earliest Earth, like Venus, was swathed in a steam atmosphere, with magma lakes over much of its surface (review by Stevenson, 1988). The long-term evolution of this atmosphere clearly was intimately tied up with near-surface tectonics, with complex regulatory feedback loops between the two; indeed, a hot atmosphere would be another factor mitigating against early continental development. Fortunately for biological evolution, perhaps because the Earth is further from the Sun, its atmosphere was at least moist, for otherwise it might have followed that of Venus forever into the oblivion of total dehydration and hostile surface temperatures. Indeed, the Earth may have lost significant water before the atmosphere rained out, substantial oceans formed, continents started to develop, and environmental conditions favoured the origin of life.

Let us now dwell on the development of the present atmosphere. Apart from oxygen, the major gases found in the atmosphere (N_2, 78%; O_2, 21%; Ar, 0.93%; CO_2, 0.03%; plus variable amounts of H_2O) are all found in volcanic gases. The ratio of argon to nitrogen in volcanic gases is the same as in the atmosphere, but water and carbon dioxide from volcanoes are considerably more abundant. The outgassed water vapour essentially has condensed into the oceans, and most of the CO_2 became dissolved in the oceans and then was deposited as calcium carbonate in limestones. Today, a small but highly significant part of the CO_2 is used in **photosynthesis**, whereby green plants use the energy of sunlight to convert H_2O and CO_2 into carbohydrates, whilst at the same time *releasing oxygen*. Most of the present atmospheric content of oxygen has been liberated by photosynthesis but, early in the Earth's history, another oxygen-releasing process, **photodissociation**, was important. This involves the breakdown of water molecules by ultraviolet light from the Sun. Lightweight hydrogen molecules tend to escape the Earth's gravity field, but oxygen is retained, and a small proportion is converted into ozone, which forms a shell in the upper atmosphere (Fig. 11.9). This ozone absorbs ultraviolet light, and so photodissociation is a self-regulating process because the oxygen produced ultimately prevents further dissociation. At least a few per cent of the present atmospheric content of oxygen may have been produced in this way, but the present high levels required extensive photosynthesis.

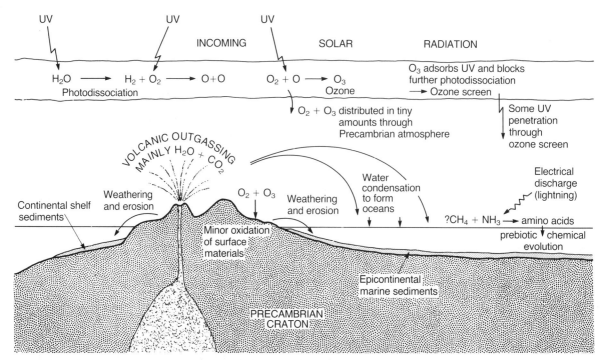

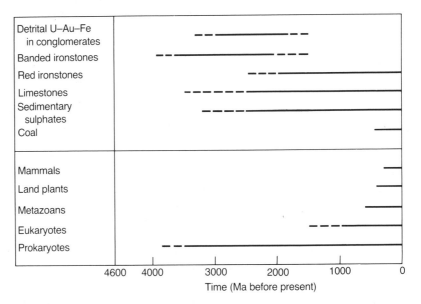

Figure 11.9 Schematic cross-section through the Earth's surface in early Archaean times illustrating the likely flow of gases in the atmosphere. (After Kershaw, 1990.)

Figure 11.10 Distribution of important environmental indicators (top) and of fossil organisms (bottom) through time. Dashed lines indicate uncertainties and/or weak extension of the relevant time-scale. Detrital U–Au–Fe conglomerates and banded ironstones are usually taken to indicate relatively reducing conditions whereas red ironstones and sedimentary sulphates require oxidizing conditions (though to a different degree).

Oxygen was built up in the Earth's atmosphere from H_2O and CO_2, and, on this theory, was virtually absent from the *early* Earth, so *reducing* conditions probably prevailed. This is supported by evidence from sedimentary rocks (Fig. 11.10); the Archaean record contains banded ironstones composed of hematite (Fe_2O_3) and chert (SiO_2), thought to have been deposited by marine organic agencies. That iron reached marine environments demonstrates that it was liberated by weathering sites in the soluble Fe^{2+} state. In contrast, modern weathering produces insoluble Fe^{3+}, which is incorporated into red, often

224

red ironstones across the interval from 1900 to 2200 Ma ago reflects increasingly oxidizing surface conditions. The common presence of detrital minerals such as uraninite (U_3O_8) and pyrite (FeS_2) in pre-1900 Ma Precambrian rocks, both of which are unstable in oxidizing conditions, and the increasing abundance of sulphate deposits since about 2500 Ma ago, are also consistent with these ideas. Interestingly, barium sulphate (baryte) evaporite deposits have now been found in 3500 Ma-old parts of the Eastern Pilbara succession, suggesting that, at least locally, conditions were mildly oxidizing. In contrast to modern conditions, the Archaean biosphere may have contained excess oxidant in its 'stagnant' pools, although the prevailing conditions were reducing.

The earliest microfossil evidence for Precambrian life is recorded in some of the oldest sediments. If reducing conditions prevailed, then the lack of a strong ozone shield would have meant that life evolved in the presence of strong ultraviolet radiation, which destroys amino acids. Thus early life-forms probably escaped the harmful effects by living in deep water. The simplest **prokaryotic** cells (single cells or chains of cells, with no nucleus) derived their food and energy by fermentation. Later, they became photosynthetic and resembled modern blue-green algae that build colonial structures and are recognized throughout the geological record as stromatolites.

Opinions differ as to where life *originated*. Two conditions seem to be necessary, an anoxic environment and protection from UV radiation. To develop proteins and nucleic acids which are built primarily of the elements C, H, N and O, we need to construct carbon chains from chemicals such as methane (CH_4) and ammonia (NH_3) and water (H_2O). A mixture of these gases in the atmosphere was synthesized into large non-biological organic molecules (amino acids) by an energy source, such as lightning (Fig. 11.9). These were incorporated into the oceans, where prebiotic molecular growth continued until the necessary organization for biological cell growth had been achieved. Given the concentration of strange bacteria around submarine hydrothermal vents today, it is conceivable that energetic shallow-water or subaerial hydrothermal systems around volcanoes were the birthplace of life. Another possibility is chemical synthesis during high-velocity impacts by planetesimals and meteorites (Mukhin *et al.*, 1989), and life may even have had an extraterrestrial origin. The answer still eludes us.

11.3   THE PROTEROZOIC

11.3.1   Geology and tectonics

Although many models of crustal evolution suggest that the majority of the continental volume had formed by 2500 Ma ago, Proterozoic rocks, surrounding and covering the ancient Archaean cratonic nuclei, are considerably more abundant at the surface (e.g. Fig. 11.11, the African picture). The shift in geological and tectonic patterns at the Archaean/Proterozoic boundary is profound; there are only rare occurrences of greenstone belts younger than 2500 Ma in age, and the early Proterozoic saw the development of extensive stable shield areas. Thick limestone–sandstone–ironstone sequences of continental-shelf type formed an extensive cover to the submerged Archaean basement. Exposed crustal rocks available for erosion seem to have become   225

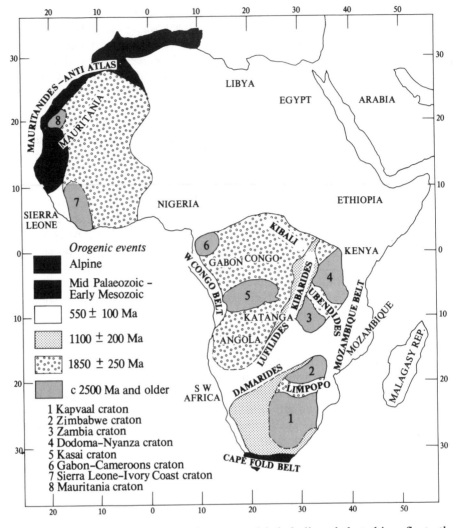

Figure 11.11 Major structural and time units in the geology of south and west Africa. Archaean cratonic nuclei are numbered and the much more areally extensive Proterozoic terrains are ornamented. Much of north and east Africa is of late Proterozoic, early Phanerozoic age (no shading) and younger peripheral belts are indicated by dark shading (see key). (After Clifford, 1970.)

Orogenic events

Alpine

Mid Palaeozoic – Early Mesozoic

550 ± 100 Ma

1100 ± 200 Ma

1850 ± 250 Ma

c 2500 Ma and older

1 Kapvaal craton
2 Zimbabwe craton
3 Zambia craton
4 Dodoma–Nyanza craton
5 Kasai craton
6 Gabon–Cameroons craton
7 Sierra Leone–Ivory Coast craton
8 Mauritania craton

much more felsic in the late Archaean, and it is believed that this reflects the increasing importance of intracrustal melting. The apparent change in granite composition discussed earlier (cf. Fig. 11.8) was not synchronous on a world scale, but was concentrated into the period 2500–3000 Ma, an interval of major growth according to curve (2), Figure 11.2. On this basis, we might expect a greater extent of crustal recycling in the early Proterozoic. However, a growing body of isotopic data from 1700–1900 Ma rocks of the northern continents (e.g. Patchett and Arndt, 1986) indicates that large volumes of the Proterozoic crust, between 50 and 90%, were newly mantle derived (further discussion in Section 11.5).

A second major feature of the early Proterozoic crust is the appearance of linear belts of trough sediments analogous to those of modern back-arc basins or closing oceans: the 'mobile belts'. One of the classic examples is the roughly N–S-orientated Labrador Trough which bisected what is now central Canada in a zone 800 km long and 100 km wide, with ancient Archaean terranes to either side. Compared with a 2 km thickness of sediments laid down on the adjacent cratons, nearly 11 km of 1800–2000-Ma-old mafic volcanics, interstratified with conglomerates, sandstones and shales were deposited in what Dimroth (1981) recognized as a subsiding basin of intracratonic origin. Later, the sequence was uplifted in the 1800 Ma Hudsonian orogeny. The mechanism proposed for the development of this **'ensialic basin'** involves fracturing of the lithosphere, followed by delamination which progressively dragged subcontinental lithosphere down into the mantle. This requires shallow mantle convection beneath the continental plate, but little horizontal motion. Similarly, the African mobile belts (for example, the Limpopo and the Ubendides in Fig. 11.11) may be sites of former ocean closure, but the alternative view is that they represent ensialic developments with limited horizontal movements. Evidence supporting the ensialic view is that some sedimentary sequences are easily traced from craton into mobile belt, that only rarely are calc-alkaline plate-margin-type lavas found in the mobile belts, and that there are few ophiolites. However, these absences may be because most mobile belts are deeply eroded; calc-alkaline intrusives and lavas *are* found in low-grade mobile belts with shallow levels of exposure.

During the late 1980s, a consensus started to emerge that plate tectonics in the contemporary style was an important determinant of Proterozoic geology. Moores (1986), for example, showed that many mobile belt basaltic complexes might be ophiolites lacking the mantle peridotite component. If oceanic crust was rather thicker than today, basal thrusts may not easily have penetrated the mantle so that only the upper layers of ocean lithosphere were obducted. Some of the best evidence for early Proterozoic (1900 Ma) plate tectonics comes from seismic reflection profiles, such as those carried out along the Baltic Sea across the boundary between an oceanic and a continental arc (Fig. 11.12). The contrasting crustal reflectors and Moho offset are extremely reminiscent of similar data from known Phanerozoic plate convergence zones.

Although true plate-tectonic processes were active during the Proterozoic, there is also evidence of **within-plate (anorogenic) magmatism**, sometimes on a grand scale, in the form of transcontinental mafic dyke swarms, enormous flood basalt provinces, massive mafic and alkaline intrusive complexes, anorthosites and kimberlites. Such magmas tend to form during periods of major lithospheric extension; for example the basic dyke swarms were developed in two major periods from 2500 to 2000 Ma and 1300 to 600 Ma ago. One interpretation of Proterozoic magmatism suggests that there were cycles of continental aggregation and dispersal. Indeed there is considerable palaeomagnetic evidence that supercontinents formed (Piper, 1987). Although accretion tends to occur over cold downwelling zones of mantle convection (subduction zones), the insulating

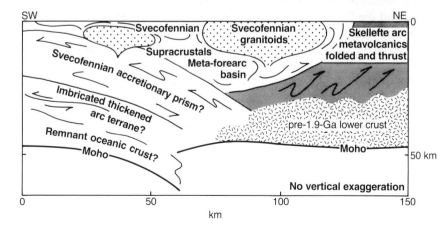

Figure 11.12 Diagrammatic interpretation of seismic reflection data from a NW–SE axial line through the Baltic Sea located at roughly 58°N to show present-day crustal structure resulting from a collision around 1900 Ma ago between an oceanic arc to the south, founded on Archaean crust, and a continental arc to the north. (After BABEL, 1990.)

227

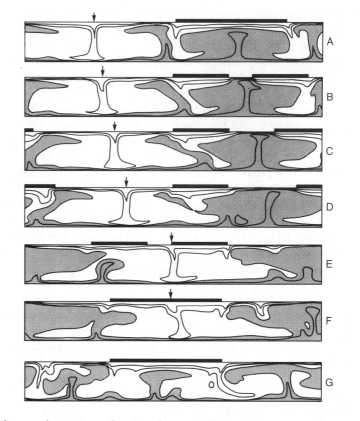

Figure 11.13 Computer simulation model showing how the upwelling of hot mantle (grey) and downwelling of cold mantle (unshaded) causes fragmentation of a supercontinent (black rectangle) in stages A–C. The continents may reassemble, perhaps 100–200 Ma later (stage F) above the main zone of downwelling (arrow). But after 400–500 Ma the new supercontinent has created its own upwelling (stage G) and the fragmentation process is repeated. (Simulation published by Gurnis, 1986.)

effect of the continent traps heat and causes the temperature to increase over several hundred million years (Fig. 11.13). Ultimately, upwelling starts, within-plate magmas rise and the supercontinent will fragment, later to be reassembled elsewhere (see Kerr, 1989 for a full analysis). When, during extension, within-plate alkali basalt magmas rose in large volumes into the cratonic crust, flotation of fractionated plagioclase would have produced anorthosite masses. In addition, the heat of the mafic magmas was sufficient to cause partial melting at lower temperatures in the crustal rocks that they rose through. These new melts mixed with the residual basalt magmas and formed the associated suite of alkali-rich diorite and granite intrusions. Similar alkaline complexes are known from younger intracontinental sites of rifting; for example, the 150-Ma-old granites of Nigeria and the 55-Ma-old granites of north-west Scotland, which heralded the incipient opening of the modern North Atlantic.

The surface evidence for plate margin developments around the cratonic nuclei becomes much clearer in later Proterozoic rocks. Examples are the Grenville Belt in eastern Canada and the Pan-African Belts of Arabia, NE Africa and elsewhere. Detailed studies have revealed calc-alkaline intrusive/extrusive lineaments separated by mafic-ultramafic ophiolite zones, again indicating horizontal continental aggregation. The geochemical and isotopic characteristics of igneous rocks from the Arabian shield led to the proposition (e.g. Gass, 1981) that the shield developed by magmatism over a series of adjacent island arcs much as those found today in the south-west Pacific. There is no indication that there was any continental crust in Arabia before 1000 Ma, and the entire shield formed during the Pan-African period – a process now termed **cratonization**.

Taking stock, it appears that Proterozoic tectonics may be relatively easily related to modern processes. This would be expected from the more stable thermal state of the Earth's interior where, with declining heat production, larger convection cells would have developed. Linear plate margin activity, mantle plumes and the familiar Phanerozoic pattern of plate fragmentation and reassembly are natural consequences of these conditions. But if the Earth's *interior* was settling down, there was still a long way to go in the evolution of *surface* environments.

11.3.2 Biological and atmospheric conditions

It is sobering to recall that not until about 400 Ma ago, within the last 10% of the Earth's history, were there any land plants or animals. The Proterozoic landscape was one of barren, rapidly eroding continents, covering 10–20% of the surface. Shallow continental seas and lagoons occupied perhaps another 20%, and the rest, as today, was deeper ocean basins. The vast shallow sedimentary basins produced thick successions of sandstones, algal limestones and, in the early Proterozoic, banded ironstones. Notable examples are the algal Gunflint limestones and siliceous cherts of Ontario and the Hammersley ironstones of Western Australia. Later, red ironstones become more widespread; for example, the thick red sandstone deposits of north-west Scotland and the Baltic shield. Sedimentary sulphates and carbonates (limestones) were becoming more common (Fig. 11.10), probably because of progressive increases in the amounts of dissolved CO_2 and SO_2 (volcanic gases) in sea-water. Like the banded ironstones, most of the early limestones were precipitated biochemically and they contain impressive numbers of prokaryotic microfossils.

Photosynthetic organisms had multiplied and so oxygen was being released increasingly rapidly throughout the Proterozoic. We read from the rocks that a major increase in atmospheric oxygen levels started around 2000 Ma ago. Until then oxygen released by photosynthesis was used up by oxidation of reduced surface sedimentary materials; once this was complete, the excess oxygen accumulated in the atmosphere. By 1500 Ma ago, atmospheric oxygen may have risen to 15% of its present level. Prokaryotic bacteria were no longer the only life-forms on Earth, and had been joined by simple **eukaryotes**. These are cellular micro-organisms with a nucleus; the earliest forms were single-celled (protozoa), like modern amoeba, but some later ones were multicellular (metazoa). It seems that the early eukaryotes were able to reproduce either asexually or sexually, but once the advantages of sexual reproduction for maintaining a healthy, shifting gene pool was hit upon, this became the norm as biological evolution progressed.

As the atmospheric oxygen levels increased, so a more extensive ozone shield would have developed (Fig. 11.9). A smaller depth of water would then be needed to protect organisms from ultraviolet radiation, so opening up a wider range of ecological environments. Increased oxygen levels also allowed the development of animals that could breathe oxygen. The first traces of **metazoan** animal life are recorded in the later Proterozoic; for example, the 670-Ma-old Ediacara fauna from sandstones of central Australia (Fig. 11.14). These are the soft-bodied ancestors of a wide range of animals that started secreting skeletons around the Cambrian–Precambrian boundary. Many of them were filter feeders and scavengers of the sea bottom; others were floating micro-carnivores. They included primitive arthropods, sea-pens, jellyfish, worms and sea-urchins. Thus,

229

Figure 11.14 A reconstruction of the late Proterozoic Ediacaran fauna from central Australia based on soft-bodied impressions in sandstones. (After McAlester, 1977.)

the dawn of major biological diversity appears in the record of the late Proterozoic. The arrival of diverse eukaryotes, both consumers and producers of oxygen, had a major effect on atmospheric evolution, for they took over the regulation and management of atmospheric gases. By now, the atmosphere was mainly nitrogen with several per cent oxygen, and small amounts of carbon dioxide and water vapour. With a significant time lag, the Earth's surface environments were beginning to follow internal environments in resembling their modern state.

11.4   THE PHANEROZOIC

11.4.1   Geology and tectonics

There is little doubt that lithosphere dynamics in the last 570 Ma have been dominated by plate tectonics. However, only the youngest island, 'Andean' and 'Alpine' arcs were described in Chapter 10. There are numerous examples of Phanerozoic arcs and sutures now recognized in the geological record; for example the Urals of central Siberia and the Variscan mountains of southern Europe are both sutures developed about 250–300 Ma ago. Arguably the best studied Phanerozoic orogenic belt, representing a complex closure between four continental masses and innumerable smaller terrane blocks (Fig. 11.15) is the 400–500-Ma-old Caledonian–Appalachian system whose tectonic imprint is found in most countries bordering the North Atlantic. The present Atlantic Ocean is only some 100 Ma old and, before it opened, Europe and America were a united continent since Caledonian times. However, the evidence for a former Atlantic Ocean of early Palaeozoic times across central Britain and eastern North America, known as Iapetus, is convincing. It embraces faunal, stratigraphical and geochemical differences between the north-west and south-east forelands, and the position of the north-dipping suture zone has been located to a high precision with offshore reflection seismic techniques (Fig. 11.16). Indeed, reflection profiling has found wide application during the last decade in unravelling deep-crustal and even upper-mantle structures (Klemperer and Peddy, 1992). Strongly reflective lower crust (e.g. south end of Fig.

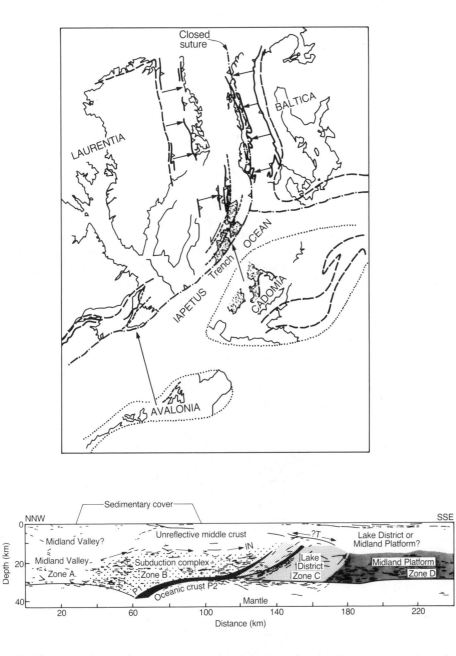

Figure 11.15 Reconstruction of the major Caledonian–Appalachian terranes and the Iapetus Ocean in mid-Silurian times. Collision has already taken place between Baltica and Laurentia, to the north, but within-plate deformation is continuing as indicated by the arrows. Cadomia has yet to collide with northern Britain–southern Baltica and Avalonia with the Appalachians. A 2 cm per year convergence rate has been assumed.

Figure 11.16 Line drawing of migrated seismic reflection profile run offshore east of northern England–southern Scotland, known as the NEC (North-East Coast) profile. Data were acquired and processed by the British Institutions Reflection Profiling Syndicate (BIRPS). Geological interpretation is indicated by shading to illustrate the four principal terrane components; the stable Midland Valley block to the north is a Proterozoic terrane and the Midland Platform is a late-Precambrian–Palaeozoic volcaniclastic terrane. Dotted ornament indicates imbricated ocean-floor sediments trapped in the suture zone. Arrows indicate probable directions of thrusting during closure. (After Freeman *et al.*, 1988.)

11.16) occurs beneath many terranes and the layering is often younger than the upper-crustal tectonic age. This points to the importance of crustal delamination and underplating processes for establishing and maintaining the typical 35–40 km thickness of continental crust. Gravity and magnetic data have been similarly influential in large-scale tectonic studies (e.g. Lee *et al.*, 1990).

Returning to the closure of Iapetus, thick Lower Palaeozoic greywacke sandstone–shale sequences developed in the closing ocean basin. As the ocean closed about 450 Ma ago across Britain, an interthrust sequence of ocean-floor lavas and sediments was forced both down into the mantle and across parts of the upper crustal forelands (Fig. 11.15) Ophiolite sequences were emplaced along the final suture and along zones where former island-arc terranes of the

231

Iapetus margins were accreted during closure of their marginal basins (e.g. in Anglesey, North Wales). Major strike-slip movements of up to 1500 km between different terrane blocks have been identified from sediment provenance studies on the northern Iapetus margin. Finally, Cordilleran calc-alkaline intrusive (granite, granodiorite, etc.). and extrusive (andesite) activity occurred on both flanks of Iapetus.

If plate *closure* events are recognized in the geological record, they must be complemented by extensional, rifting events that led to plate *separation*. As with the Proterozoic record, it seems that large-scale mantle convection has been associated with the aggregation and dispersal of supercontinents (Fig. 11.13). The last such supercontinent clearly to be identified, Pangaea (Fig. 11.17), aggregated in late Palaeozoic times, stayed united for 100 Ma or more, and dispersed during the Mesozoic. The evidence comes from palaeoclimatic and palaeomagnetic indicators, the latter in the form of a succession of apparent pole positions indicating the palaeolatitude of the continent at the times the relevant sample formed; these are **apparent polar wander curves**. Figure 11.18, for example, indicates that Europe and North America were in southerly latitudes in Cambrian times, converged at the end of the Silurian (Iapetus closure), moved together northwards as part of Pangaea, and then separated at the end of the Cretaceous as the modern North Atlantic formed. High-resolution seismic

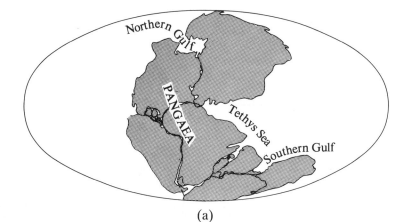

(a)

Figure 11.17 The distribution of modern continents (a) 200 Ma ago and (b) 180 Ma ago, just after fragmentation started to break up the supercontinent Pangaea into the northern and southern landmasses, Laurasia and Gondwanaland. Arrows in (b) indicate the direction of continental movements. Evidence is based on a combination of palaeoclimatic and palaeomagnetic data (see text).

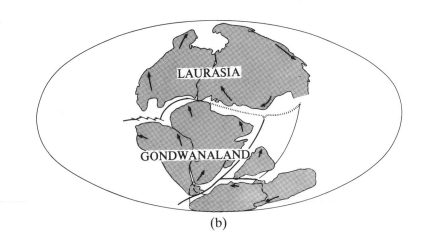

(b)

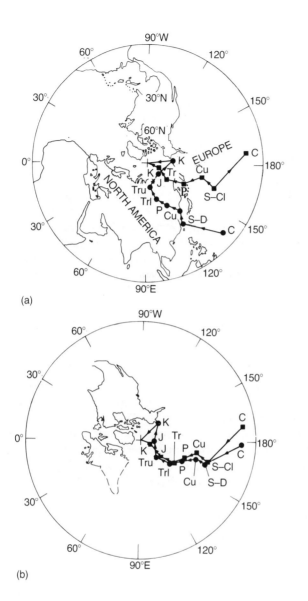

(a)

(b)

Figure 11.18 Polar wander curves for Europe (squares) and North America (circles) for Phanerozoic times with the continents (a) in their present-day positions, and (b) in their best-fitting positions as following Iapetus closure and prior to formation of the modern North Atlantic. Notice that both continents had similar curves for most of Phanerozoic time in (b), suggesting this was the appropriate relationship between them for the Silurian–Cretaceous interval. Note also that the Cambrian poles for both continents lie near the equator, and to bring these back to the actual North Pole we would have to move the continents south, beyond the equator into the Southern Hemisphere. They have drifted north by nearly 90° on a great circle since that time. Ages are: C= Cambrian, 550 Ma; S–D = Siluro – Devonian, 390 Ma; S–Cl = Silurian–Lower Carboniferous, *c.* 350 Ma; Cu = Upper Carboniferous, 300 Ma; P = Permian, 250 Ma; Trl = Lower Triassic, 220 Ma; Tru = Upper Triassic, 200 Ma; J= Jurassic, 150 Ma; K= Cretaceous, 100 Ma. (Redrawn from McElhinny, 1973.)

tomographic data for the upper mantle (Anderson *et al.*, 1992) show that we are still living in a period of plate dispersal, for there are no large continental areas overlying low-velocity upper mantle, which is concentrated instead beneath the Pacific and Indian oceans.

Anderson *et al.* (1992) emphasized, however, that moving plates overriding a hot region of the upper mantle will be put into tension and will behave as if impacted by a giant mantle plume. This is raising again (cf. Chapter 9) the question of whether there are, indeed, thermal plumes penetrating upwards from the lower mantle or whether, instead, there is simply hot upper mantle capable of plume-like behaviour that has either:

1. been insulated by continental cover; or
2. not recently been affected by cold, downgoing oceanic plates.

233

Although this remains unresolved, what is clear is that hot upper mantle was in some way responsible for the major Phanerozoic continental flood-basalt provinces. Over 1.5×10^6 km^3 was erupted in each of the following events at rates exceeding 1 km^3 per year; eastern Greenland/north-west Europe (erupted 60 Ma ago), the Deccan Traps of India (65 Ma), the Parana of Brazil and Etendeka of West Africa (130 Ma), and the Karoo of South Africa (180 Ma). Many of these flood-basalt provinces do lie along continental margins, where they were associated with incipient rifting. Richards *et al.* (1989) proposed a model in which the flood basalts themselves represent plume 'heads', and that the hot spots, which continue to be active after rifting, represent the plume 'tails' where magma is still rising through the activated conduit. Thus for our four provinces, hot-spot magmatism is seen in the ocean islands, respectively, of Iceland, Réunion, Tristan de Cunha and Marion–Prince Edward (Crozet) (Fig. 9.1).

Continental flood basalts, the surface expression of vigorous Phanerozoic extensional tectonics, remind us that extensional magmatism elsewhere may also be contributing to crustal growth by underplating, and without such obvious surface expression. From Phanerozoic evidence we are now beginning to understand the relationship between magmatism, plate tectonics and mantle convection. The continents move principally because they are attached (via oceanic plates) to subduction zones (Chapters 8 and 9), but they may break up when this motion brings them over hot, penetrative upper mantle.

11.4.2 Life in the Phanerozoic

The preserved hard parts of animal fossils first occur in Cambrian rocks, and for this reason the Proterozoic/Phanerozoic boundary traditionally has been placed at this stage in palaeontological evolution. There was a dramatic increase in the number and diversity of marine species; both the ecology and environmental conditions of the middle Cambrian have been illuminated by recent studies of the Burgess shales of the Canadian Rockies (Conway Morris, 1992). Here both the soft and hard parts of a very diverse fauna are beautifully preserved in muddy sediments at the foot of an algal reef. Land plants did not develop until about 460 Ma ago, at which point greening of the land surface commenced. The earliest plants were probably woody, and grew in marshy conditions, but by early Devonian times (380 Ma ago), dense plant-life was spreading out over the land surface. Significant quantities of coal, produced by decaying vegetation, appear first in the Devonian geological record (Fig. 11.10). The late Carboniferous saw the maximum development of massive forests and this means that by 270 Ma ago, atmospheric oxygen had probably reached its present level. Meanwhile, in the late Cambrian and Ordovician the first fish evolved; initially jawless filter feeders, by Devonian times they had developed into highly successful predatory animals. Amphibious creatures also appeared in the Devonian as an evolutionary branch of the fish kingdom. As ecological niches on land diversified, amphibians gave rise to reptiles and eventually, by 250 Ma ago, to the mammals (Fig. 11.10) – though the latter did not become widespread until about 65 Ma ago.

In the context of a book concerned primarily with the evolution of the Earth's interior, inevitably this is very much a thumbnail sketch of organic evolution. There is a view that life and the environment are part of a single self-regulating system: this is the **Gaia hypothesis** (Lovelock, 1988; Watson, 1991). The system as a whole appears to be nearly immortal, capable of surviving major crises of external origin, including **mass extinctions**. The last such event occurred at the Cretaceous–Tertiary (or K–T) boundary, and probably resulted from the impact of a comet or asteroid capable of raising global forest fires, perhaps inducing major volcanism, and temporarily changing surface environments. Large flood-basalt eruptions, releasing acid aerosols that promote global cooling, may have a similar capacity (Rampino, 1992). Despite occasional natural catastrophes, the composition of the atmosphere has probably changed very little in the past 250 Ma and, today, there is a complex equilibrium in the atmosphere–hydrosphere–biosphere system dominated by:

1. the production of oxygen from CO_2 by plants; and
2. the consumption of oxygen and regeneration of CO_2 by respiration of animals.

In the last century, particularly, human activities have also increased atmospheric CO_2 contents through the burning of fossil fuels that took hundreds of millions of years to form. But the debate about the greenhouse effect and contemporary environmental change is quite another story.

11.5 ISOTOPES AND CRUSTAL EVOLUTION

We are now in a position to continue the discussion of Section 11.1 concerning crustal growth and recycling. Isotope geochemistry is a particularly fruitful approach, for there are marked differences between the isotope ratios produced by magmas derived from the crust and the upper mantle (Fig. 9.3). To illustrate this problem, Figure 11.19(a) shows that the oldest Amîtsoq gneisses of western Greenland formed 3800 Ma ago, whereas the voluminous Nûk gneisses of the same area are 2850 Ma old. These were once interpreted as having formed by the remobilization of older, Amîtsoq-like Archaean crust, but Figure 11.19(a) shows that by the time the Nûk gneiss formed the Amîtsoq gneiss would have evolved to a $^{87}Sr/^{86}Sr$ ratio higher than that found in the Nûk gneiss. Thus the Nûk gneisses represent a new major addition to the continental crust, rather than being the product of reworking. Conversely, isotopic data from some late K-rich granites, such as the 2650-Ma-old Qôrqut intrusion of Greenland (Fig. 11.19(b)), show that crustal remelts can be distinguished from new additions to the crust.

A similar approach has been applied to many continental intrusive rocks of different ages. There are two assumptions, first that new crustal additions always have initial isotope ratios (Note 4) reflecting the upper mantle growth curve (Fig. 11.19(b)), and second, that the Rb/Sr ratio of the source region is known. Thus, suppose the Qôrqut granite (C in Fig. 11.19(b)) was derived by remelting Amîtsoq gneisses with the lowest Rb/Sr ratio. Its initial ratio of 0·709 would indicate a crustal residence age *at the time of its formation* of 1150 Ma, the difference between the source and its daughter intrusion, i.e. 3800 Ma minus 2650 Ma. Seen from the present day when we are doing the analysis, the age of intrusion is 2650 Ma and the crustal residence age is 3800 Ma. Notice one more assumption, that the intrusion is derived from just one source, namely the ancient basement. In fact most major granitic (*sensu lato*) intrusions have mixed sources (cf. Section 10.3.2), often including both new mantle and older crustal components. The mantle contribution reduces the residence age, so really we

235

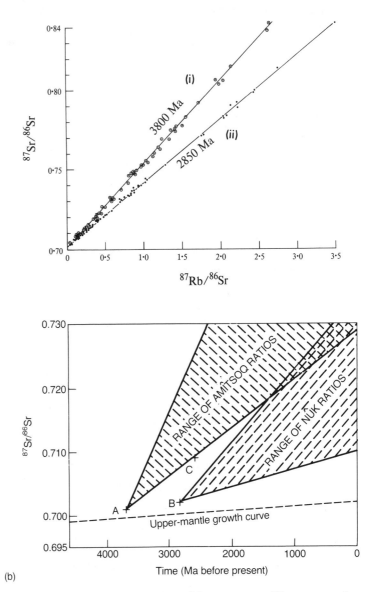

Figure 11.19 (a) Rb–Sr whole-rock isochrons for (i) the older Amîtsoq, and (ii) the younger Nûk gneisses of western Greenland. (b) Strontium isotope evolution diagram for Amîtsoq (A) and Nûk (B) gneisses, showing the possible range of $^{87}Sr/^{86}Sr$ that could evolve as ^{87}Rb decays to ^{87}Sr. Point C is the later Qôrqut granite of western Greenland. (After Moorbath, 1977.)

are speaking of *apparent* crustal residence ages. However, whenever the residence age exceeds the age of intrusion, clearly some *crustal recycling* has been involved in magma genesis.

It turns out that Sm–Nd analysis (Note 4) is generally more suitable for this approach because the major continental processes of rock alteration, erosion, sedimentation and metamorphism have a much smaller effect on Sm/Nd than on Rb/Sr ratios. A compilation of granite data (Fig. 11.20(a)) illustrates that crustal recycling was relatively minor in Archaean times, but then became much more significant. Indeed, about half the Phanerozoic granites apparently are derived from basement with residence ages greater than 2000 Ma. Other recent granites have shorter residence times, and some arise from new crustal sources. So, some totally new additions to the crust are coming from the mantle, as anticipated in Chapter 10: of course, these additions are basaltic. Of course the isotopic evidence in Figure 11.20(a) therefore supports the view that the crust is 'growing

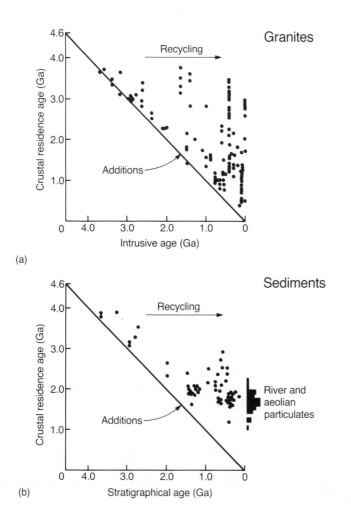

(a)

(b)

Figure 11.20 (a) Plots of crustal residence age based on the Sm–Nd isotopic method against (a) intrusive age for granites (*sensu lato*) from the continents, and (b) stratigraphical age for sedimentary continental samples. Note that new additions to the crust will be characterized by equal ages on both axes, whereas departures towards higher crustal residence ages represent greater degrees of crustal recycling. (See text for further discussion; after O'Nions, 1992.)

from below' (Rudnick, 1990) mainly by basaltic underplating. More granitic magmas develop by partial melting of crust with a variety of residence ages, often with assimilation, magma mixing and homogenization taking place before emplacement.

A complementary approach comes from the isotopic analysis of sedimentary rocks (Fig. 11.20(b)). These reflect the average isotopic composition and hence residence age of the continental material being eroded. By comparing the best-fit stratigraphical age for individual samples with the isotopic residence age (Fig. 11.20(b)) we find a smoothed version of the granite picture. This is not surprising because most sedimentary depositional environments are sampling a wide range of different source regions. It is this evidence that leads to the conclusion that the *modern* sedimentary mass, as represented by rivers and aeolian particulates, has a *mean* residence age of 1700 Ma. As with the granite data, this emphasizes the progressive nature of additions to the continental crust.

Returning to the continental growth models of Figure 11.2, at first sight, evidence from crustal residence ages strongly favours the slow-growth model (3). However, we need to account for loss of material from the continents, which we know is happening by basal delamination and, in some areas, by sediment subduction: this loss must be more than balanced by the addition of new material if growth is occurring. As O'Nions (1992) pointed out, once replacement is allowed, the isotopic record tells us merely about the continental *survival* of material rather than growth. Nevertheless, loss of continental crust is widely considered to be much smaller volumetrically than the typical additions occurring today (*c.* 0.8 km^3 per year, Section 10.3), thus favouring a growth scenario somewhere between models (2) and (3). Having reached this conclusion, however, it is only fair to reflect that some researchers believe there is good evidence for episodic crustal growth, with a major contribution around the Archaean–Proterozoic boundary.

A remaining problem is how and when the upper mantle came to be depleted of incompatible elements. An analysis of some of the oldest (3800 Ma) Isua metasediments from West Greenland has generated a residence age of 400 Ma at the time of their formation. Remarkably, the precursor crust was derived from *already* depleted mantle sources. So where is the corresponding isotopically enriched component if the upper mantle was already depleted by 4200 Ma ago? One possibility is that a primitive 'pre-Archaean' *basaltic* crust may have effectively removed incompatible elements and later have provided the foundation on which tonalite–granite crust could develop. We would not see this primitive basaltic crust because of later reworking and vertical growth. Alternatively, the early depletion event might be connected with the terrestrial magma ocean, in which case the enriched (magmatic) zone may have migrated downwards as melts were trapped in the mantle. At present, this problem remains much debated and unresolved.

We conclude that the isotopic evidence supports a model of progressive crustal growth, with some recycling, throughout the last *c.* 4000 Ma of the Earth's history. There may have been periods of above-average growth, particularly when the continents were in the early stages of disaggregation. The isotopic arguments presented above are based on an assumed mantle growth curve that became depleted relative to Bulk Earth before the continents formed. The enriched component, formed by low degrees of melting to concentrate the incompatible elements, may have produced a primitive basaltic crust and/or be trapped at depth in the Earth's mantle.

11.6 A HISTORY OF THE EARTH AND A FINAL SUMMARY

This final section brings together the available evidence concerning the evolution of the Earth as a whole. We start with interstellar gas and dust previously processed inside stars, most of which was generated in supernovae explosions. The Solar System probably formed about 4550 Ma ago from a nebula, or cloud of this material, which flattened under gravity to form a disc rotating about the proto-Sun. Chemically, the gas and dust mixture consisted mainly of hydrogen (70%) and helium (28%), with the small balance comprising all the heavier elements. The bulk of this matter, at the centre of the disc, contracted and heated up until it became hot enough to start hydrogen-burning and be a typical star: the Sun. The mixture of gas and dust in the remainder of the nebula was chemically fractionated by strong radial temperature and pressure gradients so that, for example, material accreting to form the planets tended to contain progressively more low-temperature condensates further from the Sun. Within the first 10^5 years or so, first coagulation of grains had occurred to form metre-sized planetesimals, and then collisions between these had led to the growth of planetary embryos, ranging up to 10% of the Earth's present mass. This rapid development of the Solar Nebula produced the meteorite parent bodies and established different U/Pb ratios within them. As these events all occurred within a small interval about 4550 Ma ago, it is to this stage of development of the Solar System that the 'age of the Earth' refers. Subsequently, a series of giant collisions over a period of about 100 Ma between most of these embryos produced the planets.

A starting composition for the primitive Earth is provided by the least-differentiated chondritic meteorites, which are believed to be fossils of the 'terrestrial' part of the nebula, largely unaffected by later events. But the largest planetary embryos, being more massive than the chondritic meteorite parent planets, became strongly heated in the late stages of accretion as kinetic energy was converted to heat. Internal chemical segregation started when temperatures became high enough for melting to occur. Close to the surface, an internal magma ocean developed and a eutectic mixture of iron, nickel, sulphur and/or oxygen – together with siderophile trace elements – collected and percolated down through the more abundant 'mantle' silicates, because of its higher density, to form a core. Some of these differentiated planetary embryos collided to build the Earth, and the last (oblique) impact, by a relatively oxidized embryo, formed the Earth–Moon system. The Earth's mantle was strongly heated by the impact, probably forming a surface magma ocean, and the Earth's chemistry was slightly modified by the admixture of the oxidized embryo.

Formation of the core was mostly complete within about 100 Ma of the accretion of the Earth, but development of the crust – and also the atmosphere, hydrosphere, and biosphere – took far longer, and indeed still continues. Some time within the first 500 Ma of the Earth's existence, the magma ocean froze and a thin, rigid lithosphere developed. But the interior remained very hot and so convection occurred throughout, removing the heat released by radioactive decay, by cooling of the Earth and, later, when the solid inner core had begun to form, by crystallization. Since convection, even by solid-state creep in the mantle, is such an efficient transporter of heat, the temperature gradient has never been much above the adiabat (except in the thermal boundary layers). Instead, as the production of heat by radioactive decay declined, the vigour of convection also declined. Only a moderate decrease of temperature has occurred, but the pattern of convection may have changed from mainly plumes to today's broad convection cells, plus a few plumes.

The mineral structures of mantle silicates change across the Transition Zone (410–660 km) towards more dense forms, and it appears that the 660-km discontinuity acts as a partial barrier to convection. This means that the trace-element and isotope geochemistry of the upper and lower mantle has evolved differently, with the upper mantle being rapidly depleted of incompatible elements early in the Earth's history. A two-layer convecting mantle seems the most appropriate physical description, though plumes may rise from the D'' thermal boundary layer at the base of the mantle and some rapidly descending subducted plates may also penetrate the 660-km discontinuity.

The base of the lithosphere is defined by the temperature at which mantle silicates deform inelastically, close to their melting temperature. At the very beginning of the Earth's history, this lithosphere was thin, unstable and easily resorbed into the mantle. But by 4000 Ma ago, the first light, unsinkable granitic crust had formed as the uppermost and most chemically fractionated constituent

239

of the lithosphere. For the next *c.* 1500 Ma the thermal regime favoured the development of microcontinental plates, and horizontal lithospheric forces ensured that these accumulated into cratons. Once this had occurred, the whole continental lithosphere thickened rapidly because chemical depletion produced refractory mantle lithosphere. Geologically, the small nuclei of Archaean gneisses, separated by folded low-grade greenstone sequences, contain high-pressure metamorphic rocks, indicating a thick crust and so favouring a thick lithosphere by the end of the Archaean.

Some hundreds of millions of years after the Earth had formed, its very slow cooling caused a solid inner core to form and grow at its centre. An iron-rich Fe–Ni alloy crystallized from the molten Fe–Ni–S (or O). This leaves a layer at the base of the outer core depleted in iron and so lighter, and its buoyancy probably drives the geodynamo, enhanced by the heat released by cooling and the latent heat of crystallization that inevitably accompany the inner core growth.

The atmosphere and oceans were produced by prodigious volcanic outgassing and cooling; the early atmosphere contained large amounts of CO_2, N_2 and H_2O. The oceans provided a suitable environment in which early organisms developed about 3500 Ma ago, shielded from harmful ultraviolet solar radiation. There was little oxygen in the atmosphere at this time, but it was produced, first, by the photochemical dissociation of water and then by the photosynthetic activities of simple prokaryotic organisms, such as blue-green algae.

A marked change in geological and tectonic patterns across the Archaean/Proterozoic boundary reflects the long-term cooling of the mantle. Ocean crust was getting thinner, continental plates were now thick and broad, and the magmas above subduction zones rose from the mantle wedge rather than directly from the downgoing ocean crust. Thus, Archaean bimodal basalt–rhyolite suites, where mafic magmas penetrated easily up through the crust, were replaced by calc-alkaline andesite (diorite)–dacite (granodiorite) magmatic arc series. Incompatible trace-element data demonstrate that the main additions to the crust were, and still are, mafic magmas, representing small mantle melt fractions. They are emplaced at depth, but are responsible for triggering crustal melting (large melt fractions) and magma mixing, producing intermediate–felsic magmas. Isotopic data from both intrusive complexes and sedimentary rocks of all ages demonstrate that new magmatic material is still being incorporated into the continents and, moreover, suggest that continental growth is still continuing. An important caveat is that continental mass is lost back into the mantle, particularly by delamination of the lowermost, high-density, refractory crust. Nevertheless, progressive continental growth throughout the last 4000 Ma, but at a rate declining generally in sympathy with the exponential decay of long-lived radio-isotopes, is our preferred model.

Linear zones of within-plate, usually alkaline, magmatism became more widespread in Proterozoic times and may mark the sites of lithospheric extension, sometimes leading to plate fragmentation and dispersal. Indeed, cycles of plate aggregation, to form supercontinents, and subsequent dispersal have occurred throughout the last 2500 Ma. The continents move principally because they are attached via oceanic plates to subduction zones, and they fragment when this motion brings them over hot, penetrative upper mantle. Phanerozoic *collisional* sutures are well documented by virtue of their geological similarities with contemporary continental convergence zones; older, more deeply eroded examples are Proterozoic mobile belts. In contrast, Phanerozoic continental *extension* zones occur either where tectonic stresses induce deformation and passive melting (due to pressure release) or because an active mantle plume penetrates the lithosphere. In both cases underplating by basalt magmas contributes to the crustal volume, but fragmentation, usually accompanied by flood-basalt eruption, also requires high strain rates so that the plate weakens during extension. Coming full cycle, early Archaean continental plates were much weaker than today, hence the abundance of mantle-derived mafic magmas in the eruptive suites.

The increase of oxygen in the atmosphere during the Proterozoic stimulated the evolution of and increased diversity of life-forms. The first multi-cellular animals appeared in the late Precambrian while, in the Phanerozoic, life was able to leave the seas and exploit the various evolutionary opportunities offered on land. But this aspect of the Earth's evolution is not entirely dependent on crustal processes for its future development. One day, an unpredictably long time in the future, the Earth's interior will inevitably cool so much that convection, continental movements and hence mountain building, volcanoes and earthquakes, will all decline and cease. Erosion to a base level may cause the surface to be covered with water. But will the evolution of the biosphere and atmosphere also come to a halt? Hardly likely, we suggest, though there will be radical changes by way of response to the new conditions. It is a sobering thought that human activities, which may themselves be causing irreversible changes to the environment, are no more than transient geological phenomena.

FURTHER READING

General journal:
   Kershaw (1990): early evolution of the atmosphere and biosphere.
General books:
   Brown *et al.* (1992): various chapters of 'Understanding the Earth', particularly those
      by O'Nions on the isotopic evidence for crustal growth, by Klemperer and Peddy on
      seismic reflection studies of continental sutures, and by Conway Morris on palaeon-
      tological evidence.
   Nisbet (1991): entertaining account of crustal evolution, with main emphasis on
      evolution of the atmosphere and biosphere.
   Windley (1984): comprehensive account of geological evidence for history of the
      continents.
Advanced journals:
   Bickle *et al.* (1980): geology and tectonics of the Pilbara (Australia).
   Bowring (1992): evidence bearing on the Earth's oldest rocks.
   Gurnis (1986): supercontinent aggregation and dispersal models.
   Hargraves (1986): speculations on spreading rates and ridge lengths in the Archaean.
   Jacobson and Dymek (1988): crustal residence age for the Isua metasediments.
   Martin (1986): changing compositions of subduction-related magmas with time.
   Riding (1992): early life-forms and oxygen levels.
   Schubert (1991): pre-continental evolution of the crust.
Advanced books:
   Piper (1987): palaeomagnetism, polar wander curves and continental movements.
   Taylor and McLennan (1985): geochemical aspects of crustal evolution.

241

Postscript: the new state of ignorance

The first edition of this book ended with a list of outstanding problems on which it was hoped progress would soon be made, and there has been significant progress on most of them. This second edition has therefore advanced to a new state of ignorance (which is what science is all about). Here, then, is a new list of topics needing scientific progress in the coming years:

1. whether other stars have planetary systems;
2. further understanding of the accretionary processes that led to the formation of the Solar System, including how chondrules formed and how meteorites are related to asteroids and comets;
3. what elements occur in the outer core, in addition to iron;
4. details of the geodynamo and its energy source;
5. improved geochemical constraints on modelling mantle-source regions of basalts, and development of geophysical evidence for mantle melting;
6. further research into the implications, for the migration of chemical elements, of mantle melting at small melt fractions;
7. better knowledge of the distribution of radioactivity within the Earth and what proportion of the heat outflow it generates;
8. better knowledge of the temperature distribution within the Earth;
9. better knowledge of the rheology of the mantle;
10. what form convection takes, and particularly to what extent – if any – material crosses the 660-km discontinuity;
11. whether hot spots really are the expression of plumes, and if so, at what depth(s) plumes originate;
12. the nature of the D'' layer;
13. improved interpretation of early crustal processes using modern rheological studies of tectonically active zones.

242

Notes

NOTE 1 MOMENT OF INERTIA, ANGULAR MOMENTUM, ETC.

If Newton's laws of motion are extended from linear motion to rotary motion, it is natural to replace distance moved by the angle though which a body rotates, velocity by angular velocity, force by torque, and so on. The relation:

Force = mass × acceleration

can be replaced by a corresponding one:

torque = I × angular acceleration   (N1.1)

where I is a new quantity chosen to make the relation hold, and is called the moment of inertia (Fig. N1.1).

axis of rotation

r

body

small mass δm

Figure N1.1 Moment of inertia of body rotating about any axis.

The moment of inertia of a small (point) mass δm about an axis at a distance r away turns out to be $r^2 \delta m$, and so for an extended body it is:

$$I = \Sigma r^2 \delta m = \int r^2 \rho dv$$ (N1.2)

The summation is summed over all component δms and the corresponding integral is over the whole volume of the body.

Note that I depends upon the position of the axis as well as upon the size and density of the body. The I is analogous to mass (see Eq. (N1.1)).

Table N1.1 Linear and rotary motion.

Linear quantities		Angular quantities	
Linear displacement	x	θ	Angular displacement
Linear velocity	$V = dx/dt$	$\omega = d\theta/dt$	Angular velocity
Linear acceleration	$a = dV/dt$	$\dot{\omega} = d\omega/dt$	Angular acceleration
Force	F	Γ	Couple or torque
Mass	M	$I = \Sigma mr^2$	Moment of inertia
Momentum	mV	$I\omega$	Angular momentum
	$F = ma = \dfrac{d}{dt}(mV)$	$\Gamma = I\dot{\omega} = \dfrac{d}{dt}(I\omega)$	

The last line of the table shows that momentum, linear or angular, remains constant or is conserved if no external forces act upon a body. Thus, if a body contracts due to internal forces, such as its own gravitational attraction, angular momentum is conserved, and since I must get smaller, ω must increase. This fact is used by a skater spinning on one toe: stretching out the arms slows the rate of spinning; bringing them to the sides speeds it up again. The conservation of angular momentum is important when considering the formation of the Solar System in Chapter 4. It shows that a contracting nebula of gas tends to rotate more rapidly as it grows smaller.

NOTE 2   MEAN ATOMIC WEIGHT ($\bar{m}$)

This parameter was introduced by Birch in 1961. As shown in Chapter 3 seismic velocities provide only two equations involving the three variables κ, μ and ρ, and, to get a unique solution, a third relation is needed. The attempt by Adams and Williamson to provide this by assuming that density depends only on self-compression failed when it was found to be untrue for parts of the Earth. A different approach is to find a direct relationship between seismic velocity and composition.

Birch (1961) measured the seismic P-wave velocity of a large number of minerals and found the empirical relation:

$$V_p = a\bar{m} + b\rho \tag{N2.1}$$

where $\bar{m}$ and ρ are the mean atomic weight and density, respectively, and a and b are constants.

This relation shows – to the extent that the equation holds – that the seismic velocity does not depend on the chemical bonds between the atoms that make up a mineral, but only upon their mean weight. Neither does the detailed structure of the mineral lattice matter except in so far as it affects the separation of the atoms, which in turn affects the density. Thus, changes in volume caused by pressure, temperature or phase changes are all taken into account by the density term, while composition differences affect both terms.

Although it has been shown that Birch's law holds quite well for nearly all minerals examined, the underlying physical reasons for this have not been well understood. A number of persons have suggested that, for example, at very high pressures, a power law would be a better description. In particular, Chung (1972) and Davies (1977) argued that the linear approximation would begin to break down at the higher densities produced by very high pressures. This has been found, for instance, by Liebermann and Ringwood (1973), who have shown that the law is not fully obeyed during polymorphic transitions like those discussed in Chapter 7. Anderson (1973) has suggested a form of the power law that encompasses these high-pressure transitions.

The use of Birch's law has been largely superseded by the results of experimental petrology at increasingly high pressures, but it is still useful in the lower mantle and is employed in PREM (Section 3.6). The value of the parameter $\bar{m}$ is that, being relatively insensitive to chemical composition, it provides a generalized description of the composition; conversely, it follows that even an exact knowledge of $\bar{m}$ would not yield details of what mineral or elements are present.

NOTE 3   ELECTROMAGNETIC STIFFENING, AND HEATING

Consider a loop of wire with a small gap, AB, placed in a magnetic field (Fig. N3.1). There is a magnetic flux through the loop, which is simply the number of lines of magnetic force, or induction, which thread through the loop. Suppose that the flux changes, either because the field changes or because the loop is moved, then an electromotive force (e.m.f.), or voltage, is induced between A and B, proportional to the rate of change of flux (Faraday's law):

$$\epsilon = -\frac{d\phi}{dt}. \tag{N3.1}$$

If A and B are joined, a current flows proportional to the e.m.f., and this is the basis of the generation of electricity in dynamos and generators.

The electrical power cannot be had for nothing, and work has to be done in rotating the loop. This arises because there are forces upon a wire carrying a current when it is in a magnetic field, and these forces oppose the motion. An alternative way of regarding it is that the current in the loop produces a field like that of a magnet (Fig. N3.2) and will try to twist until it is aligned along the lines of force: this twist is opposite to the rotation of the loop.

Note that the magnetic field, and hence flux through the loop, *produced by the current in the loop*, is greatest when the external flux is *changing* most rapidly, and is in such a sense that the total flux change through the loop is less than would be the case if the current did not flow. If the loop has no electrical resistance, then this compensation will be perfect. (This can be deduced because Ohm's law tells us that the e.m.f. is equal to the product of the resistance and the current: if the resistance is zero, then so must be the e.m.f. But Faraday's law, above, states that the flux change is proportional to the e.m.f., so it too must be zero.) Therefore, in a perfect conductor, a current is

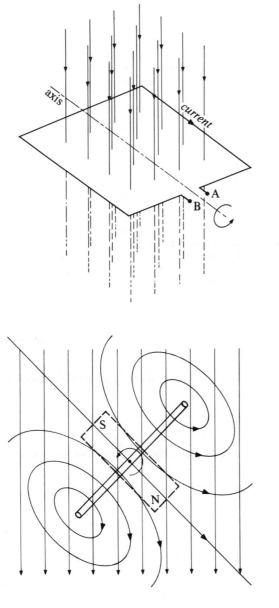

Figure N3.1 Magnetic flux
threading a loop of wire.

Figure N3.2 This uniform flux
induces a current in the loop;
in turn, this current produces
a magnetic field – shown by
the curved lines – which is
similar to that in a bar
magnet.

induced whose associated flux through the loop exactly equals the change of external flux, so that the
total flux remains constant.

If the simple loop were replaced by an arrangement of interconnected conducting loops of any
shape, this effect would ensure that the flux threading every one of them would be unchanging. This
can be generalized to a continuous, perfect conductor, such as a block; in it there can be no flux
change, appropriate currents being induced to ensure this.

Imagine that we try to rotate part of a perfect conductor, say a cylinder, within the rest. Currents
are induced to keep constant the flux within the cylinder, and forces are produced opposing the
rotation, just as with the loop (Fig. N3.2). Thus, in general, the presence of a magnetic field in a
conductor tends to prevent relative movements within it, and therefore effectively stiffens it.

These conclusions are applicable to an interstellar cloud or other gas containing ions, for such an
ionized gas is a very good conductor. This may seem surprising for such a tenuous material, but it
arises because the ions can move easily, with few collisions; as a result, even a very small electric
field, or voltage difference, will cause them to accelerate and reach high velocities, and the flow of
charges forms a current. To the extent that some collisions do occur, the cloud is not a perfect

245

conductor and so flux changes can occur, albeit slowly. The conclusion remains that a magnetic field in a fluid conductor stiffens it, and opposes relative motions within it.

One important example of a plasma is the solar wind, which is stiffened as it blows away from the Sun, and thereby transfers angular momentum from the Sun, as described in Section 4.3.2. A plasma opposes changes of magnetic flux through it, and so it follows that a moving plasma tends to carry with it any flux threading it, and so acts as a moving magnet. When the plasma passes near a body, this moving field can induce currents in the body. If the field is sufficiently strong and if the body is sufficiently conductive, the resulting currents may be large, and their resistive dissipation in the body will heat it. It is thought that the very strong solar wind produced by the early Sun contributed to the heating of planetesimals.

NOTE 4   RADIOMETRIC DATING

Some nuclei are unstable and spontaneously convert to another isotope by emission of a nuclear particle, such as an α-particle or β-particle (see Note 5). Remarkably, the chance of this happening in any given interval is quite independent of the time the nucleus has already existed, but depends upon the particular nucleus. Thus the rate of decay – the number of decays in, say, a year – depends upon the number of atoms present:

$$\frac{dN_P}{dt} = -\lambda N_P \tag{N4.1}$$

λ, the decay constant, depends upon the isotope decaying. Integrated, this gives:

$$N_{P,t} = N_{P,0}\, e^{-\lambda t} \tag{N4.2}$$

where $N_{P,0}$ is the number of atoms present initially, and t is the time elapsed. Suppose one parent atom decays into one daughter atom:

$$N_{D,t} = N_{P,0} - N_{P,t}$$
$$= N_{P,0}\,(1 - e^{-\lambda t}). \tag{N4.3}$$

Using Eq. (N4.2) to eliminate $N_{P,0}$ from Eq. (N4.3) (since we were not there to measure it), but retaining present quantities:

$$N_{D,t} = N_{P,t}\, e^{-\lambda t}\,(1 - e^{-\lambda t})$$
$$= N_{P,t}\,(e^{\lambda t} - 1). \tag{N4.4}$$

Rearranging:

$$\left(\frac{N_D}{N_P}\right)_{now} = (e^{\lambda t} - 1). \tag{N4.5}$$

This is the **Basic Dating Equation,** i.e. the present ratio of the numbers of daughter and parent atoms can be used to deduce 't', the time since the daughter atoms began accumulating (see Fig. N4.1).

Figure N4.1 As the parent isotope decays away, it turns into the daughter isotope: if one parent atom produces one daughter atom the curves are as shown, such that their sum at any instant is constant, equal to $N_{P,0}$. The ratio of numbers of atoms of daughter to parent depends uniquely upon the time elapsed and so may be used to date the time since closure. (The half life, $t\frac{1}{2}$ is shown for comparison.)

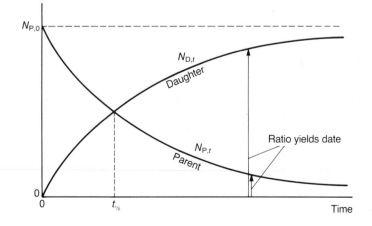

246

This equation will give a correct date provided that:

1. the system has remained closed since the event to be dated, i.e. there was no loss or gain of parent or daughter atoms, except by decay;
2. there were no daughter atoms present at closure, and
3. a parent atoms decays into one daughter.

The simplest example is the decay of uranium into lead. There are two isotopes of uranium, which decay:

$$^{238}U \rightarrow {}^{206}Pb$$
$$^{235}U \rightarrow {}^{207}Pb.$$

The ratio of either daughter/parent pair can be used in Eq. (N4.5). They should give the same date, which provides a check.

The **uranium–lead, U–Pb, dating method** can be used only on a few minerals which, when formed, contain appreciable uranium but negligible lead. Zircon is the principal mineral used.

Another widely used decay is ^{40}K into ^{40}Ar, giving rise to the **potassium–argon, K–Ar, dating method**. The ^{40}K forms only about 0·01% of potassium, but potassium is so abundant and ubiquitous that the method has wide applicability. Only 11% of the total decays of ^{40}K form ^{40}Ar (the rest forming ^{40}Ca) and so Eq. (N4.5) is modified by inserting the factor 0·11 on the right-hand side.

The **rubidium–strontium, Rb–Sr, dating method** depends upon the decay of ^{87}Rb into ^{87}Sr, but allowance has to be made for the presence of strontium at closure. This can be done because:

1. strontium also contains the isotope ^{86}Sr, which neither decays nor is formed by decay, and so can be used as a reference isotope; and
2. different minerals take up Rb and Sr in different ratios.

If a well-mixed magma crystallizes into an assemblage of different minerals, initially these minerals have different $^{87}Rb/^{86}Sr$ ratios, but the same $^{87}Sr/^{86}Sr$ ratio, since isotopes of a single element behave the same chemically. Thus initially, the compositions lie along a horizontal line in Figure N4.2.

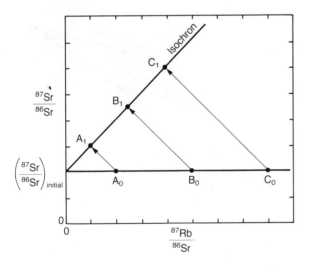

Figure N4.2 At the time of closure, all minerals A, B, C . . . have the same isotopic composition of strontium (and other elements) $(^{87}Sr/^{86}Sr)_{init}$, but different ratios of rubidium to strontium, given by the $^{87}Rb/^{86}Sr$ ratio. Subsequently, ^{87}Rb decays into ^{87}Sr and the compositions evolve along growth lines up to the left at an angle of 45°. Since the rate of decay is proportional to the amount of ^{87}Rb present, at any instant the compositions lie on a straight line, an isochron, which rotates anticlockwise with time. Its slope yields the time since closure.

Thereafter, as ^{87}Rb decays into ^{87}Sr, one for one, the $^{87}Rb/^{86}Sr$ ratio decreases and the $^{87}Sr/^{86}Sr$ ratio increases, and isotopic compositions evolve up to the left (at 45°, if the axes have the same scales). A mineral, B, whose initial $^{87}Rb/^{86}Sr$ ratio, B_0, is twice that of another, A_0, will evolve at twice the rate of A, so at some subsequent time all isotopic compositions lie on a straight line $IA_1 B_1$ . . . . The equation of this line is:

$$\left(\frac{^{87}Sr}{^{86}Sr}\right)_{now} = \left(\frac{^{87}Sr}{^{86}Sr}\right)_{init} + \left(\frac{^{87}Rb}{^{86}Sr}\right)_{now} (e^{\lambda t} - 1). \qquad (N4.6)$$

$$y \quad = \quad c \quad + \quad x \quad \quad m$$

(It derives from Eq. (N4.5) by adding the amount of ^{87}Sr present initially and dividing throughout by the reference isotope ^{86}Sr, which, since it does not change, can be given the subscripts of either 'init' or 'now'.)

The line is called an **isochron** (same age) and has a slope equal to $(e^{\lambda t} - 1)$, whence t, the age, may be calculated. Where it cuts the ^{87}Sr/^{86}Sr axis gives the initial ratio, though this is not needed to find the date.

An analogous method is the **samarium–neodymium, Sm–Nd, dating method,** involving the decay of ^{147}Sm into ^{143}Nd. In this case, the isotope ^{144}Nd is used for reference purposes, and the isochron equation becomes:

$$\left(\frac{^{143}\mathrm{Nd}}{^{144}\mathrm{Nd}}\right)_{\mathrm{now}} = \left(\frac{^{143}\mathrm{Nd}}{^{144}\mathrm{Nd}}\right)_{\mathrm{init}} + \left(\frac{^{147}\mathrm{Sm}}{^{144}\mathrm{Nd}}\right)_{\mathrm{now}} (e^{\lambda t} - 1). \tag{N4.7}$$

To use these methods, a block of igneous rock is crushed and different minerals are separated from it, and their isotopic compositions are analysed separately and then plotted as above. The compositions should fit a single line within their experimental errors. Alternatively, blocks from different parts of a single igneous body may differ sufficiently in parent/daughter ratio to employ the same theory. Blocks are selected, ground up, and a representative portion of each is taken for analysis. The two approaches are referred to as mineral and whole-rock isochron dating, respectively.

The last method to be described is the **lead–lead, Pb–Pb, isochron method.** It is applicable when lead is present initially. As with the Rb–Sr and Sm–Nd methods, a reference isotope, ^{204}Pb, is used. An equation analogous to Eq. (N4.6) is written for each of the uranium decays described above:

$$\left(\frac{^{206}\mathrm{Pb}}{^{204}\mathrm{Pb}}\right)_{\mathrm{now}} = \left(\frac{^{206}\mathrm{Pb}}{^{204}\mathrm{Pb}}\right)_{\mathrm{init}} + \left(\frac{^{238}\mathrm{U}}{^{204}\mathrm{Pb}}\right)_{\mathrm{now}} [\exp(\lambda_{238}t) - 1] \tag{N4.8}$$

$$\left(\frac{^{207}\mathrm{Pb}}{^{204}\mathrm{Pb}}\right)_{\mathrm{now}} = \left(\frac{^{207}\mathrm{Pb}}{^{204}\mathrm{Pb}}\right)_{\mathrm{init}} + \left(\frac{^{235}\mathrm{U}}{^{204}\mathrm{Pb}}\right)_{\mathrm{now}} [\exp(\lambda_{235}t) - 1].$$

Figure N4.3 On this diagram the initial lead composition – the same for all samples – is a point, P. As ^{235}U and ^{238}U decay into ^{207}Pb and ^{206}Pb, respectively, but at different rates, the lead isotopic compositions evolve along growth curves up to the right. The higher the U/Pb ratio the faster the curves evolve, but at any instant the compositions lie on a straight line, the isochron, which rotates clockwise with time and whose slope yields the time since closure. A sample without any uranium remains unchanged at P; for the Solar System this is termed primeval lead.

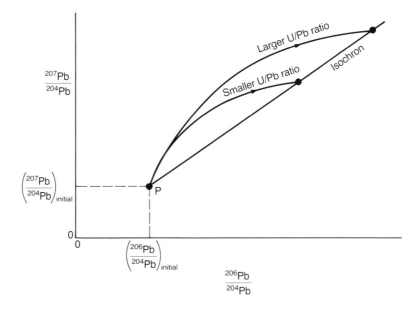

Since both parents are the same element, samples of uranium contain them in the same ratio, which at the present day is ^{238}U/^{235}U = 137·9. Therefore, ^{235}U can be replaced by ^{238}U/137·9 so that both equations have ^{238}U. Next, ^{238}U is eliminated by dividing one equation by the other. After rearrangement, we have:

$$\left(\frac{^{207}\mathrm{Pb}}{^{204}\mathrm{Pb}}\right)_{\mathrm{now}} - \left(\frac{^{207}\mathrm{Pb}}{^{204}\mathrm{Pb}}\right)_{\mathrm{init}} = \left[\left(\frac{^{206}\mathrm{Pb}}{^{204}\mathrm{Pb}}\right)_{\mathrm{now}} - \left(\frac{^{206}\mathrm{Pb}}{^{204}\mathrm{Pb}}\right)_{\mathrm{init}}\right] \times \left\{\frac{1}{137·9}\frac{\exp(\lambda_{238}t-1)}{\exp(\lambda_{235}t-1)}\right\}. \tag{N4.9}$$

This is the equation of a straight line with slope equal to the term in the furthest right bracket, so it too is an isochron (see Fig. N4.3). The isochron does not originate from a point on the $^{207}Pb/^{204}Pb$ axis, but from a point $(^{207}Pb/^{204}Pb)_{init}$, $(^{206}Pb/^{204}Pb)_{init}$, within the axes.

By combining the two independent isochron equations, we have lost the opportunity of a check similar to the two U–Pb dates, above, but it allows samples with no uranium to be used. Troilite in iron meteorites contains uranium but no lead (Section 4.5.4), and so cannot be dated, but the lead-isotopic composition lies on a Pb–Pb isochron for chondrites; thus iron meteorites have an origin in common with chondrites, and the troilite lead is identified as initial, or primeval, lead.

The pros and cons of the various dating methods. Which dating method could be useful for a given sample depends upon several factors. The parent must be present in measurable amounts, and the time elapsed since the event to be dated must have been long enough for the daughter/parent ratio to have evolved by a measurable amount. In addition, samples and methods vary in their resistance to metamorphism or alteration. Thus oceanic basalts are too young and low in Rb/Sr ratio to be datable by the Rb–Sr method, but the K–Ar method can be used. However, the K–Ar method is fairly easily disturbed by metamorphism, and so is of little use for determining the crystallization age of a rock subsequently metamorphosed, and U–Pb on zircons, or the Rb–Sr and Sm–Nd isochron methods on whole rocks would be superior; the Sm–Nd method is especially resistant to weathering and metamorphism, and is applicable to mafic rocks, which have little Rb. However, several methods may reveal more of a history than any one alone. For example, if U–Pb, Rb–Sr, Sm–Nd and K–Ar dates are the same, the sample has not been metamorphosed (unless so severely that it has been completely 'reset'). Resistance to reheating is expressed by the closure temperature, the temperature below which a particular mineral is closed for a particular dating method. Since different mineral/dating method combinations have different closure temperatures, a decrease in dates obtained from higher to lower closure temperature combinations reveals the cooling history.

Dalrymple (1991) gives an account of radiometric dating, and its application to meteorites and the age of the Earth, that is thorough but assumes no expertise. The subject can be further explored in *Isotope Geology* (Faure, 1986), while its applications to meteorites are discussed in Tilton (1988) and subsequent chapters in the same volume (Kerridge and Matthews, 1988).

NOTE 5   THE ATOM AND THE NUCLEUS

An atom consists of a tiny nucleus containing a number of protons, each having unit positive electric charge, surrounding which are an equal number of negatively charged electrons, making the atom electrically neutral. The electrons form a vague cloud about the nucleus, but it is convenient to think of them as belonging to concentric shells, those in the outer shells being more loosely attracted to the central charge than those in the inner ones. Chemical reactions between atoms depend upon interactions between the electrons in their outermost shells but, as the number of electrons depends upon the number of protons in the nucleus, determining the nuclear charge effectively determines which element an atom is. We shall be concerned only with nuclei.

In addition to the protons, there are neutrons in the nucleus. They have no charge, so their presence does not affect the number of electrons in the shells and hence which element it is; different numbers of neutrons but the same number of protons form a series of isotopes of an element. But neutrons are similar to protons in many ways and can convert from one to the other; neutrons and protons are referred to collectively as **nucleons**. The number of neutrons in a nucleus does affect its properties.

The protons, all having the same positive charge, repel each other, so to keep the nucleus together there must be an attractive force, known as the strong nuclear force. The electric repulsive force which operates between any two protons decreases as the inverse square of the separation, and so all protons repel all other protons, but the strong nuclear force acts only over very short distances and so bonds only between adjacent nucleons. Consider the number of bonds in a nucleus as nucleons are added (Fig. N5.1(a)). With two nucleons there is but one bond, with three there are three and with four there are six. So the number of bonds *per nucleon* has increased in the order: half, one, one and a half. But it does not keep increasing indefinitely because the force operates only between nearest neighbours, and a nucleon can have no more nearest neighbours than the number that surround it (Fig. N5.1(b)); therefore the number of bonds per nucleon will tend towards a constant value. But the electrical repulsion, though much weaker between neighbours than the attractive nuclear force, does keep increasing, because every additional proton repels every other proton, albeit with a force

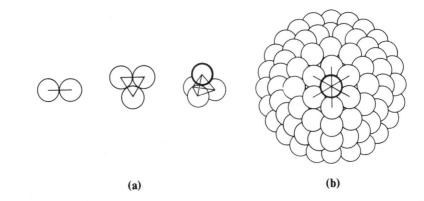

Figure N5.1 Nuclear bonds. As nucleons are added to a nucleus the number of bonds per nucleon at first increases rapidly (a), but then becomes constant because they are short range and only affect nearest neighbours (b).

(a)   **(b)**

that decreases as the square of their separations. Therefore there comes a point at which the repulsive force on an additional proton is greater than the attractive force, and then it cannot be added. Since attractive bonds release energy, while repulsive ones require input of energy to hold particles together, at some point between one and this limiting number there must be a number that releases the maximum amount of energy *per nucleon*. This is the most stable nucleus.

This account explains the general shape of the binding energy per nucleon curve (Fig. 4.7, upper curve), a rise, initially steep, that shallows progressively to give a peak value (iron), followed by a slow decline as the electric repulsive energy builds up. It does not account for the details of the curve. This is because it has ignored additional forces peculiar to atomic particles (quantum forces). One such force favours pairs of protons and neutrons, which is why there are equal number of protons and neutrons in the most stable light nuclei ($^4_2$He, $^{12}_6$C, $^{16}_8$O) (and a multiple of two protons plus two neutrons is usually particularly stable). However, as electric repulsion builds up, neutrons – having no charge – are slightly favoured over protons, giving an excess of neutrons (Fig. 4.6). Just as there is a force among electrons that favours certain numbers – the electron shells that contain successively 2, 8, 8 . . . electrons, giving rise to the inert gases, helium, neon, argon – so there are favoured numbers of nucleons. These are 50, 82 and 126 protons or neutrons (the so-called magic numbers); their effect can be seen in Figure 4.7.

A nucleus becomes unstable long before it reaches the point of repelling a further proton, because of a further quantum effect, and the nucleus adjusts to more stable form in a number of ways: radioactive decay by emission of an α-particle (helium nucleus), β-particle (electron) or γ-ray (X-ray of shorter wavelength and more energy), or fission, in which the nucleus divides into two roughly equal parts, plus some particles.

The above account discusses the stability of nuclei, assuming that the protons can be brought close enough to other nucleons for the attractive nuclear force to be able to operate. To achieve this, the electrical repulsion must be overcome, and this requires a very high speed, produced either by temperatures of millions of degrees, or by atomic particle accelerators.

NOTE 6   ELECTRONEGATIVITY AND THE ELECTRONIC STRUCTURE OF ATOMS

Figure 5.5 shows the Periodic Table, which is used to classify chemical elements according to their electronic structure, essentially a series of electron shells (negatively charged) surrounding a nucleus (positively charged). Values of electronegativity, which is a measure of the ability of a neutrally charged atom to attract an electron and so become an anion, increase from left to right across the table. This is because the outer electron shell is approaching completion and has a smaller number of vacancies towards the right. Additional electrons are attracted to complete the shell, a particularly stable configuration, resulting in negatively charged anions. Conversely, the nearer an atom is to having an empty outer electron shell, towards the left, the more easily it will *lose* an electron to form a stable *cation*. Thus electronegativity, anion-forming ability, increases from left to right. Going down the table, towards more massive and large atoms, the distance between the positively charged atomic nucleus and the most remote, negatively charged electrons increases because there are more electron shells at high period numbers. Thus it becomes more difficult for the nucleus to attract an additional electron into the outermost shell to form an anion, whereas electrons may be lost easily, producing cations. Therefore electronegativity falls from top to bottom of the table as well as from right to left. The most 'electropositive' cations are at the bottom left and the most electronegative anions are at the top right.

NOTE 7   EARTHQUAKE FAULT-PLANE SOLUTIONS

When a region containing a fault is subjected to a shear stress – perhaps by the relative movement of plates – strain builds up in the surrounding area until the fault can no longer resist movement. It yields abruptly, releasing the stored energy as an earthquake, and generates seismic waves. Figure N7.1, which fits observation of the 1906 San Francisco and other earthquakes, depicts the sequence of events.

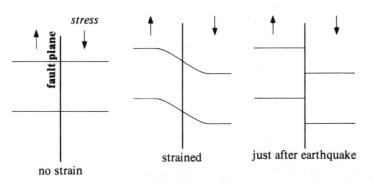

Figure N7.1 Deformation associated with an earthquake.

Any sudden movement of the ground generates seismic waves, but it is important to realize that the motion that they produce at a distant point depends partly on the position of that point (Fig. N7.2). The movement of S causes compression in the direction of A, so that the *first* effect at A is the

Figure N7.2 Motions produced about a displaced point, S.

passage of a **compression**, followed by a series of rarefactions and compressions as the rest of the wave train passes by (cf. Fig. 2.1(a)); at B the *first arrival* is a rarefaction or **dilatation**. Thus A and B both receive P-waves, though of opposite sign. Since there is no movement in the direction of C, it does not receive P-waves, but it does receive an S-wave whose first motion is parallel to the movement of S. In the general case, D, both P- and S-waves are received.

Now consider the effect of movement on the fault, shown by the half-arrows (Fig. N7.3). As far as P-waves are concerned, L receives none, since seismic movement at the focus, F, is transverse to the direction FL. M does not receive P-waves either, because, though each side of the fault generates motion along FM, the resulting waves are of opposite sign and cancel at M. At other places, such as N, the effect of the far side of the fault is less than that of the near side because of the intervening fault plane, and it receives a P-wave (and also an S-wave). Similar arguments apply in other directions, and alternate quadrants receive compressive (+) and dilational (−) first arrivals. The length of the arrows within the lobes indicate the relative *magnitude* of the first arrivals in the direction of the arrows. (S-waves have a similar distribution, though rotated through 45° with respect to that for the P-wave.)

By observing the sense of first movements at seismic stations distributed around the world, it is

251

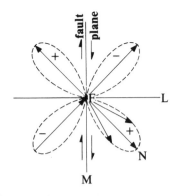

Figure N7.3 Magnitude and sign of first motions (P-waves) due to an earthquake at F.

possible – after making allowances for curvature of ray paths in the Earth – to deduce the pattern. Note that there is an inherent ambiguity: the fault plane could be as shown with dextral movement (as observed of one side of the fault from the other), but the same pattern would result from a fault plane FL with sinistral movement. In practice, it is usually easy to decide between these and so deduce the sense of movement on the fault and of the stress producing it.

Thus, by observing the first motions of the ground, due to the passage of P- and S-waves at a number of seismic stations (Fig. 2.2), it is possible to deduce not only the location of an earthquake, but also the direction of the associated fault plane and the sense of movement across it. In this way it is found, for instance, that many fault planes are not vertical. Finally, from this knowledge it is possible to estimate the directions of the stresses causing the earthquake.

NOTE 8   GRAVITY AND THE GEOID

Gravity anomalies

As explained in Section 3.2, the inverse square law of attraction between two point masses leads to the deduction that a spherical shell of material acts, at all places outside it, as if it were concentrated at its centre. Thus, to the extent that the Earth is spherically symmetrical, it is impossible to deduce anything about the internal distribution of mass. However, the Earth is not quite symmetrical, so some deductions can be made.

Gravity – more correctly, the acceleration due to gravity, g – varies over the surface of the Earth due to several effects, only some of which are important to us, although the others must be allowed for.

The largest variation occurs because of the equatorial bulge (Section 3.3) which exists because of the Earth's rotation. This affects g in two ways: because the distance from the Earth's centre increases from the poles to the equator, g decreases; and g also decreases because the centrifugal force increases from zero at the poles to a maximum at the equator. These two effects can be combined as a latitude dependence.

More local is the effect of topography. For hills (Fig. N8.1), there is a decrease in g due to the increased distance from the Earth's centre, although the pull of the mass of the hill partially offsets this decrease. The total effect is calculated in three stages, correcting to some datum level (often sea-level), that is, in effect, removing the hill. Firstly, the effect of the increased distance is equivalent to

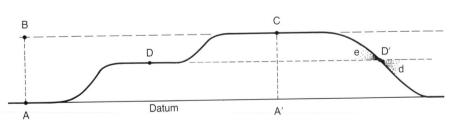

Figure N8.1 Illustration of topographical corrections. The vertical scale is greatly exaggerated.

going from A to B, and the corresponding correction is the **free-air correction**. Next, the mass of the hill is treated to a first approximation as if it were an infinitely extended slab filling the space between

A and C, and with an estimated density intended to be the same as that of the hill. A slab is used because it is not just the matter vertically below C that pulls on it, but also the matter to either side (though the more distant the matter from C the less its pull on C). This gives the **Bouguer correction**. Clearly, an infinite slab is a reasonable approximation only if the place of interest is on a plateau, as at C or D, but is not even roughly true at, say, D′. Comparing D′ to D, at the same altitude, there is an extra pull *upwards*, due to the mass *e*, and a *lack* of the downward pull of the missing mass *d*. Both of these mean that *g* at D′ is less than at D, and allowance for this is made by the **terrane correction**. The calculation of the value of the terrane correction is laborious because it depends on the densities and dispositions of masses around D′. Details of how all the corrections are made can be found in standard textbooks on applied or exploration geophysics, such as Kearey and Brooks (1991) and Robinson and Coruh (1988).

We are interested in the difference of the measured value of *g* from that expected for a uniform Earth. When all the above effects have been allowed for, any difference remaining must be due to internal variations; these differences are called **gravity anomalies**. If all the above corrections have been made, any resulting anomaly is termed a **Bouguer anomaly**, but if only those up to and including the free-air correction have been completed it is a **free-air anomaly**. A survey made on the sea surface will therefore reveal free-air anomalies, but on land, because of its variation in height, a Bouguer anomaly is needed to give the equivalent sea-level anomaly.

Anomalies can arise in two ways. One is a lateral variation of density, such as between different kinds of rocks, and gravity surveys can be used to find their approximate positions and mass excess or deficit compared to the surrounding rocks. But there is a limit on the size of such anomalies because of isostatic compensation. As we saw in Section 8.2.2, it was because of the reduced gravity anomaly of mountains that isostasy was recognized, and if compensation were perfect there would be no anomaly. Also, we saw that though relatively small masses can be supported by the strength of the lithosphere, and so are not compensated, the largest masses are almost entirely supported by buoyancy and so are compensated (given time for the adjustment), and this gives a size limit. Yet anomalies exist that are very large, both in lateral extent and magnitude, and therefore something must be acting to prevent isostatic equilibrium being attained. In the case of post-glacial rebound this is simply that equilibrium has not yet been reached, and the anomalies are temporary, but this cannot be the explanation for other large anomalies, such as those found at subduction zones. In these cases there must be long-acting forces operating, such as those associated with convection; gravity surveys can therefore help us identify where forces are acting (see Section 8.4.2).

Shape of the Earth and the geoid

The variation of gravity helps us to determine the shape of the Earth. The balance of the central pull of the Earth's mass and the outward centrifugal force due to its rotation results in a spheroidal shape (i.e. a cross-section along any line of longitude that is elliptical), which is referred to as the equatorial bulge. But because of local variations of gravity, there are small deviations from the spheroid, which are most easily observed on the sea surface where there is no topography.

The sea surface is a gravitational equipotential, which means that the pull of gravity is perpendicular to the surface; if it were not, then the water would readjust itself until it was. But *g* need not be constant along an equipotential. Figure N8.2 shows, as an example, the effect of a dense

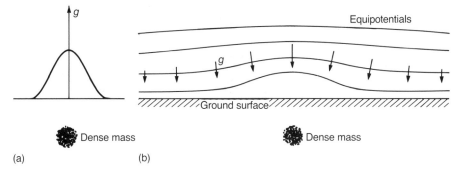

(a)　　　　　　　(b)

Figure N8.2 Deflection of equipotentials by a dense mass. (a) Shows the gravity anomaly, the variation in *g* due to the mass. (b) Shows how the equipotentials are both domed and closer together above the mass; the arrows show how the direction and magnitude of *g* vary along one equipotential (the curvature of the equipotentials is greatly exaggerated).

mass. This will produce a positive anomaly, i.e. the gravity above it will be above average. This means that, for example, firing a rocket into space will require more work above an excess mass, compared with elsewhere. Conversely, if we map surfaces where the work required is the same, they will be a little higher above the excess mass; in effect, to offset the extra pull the rocket has to start a little further from the Earth's centre. These surfaces are equipotentials, and their distortion by the mass is greatest closest to the mass, as shown. The direction of g is perpendicular to the equipotentials, and where the equipotentials are closest the value of g is greatest, showing that g is greater in the vicinity of the excess mass. It follows that the sea surface over such a mass will be a little higher than elsewhere.

The **geoid** is simply the sea surface (after allowance for tides and waves; for continents one has to imagine sea-level canals), so it follows that, compared with the spheroid, it will be high where there is a positive gravity anomaly, and low where there is a negative anomaly. The difference in height between the spheroid and the geoid gives the **geoid anomaly**. Anomalies are only tens of metres in magnitude but may be hundreds or thousands of kilometres broad. They can be measured by surveying, or by timing a radar beam reflected off the sea surface back up to a satellite and comparing the actual sea surface with that predicted from the theoretical spheroid. Geoid anomalies provide information about density variations and, by inference, mantle processes on a large scale.

References

Adams, F. D. (1938) *The Birth and Development of the Geological Sciences*, Baillière, Tindall and Cox, London.

Ahrens, T. J. (1979) Equations of state of iron sulfide and constraints on the sulfur content of the Earth. *J. Geophys. Res.*, **84**, 985–98.

Anders, E. and Grevasse, N. (1989) Abundances of the elements: meteoritic and solar. *Geochim. Cosmochim,. Acta*, **53**, 197–214.

Anderson, D. L. (1973) Comments on the power law representation of Birch's law. *J. Geophys. Res.*, **78**, 4901–14.

Anderson, D. L. (1989a) *Theory of the Earth*, Blackwell, California, 366 pp.

Anderson, D. L. (1989b) Composition of the Earth. *Science*, **243,** 367–71.

Anderson, D. L. (1992) The Earth's Interior, in *Understanding the Earth* (eds G. C. Brown, C. J. Hawkesworth and R. C. L. Wilson), Cambridge University Press, Cambridge, pp. 44–66.

Anderson, D. L. and Hart, R. S. (1976) An Earth model based on free oscillations and body waves. *J. Geophys. Res.*, **81**, 1461–75.

Anderson, D. L., Zhang, Y.-S. and Tanimoto, T. (1992) Plume heads, continental lithosphere, flood basalts and tomography. *J. Geol. Soc. Lond.*, in press.

Armstrong, R. L. (1981) Radiogenic isotopes: the case for crustal recycling on a near-steady-state no-continental growth Earth. *Phil. Trans. Roy. Soc. Lond.*, **301A**, 443–72.

Atherton, M. P. and Naggar, M. H. (1990) Granite: a symposium celebrating the 70th birthday of W. S. Pitcher. *Geol. J.*, **25**, 209–456.

BABEL Working Group (1990) Evidence for early Proterozoic plate tectonics from seismic reflection profiles in the Baltic shield. *Nature*, **348**, 34–8.

Badash, L. (1989) The age-of-the-Earth debate. *Sci. Am.*, **261** (2) (August) 78–83.

Badding, J. V., Hemley, R. K. and Mao, H. K. (1991) High-pressure chemistry of hydrogen in metals: in situ study of iron hydride. *Science*, **253**, 421–4.

Benz, W., Slattery, W. L. and Cameron, A. G. W. (1987) The origin of the moon and the single impact hypothesis II. *Icarus*, **71**, 30–45.

Bethe, H. A. and Brown, G. (1985) How a supernova explodes. *Sci. Am.*, **252** (5) (May), 40–8.

Bickle, M. J. (1986) Global thermal histories. *Nature*, **319**, 13–14.

Bickle, M. J., Bettanay, L. F., Boulter, C. A. *et al.* (1980) Horizontal tectonic interactions of an Archaean gneiss belt and greenstones, Pilbara Block, Western Australia. *Geology*, **8**, 525–9.

Binzel, R. P., Barucci, M. A. and Fulchignoni, M. (1991) The origins of the asteroids. *Sci. Am.*, **265** (4) (Oct.), 66–72.

Birch, F. (1961) Composition of the Earth's mantle. *Geophys. J. R. Astr. Soc.*, **4**, 295–311.

Bloxham, J. and Gubbins, D. (1985) Field and flow in the core. *Nature*, **317**, 777–81.

Bloxham, J. and Gubbins D. (1989) The evolution of the Earth's magnetic field. *Sci. Am.*, **261** (6) (Dec.), 30–7.

Bohlen, S. R. and Mezger, K. (1989) Origin of granulite terranes and the formation of the lowermost continental crust. *Science*, **244**, 326–9.

Bolt, B. A. (1976) *Nuclear explosions and earthquakes*, Freeman, San Francisco, pp. 309.

Bolt, M. H. P. (1982) *The Interior of the Earth*, Arnold, London, pp. 403.

Bowring, S. A. (1992) Earth's early crust. *EOS, Trans. Am. Geophys. Union*, **73**, p. 33.

Brandt, J. C. and Maran, S. P. (1972) *New Horizons in Astronomy*, W. H. Freeman, San Francisco.

References   Brown, G. C. (1985) Processes and problems in the continental lithosphere: geological history and physical implications, in *The Chronology of the Geological Record* (ed. N. J. Snelling), *Geol. Soc. Lond. Mem.*, **10**, 326–34.

Brown, G. C. and Hennessy, J. (1978) The initiation and thermal diversity of granite magmatism. *Phil. Trans. R. Soc. Lond.*, **288A**, 631–43.

Brown, G. C., Hawkesworth, C. J. and Wilson, R. C. L. (1992) *Understanding the Earth*, Cambridge University press, Cambridge.

Brush, S. G. (1983) Chemical history of the Earth's core. EOS *Trans. Am. Geophys. Union*, **63**, 1185–9.

Bullard, E. C. (1957) The density within the Earth. *Verh. Geol. Mijnb. Genoot. Ned. Kolon*, **18**, 23–41.

Bullard, E. C. (1971) The Earth's magnetic field and its origin, in *Understanding the Earth* (eds I. G. Gass, P. J. Smith and R. C. L. Wilson), Artemis Press, Sussex pp. 71–80.

Bullen, K. E. (1963). *Introduction to the Theory of Seismology*, Cambridge University Press, Cambridge.

Burbidge, E. M., Burbidge, G. R., Fowler, W. A. *et al.* (1957) Synthesis of the elements in stars. *Rev. Mod. Phys.*, **29**, 547–650.

Burchfield, J. D. (1975) *Lord Kelvin and the Age of the Earth*, Macmillan, London, pp. 260.

Burnett, M. S., Caress, D. W. and Orcutt, J. A. (1989) Tomographic image of the magma chamber at 12°50′ N on the East Pacific Rise. *Nature*, **339**, 206–8.

Busse, F. H. (1975) A model of the geodynamo. *Geophys. J.*, **42**, 437–59.

Campbell, I. H., Griffiths, R. W. and Hill, R. I. (1989) Melting in an Archaean mantle plume: heads it's basalts, tails it's komatiites. *Nature*, **339**, 697–9.

Campbell, J. A. (1970) *Chemical systems – Energetics, Dynamics, Structure*, W. H. Freeman, San Francisco, pp. 1111.

Cann, J. R. (1970) New model for the structure of ocean crust. *Nature*, **226**, 928–30.

Cattermole, P. (1989) *Planetary Volcanism*, Ellis Horwood, Chichester.

Chang, A. C. and Cleary, J. R. (1978) Precursors to PKKP. *Bull. Seism. Soc. Am.*, **68**, 1059–79.

Chevalier, R. A. (1992) Supernova 1987A at five years of age. *Nature*, **355**, p. 691.

Christiansen, G. S. and Garrett, P. M. (1960) *Structure and Change*, W. H. Freeman, San Francisco.

Chung, D. H. (1972) Birch's Law: why is it so good? *Science*, **177**, 261–3.

Clayton, R. N., Mayeda, T. K. and Molini-Velsko, C. A. (1985) Isotopic variations in Solar System material: evaporation and condensation of silicates, in *Protostars and Planets II* (eds D. C. Black and M. S. Matthews), University of Arizona Press, Tucson, pp. 775–81.

Clement, B. M. (1991) Geographical distribution of transitional VGPs: evidence for non-zonal equatorial symmetry during the Matuyama–Brunhes geomagnetic reversal. *Earth Planet. Sci. Lett.* **104**, 48–58.

Clifford, T. N. (1970) The structural framework of Africa, in *African Magmatism and Tectonics* (eds T. N. Clifford and I. G. Gass), Oliver and Boyd, London, pp. 1–26.

Cogley, J. G. (1984) Continental margins and the extent and number of the continents. *Rev. Geophys. Space Physics*, **22**, 101–22.

Cole, G. H. A. (1984) *Physics of Planetary Interiors*, Adam Hilger, Bristol, 208 pp.

Coleman, R. G. (1977) *Ophiolites: Ancient Oceanic Lithosphere?* Springer-Verlag, Heidelberg.

Coney, P. J., Jones, D. L. and Monger, J. W. H. (1980) Cordilleran suspect terranes. *Nature*, **288**, 329–33.

Conway Morris, S. (1992) The early evolution of life, in *Understanding the Earth* (eds G. C. Brown, C. J. Hawkesworth and R. C. L. Wilson), Cambridge University Press, Cambridge.

Coward, M. P., Dewey, J. F. and Hancock, P. L. (eds) (1987) Continental extensional tectonics. *Geol. Soc. Lond. Spec. Publ.*, **28**.

Crough, S. T. (1983) Hotspots swells. *Ann. Rev. Earth. Planet. Sci.*, **11**, 165–93.

Dalrymple, G. B. (1991) *The Age of the Earth*, Stanford University Press, Stanford, 474 pp.

Davies, G. F. (1977) Whole mantle convection and plate tectonics. *Geophys. J. R. Astr. Soc.*, **49**, 459–86.

Davies, G. R., Norry, M. J., Gerlach, D. C. *et al* (1989) A combined chemical and Pb–Sr–Nd isotope study of the Azores and Cape Verde hot spots: the geodynamical implications, in *Magmatism in the Ocean Basins* (eds A. D. Saunders and M. J. Norry), *Geol. Soc. Lond. Spec. Publ.*, **42**, pp. 231–55.

Detrick, R. S., Buhl, P., Vera, E. *et al.* (1987) Multi-channel seismic imaging of a crustal magma chamber along the East Pacific Rise. *Nature*, **326**, 35–41.

Dimroth, E. (1981) Labrador Geosyncline: type example of early Proterozoic cratonic reactivation, in *Precambrian Plate Tectonics* (ed. A. Kröner), Elsevier, Amsterdam, pp. 331–52.

Drury, S. A. (1978) Basic factors in Archaean geotectonics, in *Developments in Precambrian Geology I: Archaean Geochemistry* (eds B. F. Windley and S. M. Maqvi), Elsevier, Amsterdam.

Duffield, W. A. (1972) A naturally occurring model of global plate tectonics. *J. Geophys. Res.*, **77**, 2543–55.

Duffy, T. S. and Anderson, D. L. (1989) Seismic velocities in mantle minerals and the mineralogy of the upper mantle. *J. Geophys. Res.* **94**, 1895–912.

Dupré, B. and Allègre, C. J. (1983) Pb–Sr isotopic variations in Indian Ocean basalts and mixing phenomena. *Nature*, **303**, 142–6.

Dziewonski, A. M. and Anderson, D. L. (1981) Preliminary reference Earth model. *Phys. Earth Planet. Int.* **25**, 297–356.

Eldridge, C. S., Compston, W., Williams, J. S. *et al.* (1991) Isotopic evidence for the involvement of recycled sediments in diamond formations. *Nature*, **353**, 649–53.

England, P. (1992) Deformation of the continental crust, in *Understanding the Earth* (eds G. C. Brown, C. J. Hawkesworth and R. C. L. Wilson), Cambridge University Press, Cambridge, pp. 275–300.

England, P. C. and Houseman, G. A. (1986) Finite strain calculations of continental deformation 2. Comparison with the India-Asia collision zone. *J. Geophys. Res.*, **91**, 3664–76.

Faure, G. (1986) *Principles of Isotope Geology*, 2nd edn, Wiley, New York, 589 pp.

Feber, R. C., Wallace, T. C. and Libby, L. M. (1984) Uranium in the Earth's core. *EOS, Trans. Am. Geophys. Un.* **65**, No. 44.

Fisher, D. E. (1985) Radiogenic rare gases and the evolutionary history of the depleted mantle. *J. Geophys. Res.*, **90**, 1801–7.

Floyd, P. A. (1991) Ocean islands and seamounts, in *Oceanic Basalts* (ed. P. A. Floyd), Blackie, Glasgow, pp. 174–218.

Forsyth, D. and Uyeda, S. (1975) On the relative importance of the driving forces of plate motion. *Geophys. J. R. Astron. Soc.*, **43**, 163–200.

Fowler, C. M. R. (1990) *The Solid Earth: an Introduction to Global Geophysics*; Cambridge University Press, Cambridge, 472 pp.

Freeman, B., Klemperer, S. L. and Hobbs, R. W. (1988) The deep structure of northern England and Iapetus suture zone from BIRPS deep seismic reflection profiles. *J. Geol. Soc. Lond.*, **145**, 727–40.

Fukai, Y. and Suzuki, T. (1986) Iron–water reaction under high pressure and its implications in the evolution of the Earth. *J. Geophys. Res.*, **91**, 9222–30.

Fyfe, W. S. and Brown, G. C. (1972) Granites past and present. *J. Earth Sci.*, **8**, 249–60.

Garland, G. D. (1971) *Introduction to Geophysics*, Philadelphia, Saunders.

Gass, I. G. (1981) Pan-African plate tectonics of the Arabian–Nubian shield, in *Precambrian Plate Tectonics* (ed. A. Kröner), Elsevier, Amsterdam, pp. 387–405.

Gass, I. G. (1982) Ophiolites. *Sci. Am.*, **247** (2) (Sept.), 122–31.

References   Gass, I. G. and Smewing, J. D. (1981) Ophiolites: obducted oceanic lithosphere, in *The Ocean Lithosphere* (*The Sea*, Vol. 7), (ed. C. Emiliani), Wiley, New York, pp. 339–62.

Gillespie, C. C. (1951) *Genesis and Geology*, Harvard University Press, Cambridge, Mass.

Goldstein, J. I. and Short, J. M. (1967) The iron meteorites, their thermal history and parent bodies. *Geochim. Cosmochim. Acta*, **31**, 1733–70.

Griffin, W. L. and O'Reilly, S. Y. (1987) Is the continental Moho the crust–mantle boundary? *Geology*, **15**, 241–4.

Grossman, L. (1972) Condensation in the primitive solar nebula. *Geochim. Cosmochim. Acta*, **36**, 597–619.

Grossman, L. and Larimer, J. W. (1974) Early chemical history of the Solar System. *Rev. Geophys. Space Phys.*, **12**, 71–101.

Gubbins, D. (1990) *Seismology and Plate Tectonics*, Cambridge University Press, Cambridge.

Gubbins, D. and Masters, T. G. (1979) Driving mechanisms for the Earth's dynamo. *Advances in Geophysics*, **21**, 1–50.

Gurnis, M. (1986) Large-scale mantle convection and the aggregation and dispersal of supercontinents. *Nature*, **322**, 695–9.

Haber, F. C. (1959) *The Age of the World – Moses to Darwin*, Johns Hopkins, Baltimore.

Hargraves, R. B. (1986) Faster spreading or greater ridge length in the Archaean? *Geology*, **14**, 750–2.

Harland, W. B., Armstrong, R. L., Cox, A. V. *et al.* (1990) *A Geologic Time Scale 1989*, Cambridge University Press, Cambridge.

Harris, N. B. W., Holt, R. W. and Drury, S. A. (1982) Geobarometry, geothermometry and late Archaean geotherms from the granulite facies terrain of South India. *J. Geol.*, **90**, 509–27.

Hart, S. R. (1988) Heterogeneous mantle domains: signatures, genesis and mixing chronologies. *Earth. Planet. Sci. Lett.*, **90**, 273–96.

Hart, S. R. and Staudigel, H. (1989) Isotopic characterisation and identification of recycled components, in *Crust/Mantle Recycling at Convergence Zones* (eds. S. R. Hart and L. Gülen), Kluwer Academic Publishers, Dordrecht, pp. 15–28.

Hayashi, C., Nakazawa, K. and Nakagawa, Y. (1985) Formation of the Solar System, in *Protostars and Planets II* (eds D. C. Black and M. S. Matthews), University of Arizona Press, Tucson; pp. 1100–53.

Head, J. W., Campbell, D. B., Elachi, C. *et al.* (1991) Venus volcanism: initial analysis from Magellan data. *Science*, **252**, 276–88.

Heiken, G. H., Vaniman, D. T. and French B. M. (1991) *Lunar Sourcebook: a User's guide to the Moon*, Cambridge University Press, Cambridge, 736 pp.

Henbest, N. (1992) Happy birthday, Supernova 1987A. *New Sci.*, **33** (1809) (22 Feb.), 25–9.

Hergt, J. and Hawkesworth, C. J. (1992) The remobilisation of continental mantle lithosphere. *Sci. Progress*, in press.

Hildreth, W. and Moorbath, S. (1988) Crustal contributions to arc magmatism in the Andes of central Chile. *Contr. Min. Pet.*, **98**, 455–89.

Hill, R. I., Campell I. H., Davies, G. F. *et al.* (1992) Mantle plumes and continental tectonics. *Science*, **256**, 186–93.

Howell, D. G. (1985) Terranes. *Sci. Am.* **253** (5) (Nov.), 90–103.

Huppert, H. E. and Sparks, R. S. J. (1988). The generation of granitic magmas by intrusion of basalt into continental crust. *J. Petrol.*, **29**, 599–624.

Huggett, R. (1990) *Catastrophism: Systems of Earth History*, Arnold.

Irifune, T., Fujino, K. and Ohtani, E. (1991) A new high-pressure form of $MgAl_2O_4$. *Nature*, **349**, 409–11.

Ito, E., Akaogi, M., Topor, L. *et al.* (1990) Negative pressure–temperature slopes for reactions forming $MgSiO_3$ perovskite from calorimetry. *Science*, **249**, 1275–8.

Jacobs, J. A. (1975) *The Earth's Core*, Academic Press, London, pp. 253.

Jacobson, S. B. and Dymek, R. F. (1988) Nd and Sr isotope systematics of clastic
metasediments from Isua, West Greenland: Identification of pre-3.8 Ga differentiated
crustal component. *J. Geophys. Res*, **93**, 338–54.

Jeanloz, R. (1990) The nature of the Earth's core. *Ann. Rev. Earth Planet. Sci.*, **18**,
357–86.

Jeffreys, H. (1962) *The Earth*, 4th edn., Cambridge University Press, Cambridge,
pp. 438.

Jeffreys, H. and Bullen, K. E. (1940) *Seismological Tables*, Brit. Ass. Adv. Sci., London.

Jones, J. H. and Drake, M. J. (1986) Geochemical constraints on core formation in the
Earth. *Nature*, **322**, 221–8.

Kasting, J. F. (1991) CO_2 condensation and the climate of early Mars. *Icarus*, **94**, 1–13.

Kaula, W. M. (1990) Venus: A Contrast in Evolution to Earth. *Science*, **247**, 1191–6.

Kearey, P. and Brooks, M. (1991) *An Introduction to Geophysical Exploration*, 2nd edn,
Blackwell, Oxford, 254 pp.

Kearey, P. and Vine, F. J. (1990) *Global Tectonics*, Blackwell, Oxford, 302 pp.

Kennett, B. L. N. and Engdahl, E. R. (1991) Traveltimes for global earthquake location
and phase identification. *Geophys. J. Int.*, **105**, 429–65.

Kerr, R. A. (1989) Another movement in the dance of the plates. *Science*, **244**, 529–
30.

Kerridge, J. F and Matthews, M. S. (1988) *Meteorites and the early Solar System*,
University of Arizona Press, Tuscon.

Kershaw, S. (1990) Evolution of the Earth's atmosphere and its geological impact.
Geology Today, **6**, 55–60.

Klemperer, S. L. and Peddy, C. (1992) Seismic reflection profiling and the structure of
the continental lithosphere, in *Understanding the Earth* (eds G. C. Brown, C. J.
Hawkesworth and R. C. L. Wilson), Cambridge University Press, Cambridge,
pp. 251–74.

Knittle, E. and Jeanloz, R. (1991) The high-pressure phase diagram of $Fe_{0.94}O$: a possible
constituent of the Earth's core. *J. Geophys. Res.*, **96**, 16 169–180.

Knopoff, L. (1972) Observation and inversion of surface wave data. *Tectonophysics*, **13**,
497–519.

Kuskov, O. L., Fabrichnaya, O. B. and Truskinovsky, L. M. (1991) Constitution of the
mantle. 2 Petrological model of transition zone based on FMS phase diagram. *Phys.
Earth. Planet. Int.*, **69**, 72–84.

Lada, C. J. and Shu, F. H. (1990) The formation of Sun-like stars. *Science*, **248**, 564–
72.

Larimer, J. W. and Wasson, J. T. (1988) Siderophile element fractionation, in *Meteorites
and the Early Solar System* (eds J. F. Kerridge and M. S. Matthews), University of
Arizona Press, Tucson, pp. 416–35.

Larson, R. L. (1991a) Latest pulse of Earth: evidence for a mid-Cretaceous superplume.
Geology, **19**, 547–50.

Larson, R. L. (1991b) Geological consequences of superplumes. *Geology*, **19**, 963–6.

Larson, R. L. and Olson, P. (1991) Mantle plumes control magnetic reversal frequency.
Earth Planet. Sci. Lett., **107**, 437–47.

Lee, M. K., Pharoh, T. C. and Soper, N. J. (1990) Structural trends in central Britain
from images of gravity and aeromagnetic fields. *J. Geol. Soc. Lond.*, **147**, 241–58.

Le Grand, H. E. (1988) *Drifting Continents and Shifting Theories*, Cambridge University
Press, Cambridge.

Lewis, B. T. R. (1983) The process of formation of ocean crust. *Science*, **220**, 151–7.

Liebermann, R. C. and Ringwood, A. E. (1973) Birch's law and polymorphic phase
transformations. *J. Geophys. Res.*, **78**, 6926–32.

Lister, C. R. B., Slater, J. G., Davis, E. E. *et al.* (1990) Heat flow maintained in ocean
basins of great age: investigations in the north-equatorial West Pacific. *Geophys. J.
Int.*, **102**, 603–30.

Liu, L.-G. (1988) Water in the terrestrial planets and the Moon. *Icarus*, **74**, 98–107.

References Lovelock, J. E. (1988) *The Ages of Gaia: a Biography of our Living Earth*, Oxford University Press, Oxford.

Maaloe, S. and Aoki, K. (1977) The major element chemistry of the upper mantle estimated from the composition of lherzolites. *Contrib. Mineral. Petrol.*, **63**, 161–73.

Martin, H. (1986) Effect of steeper Archaean geothermal gradient on geochemistry of subduction zone magmas. *Geology*, **14**, 753–6.

Masters, T. G. and Shearer, P. M. (1990) Summary of seismological constraints on the structure of the Earth's core. *J. Geophys. Res.* **95**, 21, 691–5.

Mather, K. F. and Mason, S. L. (1939) *A Source Book in Geology*. McGraw-Hill, New York.

McAlester, A. L. (1977) *The History of Life*, 2nd edn, Prentice-Hall, Englewood Cliffs, New Jersey.

McCall, G. J. H. (1991) A return to the Earth's oldest rocks. *Geoscientist*, **1**(5), 24–7.

McGregor, A. M. (1951) Some milestones in the Precambrian of southern Africa. *Proc. Geol. Soc. S. Afr.*, **54**, 27–71.

McKenzie, D. P. (1983) The Earth's mantle. *Sci. Am.*, **249** (3) (Sept.), 51–62.

McKenzie, D. P. (1985) The extraction of magma from the crust and mantle. *Earth Planet. Sci. Lett.* **74**, 81–91.

McKenzie, D. and Bickle, M. J. (1988) The volume and composition of melt generated by extension of the lithosphere. *J. Pet.*, **29**, 625–79.

McSween, H. Y. (1987) *Meteorites and their Parent Planets*, Cambridge University Press, Cambridge.

Melchior, P. (1986) *Physics of the Earth's Core*, Pergamon Press, Oxford.

McElhinny, M. W. (1973) *Palaeomagnetism and plate tectonics*. Cambridge University Press, Cambridge, pp. 358.

Merrill, R. T. and McElhinny, M. W. (1983) *The Earth's Magnetic field, its History, Origin and Planetary Perspective*, Academic Press, London, pp. 401.

Merrill, R. T. and McFadden, P. L. (1990) Paleomagnetism and the nature of the geodynamo. *Science*, **248**, 345–50.

Minster, J. B., Jordan, T. H., Molnar, P. *et al.* (1974) Numerical modelling of instantaneous plate tectonics. *Geophys. J. R. Astr. Soc.* **36**, 541–76.

Minster, J.-F., Birck, J.-L. and Allègre, C. J. (1982) ^{87}Rb–^{87}Sr chronology of enstatite chondrites. *Earth Planet. Sci. Lett.*, **44**, 420–40.

Montagner, J.-P. and Anderson, D. L. (1989) Constrained reference mantle model. *Phys. Earth Planet. Int.*, **58**, 205–27.

Mooney, W. D. and Meissner, R. (1991) Continental crustal evolution observations. *EOS, Trans. Am. Geophys. Union*, **72**, 537–41.

Moorbath, S. (1977) Ages, isotopes and evolution of the Precambrian continental crust. *Chem. Geol.*, **20**, 151–87.

Moores, E. M. (1986) The Proterozoic ophiolite problem, continental emergence, and the Venus connection. *Science*, **234**, 65–71.

Morgan, J. W. and Anders, E. (1980) Chemical compositions of the Earth, Venus and Mercury. *Proc. Nat. Acad. Sci.*, **77**, 6973–7.

Mukhin, L. M., Gerasimov, M. V. and Safnova, E. N. (1989) Origin of precursors of organic molecules during evaporation of meteorites and mafic terrestrial rocks. *Nature*, **340**, 46–8.

Mulargia, F. and Quareni, F. (1988) Validity of the Sutherland–Lindemann law and melting temperatures in the Earth's interior. *Geophys. J*, **92**, 269–82.

Murray, B. (1983) *The Planets – readings from Scientific American*, Freeman, San Francisco, pp. 132.

Murrell, M. T. and Burnett, D. S. (1986) Partitioning of K, U and Th between sulfide and silicate liquids: implications for radioactive heating of planetary cores. *J. Geophys. Res.*, **91**, 8126–36.

Murty, V. R. and Patterson, C. C. (1962) Primary isochron of zero age for meteorites and the Earth. *J. Geophys. Res.*, **67**, 1161–7.

Nataf, H.-C. (1989) One-and-a-half layer convection? in *Crust/Mantle Recycling at Convergence Zones* (eds S. R. Hart and L. Gülen,), Kluwer Academic Publishers, Dordrecht, pp. 197–200.

Newsom, H. W. and Jones, J. H. (1990) *Origin of the Earth*, Oxford University Press, New York/London.

Newsom, H. E. and Sims, K. W. W. (1991) Core formation during early accretion of the Earth. *Science*, **252**, 926–33.

Newsom, H. E. and Taylor, S. R. (1989) Geochemical implications of the formation of the Moon by a single giant impact. *Nature*, **338**, 29–34.

Nisbet, E. G. (1987) *The Young Earth*, Unwin Hyman, Boston, 402 pp.

Nisbet, E. G. (1991) *Living Earth*. HarperCollins, London, 237 pp.

Ohtani, E. and Ringwood, A. E. (1984) Composition of the core, I. Solubility of oxygen in molten iron at high temperatures. *Earth Planet. Sci. Lett.*, **71**, 85–93.

Ohtani, E., Ringwood, A. E. and Hibberson, W. (1984) Composition of the core, II effect of high pressure on solubility of FeO in molten iron. *Earth Planet. Sci. Lett.*, **71**, 94–103.

Olson, P., Silver, P.G and Carlson, R. W. (1990) The large-scale structure of convection in the Earth's mantle. *Nature*, **344**, 209–15.

O'Nions, R. K. (1987) Relationships between chemical and convective layering in the Earth. *J. Geol. Soc. Lond.*, **144**, 259–74.

O'Nions, R. K. (1992) The Continents, in *Understanding the Earth* (eds G. C. Brown, C. J. Hawkesworth and R. C. L. Wilson), Cambridge University Press, Cambridge, pp. 145–63.

Ott, U. and Begemann, F. (1985) Are all the 'martian' meteorites from Mars? *Nature*, **317**, 509–13.

Palme, H. and Nickel, K. G. (1985) Ca/Al ratios and composition of the Earth's upper mantle. *Geochim. Cosmochim. Acta*, **49**, 2123–32.

Park, R. G. (1988) *Geological Structures and Moving Plates*, Blackie, Glasgow.

Patchett, J. and Arndt, N. T. (1986) Nd isotopes and tectonics of 1.9–1.7 Ga crustal genesis. *Earth Planet. Sci. Lett.*, **78**, 329–38.

Pauling, L. (1959) *The Nature of the Chemical Bond*, 3rd edn., Oxford University Press, Oxford.

Peltier, W. R. (1989) Mantle viscosity, in *Mantle Convection* (ed. W. R. Peltier), Gordon and Breach, New York.

Piper, J. D. A. (1987) *Palaeomagnetism and the Continental Crust*, Open University Press, Milton Keynes, 434 pp.

Pitcher, W. S., Atherton, M. P., Cobbing, E. J. *et al.* (1985) *Magmatism at a Plate Edge: The Peruvian Andes*, Blackie and Halstead Press, Glasgow.

Press, F. and Siever, R. (1986) *Earth*, 4th edn, Freeman, New York.

Price, G. D., Wall, A. and Parker, S. C. (1989) The properties and behaviour of mantle minerals: a computer simulation approach. *Phil. Trans. R. Soc. Lond.*, **328A**, 391–407.

Rama Murthy, V. (1991) Early differentiation of the Earth and the problem of mantle siderophile elements: a new approach. *Science*, **253**, 303–6.

Rampino, M. (1992) Volcanic hazards, in *Understanding the Earth* (eds G. C. Brown, C. J. Hawkesworth and R. C. L. Wilson), Cambridge University Press, Cambridge, pp. 506–22.

Ranalli, G. (1987) *Rheology of the Earth*, Allen and Unwin, Boston, 366 pp.

Richards, M. A. and Engebretson, D. C. (1992) Large-scale mantle convection and the history of subduction. *Nature*, **355**, 437–40.

Richards, M. A., Duncan, R. A. and Courtillot, V. E. (1989) Flood basalts and hot-spot tracks: plume heads and tails. *Science*, **246**, 103–7.

Richardson, S. H., Gurney, J. J., Erlank, A. J. *et al.* (1984) Origin of diamonds in old enriched mantle. *Nature*, **310**, 198–202.

Richter, C. F. (1958). *Elementary Seismology*, Freeman, San Francisco.

Riding, R. (1992) The algal breath of life. *Nature*, **359**, 13–14.

References   Ringwood, A. E. (1977) Composition of the core and implications for origin of the Earth. *Geochem. J.*, **11**, 111–35.

Ringwood, A. E. (1989) Significance of the terrestrial Mg/Si ratio. *Earth Planet. Sci. Lett.*, **95**, 1–7.

Ringwood, A. E. and Major, A. (1971) Synthesis of majorite and other high pressure garnets and perovskites. *Earth. Planet. Sci. Lett.*, **12**, 411–18.

Roberts, P. H. and Gubbins, D. (1987) Origin of the main field: kinematics, in *Geomagnetism*, Vol. 2 (ed. J. A. Jacobs), Academic Press, London, pp. 185–249.

Robinson, E. S. and Coruh, C. (1988) *Basic Exploration Geophysics*, Wiley, New York.

Rogers, G. and Hawkesworth, C. J. (1989) A geochemical traverse across the North Chilean Andes: evidence for crust generation from the mantle wedge. *Earth Planet. Sci. Lett.*, **91**, 271–85.

Romanowicz, B. (1991) Seismic tomography of the Earth's mantle. *Ann. Rev. Earth. Planet. Sci.*, **19**, 77–99.

Ronov, A. B. and Yaroshevsky, A. A. (1969) Chemical composition of the Earth's crust, in *The Earth's Crust and Upper mantle* (ed. P. J. Hart), Am. Geophys. Union, Washington, pp. 35–57.

Rothery, D. A. (1992) *Satellites of the Outer Planets*, Clarendon Press, Oxford.

Rudnick, R. (1990) Growing from below. *Nature*, **347**, 711–12.

Russell, S. S., Arden, J. W. and Pillinger, C. T. (1991) Evidence for multiple sources of diamond from primitive chondrites. *Science*, **254**, 1188–91.

Saikumar, V. and Goldstein, J. I. (1988) An evaluation of the methods to determine the cooling rates of iron meteorites. *Geochim. Cosmochim. Acta*, **52**, 715–26.

Scarfe, C. M. and Takahashi, E. (1986) Melting of garnet peridotite to 13 GPa and the early history of the upper mantle. *Nature*, **322**, 354–6.

Sclater, J. G., Jaupart, C. and Galson, D. (1980) The heat flow through oceanic and continental crust and the heat loss of the earth. *Rev. Geophys. Space Sci.*, **18**, 269–311.

Schubert, G. (1991) The lost continents. *Nature*, **354**, 358–9.

Schult, F. R. and Gordon, R. G. (1984) Root mean square velocities of the continents with respect to the hot spots since the early Jurassic. *J. Geophys. Res.*, **89**, 1789–1800.

Sears, D. W. G. and Dodd, R. T. (1988) Overview and classification of meteorites, in *Meteorites and the Early Solar System* (eds J. F. Kerridge and M. S. Matthews), University of Arizona Press, Tucson, pp. 3–31.

Seeds, M. A. (1990) *Foundations of Astronomy*, Wadsworth Pub. Co., Belmont, California, pp. 654.

Semat, H. (1954) *Introduction to Atomic and Nuclear Physics*, 3rd edn., Rinehart, New York, pp. 561.

Sharma, P. V. (1984) The Fennoscandian uplift and glacial isostasy. *Tectonophys.*, **105**, 249–62.

Shearer, P. M. (1991) Constraints on upper mantle discontinuities from observations of long-period reflected and converted phases. *J. Geophys. Res.*, **96**, 18 147–82.

Smylie, D. E. (1992) The inner core translational triplet and the density near the Earth's core. *Science*, **255**, 1678–82.

Snowdon, P. A. (1984) Non-diapiric batholiths in the north of the Zimbabwe Shield, in *Precambrian Tectonics Illustrated* (eds A. Kröner and R. Greilung), Schweizerbart, Stuttgart, pp. 135–46.

Soper, N. J. and Hutton, D. H. W. (1984) Late Caledonian sinistral displacements in Britain: implications for a three-plate collision model. *Tectonics*, **3**, 781–94.

Spohn, T. (1991) Mantle differentiation and thermal evolution of Mars, Mercury and Venus. *Icarus*, **90**, 222–36.

Spray, J. (1991) Structure of oceanic crust as deduced from ophiolites, in *Oceanic Basalts* (ed. P. A. Floyd), Blackie, Glasgow, pp. 49–62.

Stacey, F. D. (1977) *Physics of the Earth*, 2nd edn. Wiley, New York.

Stacey, F. D. and Loper, D. E. (1988) Thermal history of the Earth: a corollary concerning non-linear mantle rheology. *Phys. Earth Planet. Int.*, **53**, 167–74.

Stahler, S. W. (1991) The early life of stars. *Sci. Am.* **265** (1) (July), 28–35.

Stevenson, D. J. (1981) Models of the Earth's core. *Science*, **214**, 611–19.

Stevenson, D. J. (1988) Greenhouses and magma ocean. *Nature*, **335**, 588–9.

Stevenson, D. J. and Lunine, J. I. (1988) Rapid formation of Jupiter by diffusive redistribution of water vapour in the Solar Nebula. *Icarus*, **75**, 146–55.

Stixrude, L. and Bukowinski, M. S. T. (1990) Fundamental thermodynamic relations and silicate melting with implications of the constitution of D″. *J. Geophys. Res.*, **95**, 19 311–325.

Sun, S. S. and McDonough, W. F. (1989) Chemical and isotopic systematics of ocean basalts: implications for mantle composition and processes, In *Magmatism in the Ocean Basins* (eds A. D. Saunders and M. J. Norry), *Geol. Soc. Lond. Spec. Publ.*, **42**, pp. 313–45.

Talwani, M., Le Pichon, X. and Ewing M. (1965) Crustal structure of mid-ocean ridges. 2:Computed model from gravity and seismic refraction data. *J. Geophys. Res.*, **70**, 341–52.

Tapponier, P., Peltzer, G., Le Dain, A. Y. *et al.* (1982) Propagating extrusion tectonics in Asia: new insights from simple experiments with plasticine. *Geology*, **10**, 611–16.

Taylor, S. R. (1982) *Planetary Science: A Lunar Perspective*. Lunar and Planetary Institute, Houston.

Taylor, S. R. (1988) Planetary compositions, in *Meteorites and the Early Solar System* (eds J. F. Kerridge and M. S. Matthews), University of Arizona Press, Tucson, pp. 512–34.

Taylor, S. R. (1992) The origin of the Earth, in *Understanding the Earth* (eds G. C. Brown, C. J. Hawkesworth and R. C. L. Wilson), Cambridge University Press, Cambridge, pp. 25–42.

Taylor, S. R. and McLennan, S. M. (1985) *The Continental Crust; its Composition and Evolution*, Blackwell, Oxford.

Thompson, A. B. (1992) Metamorphism and fluids, in *Understanding the Earth*, (eds G. C. Brown, C. J. Hawkesworth and R. C. L. Wilson), Cambridge University Press, Cambridge, pp. 223–48.

Thorpe, R. S. (1982), *Andesites: Orogenic Andesites and Related rocks*, Wiley and Sons, Chichester.

Tilton, G. R. (1988) Principles of radiometric dating, in *Meteorites and the Early Solar System* (eds J. F. Kerridge and M. S. Matthews), University of Arizon Press, Tucson, pp. 249–58.

Toksöz, M. N. (1976) The subduction of lithosphere, in *Continents Adrift and Aground*, (ed. J. T. Wilson), W. H. Freeman, San Francisco, pp. 113–22.

Turcotte, D. L. (1980) On the thermal evolution of the Earth. *Earth Planet. Sci. Lett.*, **48**, 53–88.

Turcotte, D. L. and Schubert, G. (1982) *Geodynamics*. Wiley, New York.

Turner, J. S. (1973) Convection in the mantle: a laboratory model with temperature dependent viscosity. *Earth Planet. Sci. Lett.*, **17**, 369–74.

Van der Hilst, R., Engdahl, R., Spakman, W. and Nolet, G. (1991) Tomographic imaging of subducted lithosphere below northwest Pacific island arcs. *Nature*, **353**,37–43.

Van Schmus, W. R. and Wood, J. A. (1967) A chemical petrologic classification for the chondritic meteorites. *Geochim. Cosmochim. Acta*, **31**, 747–65.

Verhoogen, J. (1980) *Energetics of the Earth*. National Academy of Science, Washington.

Vidale, J. E. and Benz, H. M. (1992) Upper-mantle seismic discontinuities and the thermal structure of subduction zones. *Nature*, **356**, 678–83.

Vink, G. E., Morgan, W. J. and Vogt, P. R. (1985) The Earth's hot spots. *Sci. Am.* **252** (4) (April), 32–9.

Wasson, J. T. (1985) *Meteorites*, W. H. Freeman, New York, 267 pp.

Watson, A. (1991) Gaia. *New Scientist*, **131**, (Inside Science No. 48), 4 pp.

Weaver, H. A. and Danly, L. (1989) *The Formation and Evolution of Planetary Systems*, Cambridge University Press, Cambridge.

References   Wernicke, B. (1985) Uniform-sense normal simple shear of the continental lithosphere. *Can. J. Earth Sci.*, **22**, 108–25.

Wetherill, G. W. (1989) The formation of the Solar System: consensus, alternatives and missing factors, in *The Formation and Evolution of Planetary Systems* (eds H. A. Weaver and L. Danly), Cambridge University Press, Cambridge, pp. 1–24.

White, R. S. and McKenzie, D. P. (1989a) Volcanism at rifts. *Sci. Am.*, **262** (1) (July), 44–55.

White, R. S. and McKenzie, D. P. (1989b) Magmatism at rift zones: the generation of volcanic continental margins and flood basalts. *J. Geophys. Res.*, **94**, 7685–730.

Williamson, E. D. and Adams, L. H. (1923) Density distribution in the Earth. *J. Wash. Acad. Sci.*, **13**, 413–28.

Windley, B. F. (1984) *The Evolving Continents*, 2nd edn, Wiley, London.

Wood, J. A. (1979) *The Solar System*, Prentice-Hall, Englewood Cliffs, New Jersey.

Wood, J. A. (1981) Meteorites, in *The New Solar System* (eds J. K. Beatty, B. O'Leary and A. Chaikin), Sky Publ. Co., Cambridge, pp. 187–96.

Woosley, S. and Weaver, T. (1989) The great supernova of 1987. *Sci. Am.*, **261** (2) (Aug.), 24–32.

Wyllie, P. J. (1971) *The Dynamic Earth*, Wiley, New York.

Wyllie, P. J. (1992) Experimental petrology: Earth materials science, in *Understanding the Earth* (eds G. C. Brown, C. J. Hawkesworth and R. C. L. Wilson), Cambridge University Press, Cambridge, pp. 67–87.

Zhang, Y.-S. and Tanimoto, T. (1992) Ridges, hotspots and their interaction as observed in seismic velocity maps. *Nature*, **355**, 45–9

Zoback, M. L., Anderson, R. E. and Thompson, G. A. (1981) Cainozoic evolution of the state of stress and style of tectonism of the Basin and Range province of the western United States. *Phil. Trans. R. Soc. London*, **300A**, 407–34.

Zoback, M. L., Zoback, M. D., Adams, J. *et al.* (1989) Global patterns of tectonic stress. *Nature*, **341**, 291–8.

Index

F following a page number denotes a Figure, T a Table. Numbers in **bold** indicate where the term is defined or explained.

266